QUANTUM THEORY OF MATTER

A NOVEL INTRODUCTION

QUANTUM THEORY OF MATTER

A NOVEL INTRODUCTION

A. Modinos
National Technical University of Athens, Greece

JOHN WILEY & SONS
Chichester · New York · Brisbane · Toronto · Singapore

National 01243 779777
International (+44) 1243 779777

Other Wiley Editorial Offices

John Wiley & Sons, Inc., 605 Third Avenue,
New York, NY 10158-0012, USA

Jacaranda Wiley Ltd, 33 Park Road, Milton,
Queensland 4064, Australia

John Wiley & Sons (Canada) Ltd, 22 Worcester Road,
Rexdale, Ontario M9W 1L1, Canada

John Wiley & Sons (Asia) Pte Ltd, 2 Clementi Loop #02-01,
Jin Xing Distripark, Singapore 0512

Library of Congress Cataloging-in-Publication Data

Modinos, A., 1938–
Quantum theory of matter : a novel introduction / by A. Modinos.
p. cm.
Includes bibliographical references and index.
ISBN 0-471-96363-1 (hardback : alk. paper). – ISBN 0-471-96364-X (pbk. : alk. paper)
1. Matter–Constitution. 2. Quantum Theory. I. Title.
QC173.M564 1996
530.1'2–dc20 95-49410
CIP

British Library Cataloguing in Publication Data

A catalogue record for this book is available from the British Library

ISBN 0 471 96363 1 (cloth)
ISBN 0 471 96364 X (paper)

Typeset in 10/12pt Times by Keytec Typesetting Ltd, Bridport, Dorset
Printed and bound in Great Britain by Bookcraft (Bath) Ltd.
This book is printed on acid-free paper responsibly manufactured from sustainable forestation, for which at least two trees are planted for each one used for paper production.

CONTENTS

Preface **ix**

1 QUANTUM MECHANICS **1**

1.1 Introduction 1
1.2 The wavefunction 3
1.3 Particle in a box 5
1.3.1 Motion in one dimension 6
1.3.2 Motion in three dimensions 12
1.3.3 Degeneracy of energy-eigenvalues—the role of symmetry 13
1.3.4 Density of states 17
1.4 Potential fields 18
1.4.1 Classical picture of a particle in a potential field 18
1.4.2 Quantum-mechanical description of a particle in a potential field 20
1.5 Particle in a rectangular potential well 22
1.6 Mean values of physical quantities and measurements 25
1.7 Non-bound states 27
1.7.1 Eigenstates of a free particle 28
1.7.2 Wavepackets of a free particle 30
1.8 Scattering by a potential barrier and the phenomenon of tunnelling 32
1.9 Bound states as superpositions of plane waves 38
Problems 39

2 ATOMS **45**

2.1 The one-electron atom—introduction 45
2.2 Angular momentum 46

2.3 Scattering states in a spherically symmetric field 52
2.3.1 The free particle 52
2.3.2 Scattering by a spherically symmetric barrier 54
2.4 Bound states in a spherically symmetric field 57
2.5 Bound states of an electron in a Coulomb field 59
2.6 Hydrogen gas in thermodynamic equilibrium 64
2.7 Interaction of atoms with the electromagnetic field 65
2.7.1 Atom in an external field 65
2.7.2 De-excitation of an atom—emission of light 69
2.8 Quantum description of the electromagnetic field 70
2.9 Fine structure of the energy levels of the hydrogen atom 75
2.10 Spin 77
2.11 Description of the eigenstates of a particle with spin ($s = \frac{1}{2}$) 79
2.12 Eigenstates of a particle in a spin-independent potential field 82
2.13 Spinors of the hydrogen atom 83
2.14 Fermions and bosons 88
2.15 The N-electron atom 97
2.15.1 Self-consistent electronic structure of many-electron atoms 100
2.15.2 Energy-level diagram of a complex atom 106
2.15.3 The radii of atomic shells 111
2.15.4 Periodic system of elements 114
2.15.5 Total energy of an atom 116
Problems 118

3 MOLECULES 129

3.1 The Born–Oppenheimer approximation 129
3.2 Hydrogen molecular ion—electronic terms 131
3.3 Nuclear motion 137
3.3.1 Rotational and vibrational motion of a diatomic molecule 137
3.3.2 Harmonic oscillations 141
3.4 Diatomic molecules of N electrons 143
3.5 Absorption–emission of electromagnetic radiation 145
3.6 Polyatomic molecules 148
Problems 154

4 SOLIDS 159

4.1 Space lattices 159
4.2 Reciprocal lattice 167
4.3 One-electron eigenstates in a periodic potential field 170
4.3.1 Energy bands of copper 175
4.3.2 Energy bands of tungsten 183

4.4 Surfaces of constant energy 184
4.5 Density of one-electron states in a crystal 187
4.6 Metals, insulators and semiconductors 189
4.7 Specific heat 192
4.8 Electron transport—conductivity of metals 198
4.8.1 Mean free path 207
4.9 The free-electron model of metals 208
4.10 Conductivity of semiconductors 209
4.10.1 Doped semiconductors 211
4.11 The self-consistent field in crystals 216
4.11.1 Non-magnetic metals 218
4.11.2 Ferromagnetic metals 219
4.11.3 Total energy of the metal 223
4.12 Lattice vibrations 225
4.13 Surfaces 232
4.13.1 General properties 232
4.13.2 One-electron states of a semi-infinite crystal—surface states 237
4.13.3 Electron emission 249
4.14 Electronic excitations 251
4.15 The dielectric function 258
4.16 Plasmons 260
4.17 Superconductivity 263
4.18 Non-crystalline solids 268
4.18.1 Amorphous films and glasses 268
4.18.2 Localisation of one-electron states due to disorder 271
4.18.3 The Anderson transition 275
4.18.4 Hopping transport 277
4.18.5 Energy bands in amorphous materials 279
Problems 282

Appendix A VECTORS AND COMPLEX NUMBERS **291**

A.1 Vector quantities 291
A.2 Complex numbers 295

Appendix B FORMAL QUANTUM MECHANICS **299**

B.1 Schrödinger's equation 299
B.2 Applications of the time-independent Schrödinger equation 302
B.2.1 The rectangular potential well 302
B.2.2 Particle in a one-dimensional box 306
B.3 Momentum-eigenstates of a particle 306
B.4 Operators 308
B.5 Mean values of physical quantities 310

B.6 Angular momentum 312
B.6.1 Orbital angular momentum 312
B.6.2 Spin and spin-dependent operators 316
B.7 Motion in a spherically symmetric field 320
B.7.1 Free motion (spherical coordinates) 322
B.7.2 The spherical potential well 322
B.8 Addition of angular momenta 324
B.9 Eigenstates of an electron in a periodic potential field 332
B.10 Time-independent perturbation theory 335
B.11 Time-dependent perturbation theory 339
B.11.1 Periodic perturbation—an example 341
B.11.2 Transitions 342

Appendix C PHYSICAL CONSTANTS **345**

REFERENCES **347**

INDEX **351**

Out of confusion, as the way is,
And the wonder that man knows,
Out of the chaos would come bliss.

That, then, is loveliness, we said,
Children in wonder watching the stars,
Is the aim and the end.

***DYLAN THOMAS**, Being but men*

PREFACE

There are many books introducing the beginner to the Quantum Theory of Matter. Most of them follow the same pattern; they introduce Schrödinger's equation, apply it to simple one-dimensional problems and usually end with a treatment of the hydrogen atom and, possibly, a presentation of the Kronig–Penney model of a one-dimensional crystal. The student of physics will learn about many-electron atoms, molecules and solids from specialised texts on these topics, after she (or he) has completed a formal course of quantum mechanics and mastered the mathematical techniques necessary for working out the answers to specific problems. This, the traditional, way of getting to know the quantum world has merits but has its disadvantages as well. The difference in emphasis and style of specialised texts does not help the student to see and appreciate the unifying principles and the common thread which runs through and holds together a quantum-mechanical description of atoms, molecules and solids.

The difficulties are greater for the student of applied physics who may not have had a formal course of quantum mechanics and yet needs to know, and be familiar with, the properties of real atoms, molecules and solids. He would like to be able to read and understand the results of quantum-mechanical calculations even when his over-all knowledge, or the time at his disposal, does not permit him to follow all the mathematical steps from the statement of a problem up to its solution.

I have the conviction, born of experience, that once the student acquires a feeling for the quantum world and how it works, he finds no difficulty with the mathematical implementation of the theory, at least as far as the 'easier' problems are concerned, and then he is willing to join a guided tour of the world of atoms, molecules and solids, not insisting on 'proof' at every step along the way. The exposition of the quantum world to be had in this way need not be superfluous if it is argued properly; on the contrary it can be extremely useful. To the student of pure physics it provides an over-all view, making it easier for him to fill in the gaps in subsequent studies. To the student of applied physics it provides sufficient knowledge and the familiarity with the real world that he needs in order to make the connection with his particular field of applied research and technology.

The above defines the aim of my book: to provide the reader with an over-all view of the quantum world of atoms, molecules and solids, to teach him to 'think quantum-mechanically', and to do this using a minimum of mathematics.

The basic concepts of quantum mechanics are introduced in the first chapter of the book. The emphasis is put on the interpretation of the mathematical formulae which describe the state of a physical system, and not on their derivation. The underlying mathematical formalism, and some solutions, are described in the appendix on formal quantum mechanics (Appendix B). The same pattern is followed in the other three chapters of the book which deal, respectively, with atoms, molecules and solids. We emphasise, wherever possible, the common arguments in the treatment of the above (which results in a considerable economy of effort) and, consistently, point out the relation which exists between the properties of molecules and solids and those of the atoms of which they are made up. The last chapter of the book, on solids, is more extended than the other chapters. This is partly, no doubt, due to the fact that I know more about solids than about molecules. I have included sections on surfaces, non-crystalline solids and superconductivity to give, I hope, a modern outlook to the book.

The four chapters of the book contain no differential equations (they can be read independently of Appendix B), and to read them one requires no more than the rudiments of calculus and some familiarity with functions of more than one variable. One needs also to know something (this is summarised in Appendix A) about vectors and complex numbers. In order to read the formal part of the theory, presented in Appendix B, one needs to know, or have some familiarity with, partial differential equations. Appendix B provides the mathematical backbone of the theory presented in the four chapters of the book and the connection to the more traditional approach to the quantum theory of matter. It contains some proofs and some of the missing steps in the exposition of the theory presented in the main body of the book, but not all the missing steps; that would be beyond the scope of the present book. Some readers may find it easier to read through a chapter without reference to the mathematical appendix, and then take what they need from it to supplement the main text. Those who have already had a course in quantum mechanics may need to refer to it only occasionally; and others may choose to disregard it, relying entirely on the main text for an over-all view of the quantum world which will be incomplete in some respects, but a consistent one nevertheless.

I have included a number of problems at the end of each chapter. Some of these ask the reader to supply the 'missing steps' in the derivation of some of the formulae given in the text; others ask him (or her) to apply the given formulae to a specific physical problem; and finally there are some problems which supplement the text by pointing out applications of the theory relating to physical models and phenomena not discussed in the text. The problems marked with an asterisk are meant for those readers who have read the relevant sections of Appendix B, besides the chapter under consideration.

I hope that my book will be useful to beginners, as a complement to more traditional textbooks on quantum mechanics and its applications, and also to those who have already had a course of quantum mechanics as part of an extended course in general physics and are now embarking on courses of atomic, molecular or solid state physics.

The book may also serve as a textbook for a course, suitable for students of applied physics, on the quantum theory of atoms, molecules and solids, especially if the emphasis is to be put on the solid state.

The preparation of the manuscript in electronic form was done with care by Mrs. Sophie Letzis of COSMOSWARE.

A. Modinos

The true philosophy is written in that great book of nature which lies ever open before our eyes but which no one can read unless he has first learned to understand the language and to know the characters in which it is written. It is written in mathematical language, and the characters are triangles, circles, and other geometric figures.

GALILEO GALILEI, Saggiatore (Opere VI)

1 QUANTUM MECHANICS

1.1 Introduction

About the beginning of the twentieth century it became gradually apparent that the way we describe the motion of particles and bodies of the macroscopic world, of the stars, the ordinary wheel, the motion of a visible particle in the field of gravity etc., fails or leads to inconsistencies when applied to the motion of 'particles' of the microscopic world, the electrons and the protons and other such particles, which we can not see, but whose existence we do not doubt, because the phenomena we observe cannot be otherwise explained. We refer to the mechanics (a set of rules) which describe the motion of the macroscopic world as classical or Newtonian mechanics, and to the mechanics which describe the motion of particles of the microscopic world as quantum mechanics.[1]

Let us imagine an electron in motion around a proton, that is a hydrogen atom. We know (see Appendix C) that the electron has a negative charge ($-e$) and the proton a positive charge ($+e$). We know, also, that the mass of the proton (M_p) is much larger than the mass (m) of the electron, and this means that, to a very good approximation, the centre of mass of the atom coincides with the atomic nucleus (the proton). We are, presently, not interested in the translational motion of the atom and we therefore assume that the centre of mass of the atom remains motionless at the origin of a system of coordinates, as shown in Fig. 1.1. For the sake of completeness, let us mention that the spatial extensions of the proton and the electron are so small that, in relation to the phenomena described in this volume, we may assume that these are point-particles. We know also, and this is very important, that the electron moves around the proton at a mean distance from it: $\bar{r} \simeq 5.3 \times 10^{-9}$ cm.

The above picture of the hydrogen atom we owe to Rutherford (1910).[2] To complete this picture we must describe the motion of the electron around

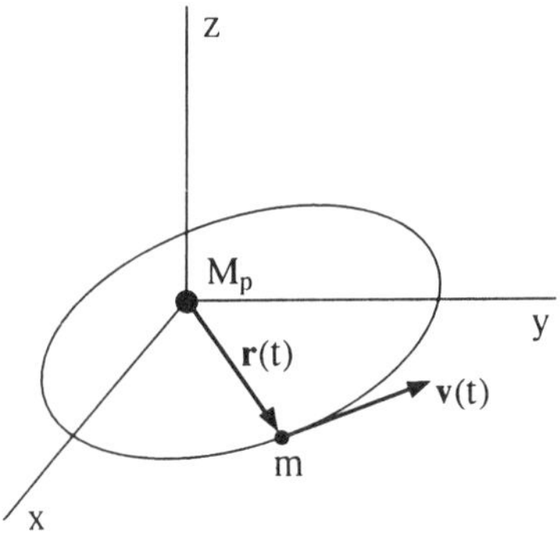

Figure 1.1. Classical orbit of an electron (m) about a proton (M_p)

the nucleus. If classical mechanics were valid in this case, we would have been able to calculate (we are not concerned with how difficult such a calculation might be as long as it is, in principle, possible) the position $\boldsymbol{r}(t) \equiv (x(t), y(t), z(t))$ and the velocity $\boldsymbol{v}(t) \equiv (\mathrm{d}x/\mathrm{d}t, \mathrm{d}y/\mathrm{d}t, \mathrm{d}z/\mathrm{d}t)$ of the electron at any time t, if by some means one could determine (measure) for us the values of these quantities at a given moment, which we may choose arbitrarily. We often call this the initial moment of time and denote it by $t = 0$ (one may think of it as the moment when our interest in the given motion began). And here we meet, already, a great difficulty: however hard we try, whatever the method and the quality of our instruments, we find that it is impossible to determine accurately the position *and* the velocity of a particle at any given moment of time. If Δx is the uncertainty[3] in our knowledge of the position coordinate x, and Δp_x the uncertainty about the momentum p_x of the particle along the same direction[4], then, at best, we have

$$\Delta x \Delta p_x = \hbar/2 \tag{1.1}$$

Identical equations hold for the y and z directions. The constant $\hbar \equiv h/2\pi \equiv (1.054\,59 \pm 0.000\,001) \times 10^{-34}$ J s, known as Planck's constant, is a universal constant of primary importance. The relation (1.1), formulated by Heisenberg, constitutes the basis and the springboard of quantum mechanics. It is known as the uncertainty principle. It forces us to abandon for good the classical concept of the orbit of a particle as shown in Fig. 1.1. Knowing the orbit of a particle means that we know the position $\boldsymbol{r}(t)$ of the particle at any and every moment of time, and this implies the simultaneous knowledge of the velocity $\boldsymbol{v}(t)$ of the particle, and this is impossible. Now the question arises: What can we know? What is there that we can calculate, which bears meaningful comparison with experimental observations?

Let us assume that the electron of a hydrogen atom executes some kind of 'periodic motion' around the proton, and let us estimate the order of magnitude of the frequency of this motion from the formula:

$$f \simeq \Delta E/h$$

where ΔE is a typical excitation energy (say, the least amount of energy that the atom can absorb, as determined experimentally), and h is Planck's constant which enters the formula with the observation that in this way f has the correct dimensions (inverse time). We have: $\Delta E \simeq 1$ eV and $h = 4.14 \times 10^{-15}$ eV s, therefore $f \simeq 10^{14}\ \mathrm{s}^{-1}$: in one second the electron goes around the proton 10^{14} times. It is evident that, even if we were in a position to determine such a quantity, it would not be of much interest to us to know the position of the electron at every moment of time. We would rather know how often the electron is to be found at a point $\boldsymbol{r} = (x, y, z)$ and how long it stays in the vicinity of this point. We can put the question in a different way, as follows: What is the probability of finding the electron, at time t, within a very small volume $\mathrm{d}V = \mathrm{d}x\,\mathrm{d}y\,\mathrm{d}z$ about the point $\boldsymbol{r}$? It turns out that this question is admissible (answerable) within the framework of quantum mechanics.

1.2 The wavefunction

Let us consider a particle. This may be free, which means that no force is acting upon it; or it may be moving under the influence of a known force; for example, in the hydrogen atom the electron moves under the influence of the electrostatic force which the proton exerts upon it. In every case the answer to the question 'What is the probability of finding the particle, at time t, within a volume $\mathrm{d}V$ about $\boldsymbol{r}$?', is encoded in a wavefunction $\psi(\boldsymbol{r}, t)$, which has the form of a complex function:

$$\psi(\boldsymbol{r}, t) = \psi_{\mathrm{R}}(\boldsymbol{r}, t) + \mathrm{i}\psi_{\mathrm{I}}(\boldsymbol{r}, t) \tag{1.2}$$

where $\mathrm{i} \equiv \sqrt{-1}$. In Appendix A we describe how one adds, subtracts, multiplies, etc., complex quantities. Here we simply note that the complex function $\psi(\boldsymbol{r}, t)$ is determined by the two (real) functions $\psi_{\mathrm{R}}(\boldsymbol{r}, t)$ and $\psi_{\mathrm{I}}(\boldsymbol{r}, t)$. We refer to ψ_{R} as the real part of ψ, and to ψ_{I} as the imaginary part of ψ[5]. Of particular importance to us is the absolute value of the complex function, which is defined as follows

$$|\psi(\boldsymbol{r}, t)| = ((\psi_{\mathrm{R}}(\boldsymbol{r}, t))^2 + (\psi_{\mathrm{I}}(\boldsymbol{r}, t))^2)^{1/2} \tag{1.3}$$

and which is always a positive number.

Before we deal with quantum-mechanical wavefunctions, it would be useful to look at similar wavefunctions which describe a well known phenomenon of classical (macroscopic physics).

In the top diagram of Fig. 1.2 we present snapshots of the first harmonic

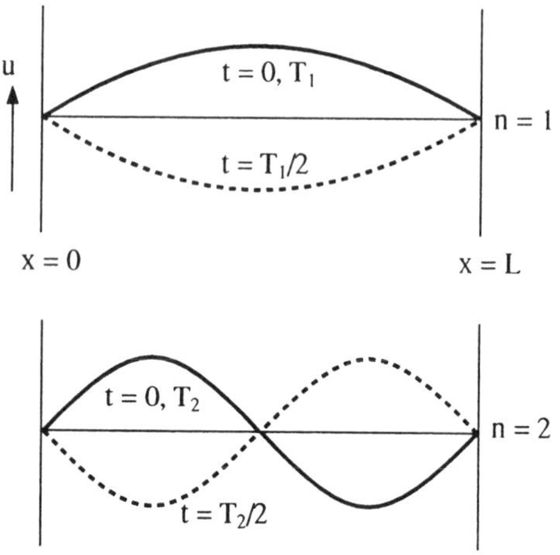

Figure 1.2. Vibrating string with its endpoints fixed at $x = 0$ and $x = L$. $n = 1$, first harmonic; $n = 2$, second harmonic

motion ($n = 1$) of a string. The displacement of the string at point x, at time t, is given by a wavefunction:

$$u_1(x, t) = A_1 \sin \frac{\pi x}{L} \cos \frac{\pi \alpha t}{L} \tag{1.4}$$

where L denotes the length of the string, and α is a constant determined by its elastic properties. The multiplicative constant A_1 gives the maximum displacement of the string (at $x = L/2$) and is determined by the initial state (stretching) of the string. The higher harmonics ($n = 2, 3, 4, \ldots$) of the string are described in similar manner by the wavefunctions:

$$u_n(x, t) = A_n \sin \frac{n\pi x}{L} \cos \frac{n\pi \alpha t}{L} \tag{1.5}$$

Snapshots of the second harmonic motion ($n = 2$) are shown in the second diagram of Fig. 1.2. We observe that the harmonics of the string correspond to a discrete set of (angular) frequencies

$$\omega_n = n\pi\alpha/L, \quad n = 1, 2, 3 \ldots \tag{1.6}$$

with corresponding periods given by

$$T_n = 2\pi/\omega_n \tag{1.7}$$

We call the frequencies (1.6) eigenfrequencies (or normal frequencies) of the string, and the corresponding states of motion (1.5) eigenstates (or normal modes) of the vibrating string. We know, and this is very important, that any motion of the string can be described as a linear sum[6] of harmonic motions. The most general displacement $u(x, t)$ of the string is described by the following sum

$$u(x, t) = A_1u_1(x, t) + A_2u_2(x, t) + A_3u_3(x, t) + \cdots = \sum_{n=1}^{\infty} A_nu_n(x, t) \tag{1.8}$$

where $u_n(x, t)$ are the eigenstates (1.5) and A_n are coefficients determined by the initial state $u(x, 0)$ of the string. The method of calculation of these coefficients does not concern us here. We simply note that, in actual situations the infinite sum of (1.8) is replaced by a finite sum. A limited number of terms is usually sufficient for the description of a real motion of the string.

We shall see, in what follows, that the mathematical language of the vibrating string is particularly useful in quantum mechanics. We conclude this section by noting that the eigenstates (1.5) of the string can also be written in the following form

$$u_n(x, t) = \text{Re}\left\{A_n \sin\frac{n\pi x}{L}\, \mathrm{e}^{\mathrm{i}\omega_n t}\right\} \tag{1.9}$$

which follows from the mathematical identity (see Appendix A)

$$\exp(\mathrm{i}\omega_n t) \equiv \cos\omega_n t + \mathrm{i}\sin\omega_n t \tag{1.10}$$

and the fact that A_n are real quantities. We refer to the expression inside the curly brackets of (1.9) as the complex amplitude of the harmonic.

1.3 Particle in a box

Let us imagine a particle, of mass m, moving inside a box (the rectangular parallelepiped of Fig. 1.3). No force is acting on the particle, except when it collides with one of the walls of the box. There it is reflected, continuing its motion with constant kinetic energy. We observe that the motion of the particle is a combination of three component motions parallel to the axes x, y, z with kinetic energies given, respectively, by $E_x = mv_x^2/2$, $E_y = mv_y^2/2$, $E_z = mv_z^2/2$. The motion parallel to one of the axes is not affected by the motions parallel to the other two axes and it can, therefore, be studied independently. This means that E_x, E_y, E_z are constant quantities: no transfer of energy from the motion along one direction to the motion along another occurs when the particle is reflected at one of the walls of the box. When the particle collides with a wall, the velocity of the particle normal to this wall changes sign; nothing else happens. Finally, in the world of classical mechanics the kinetic energy of the particle associated with any of the three directions can have any value between zero and infinity.

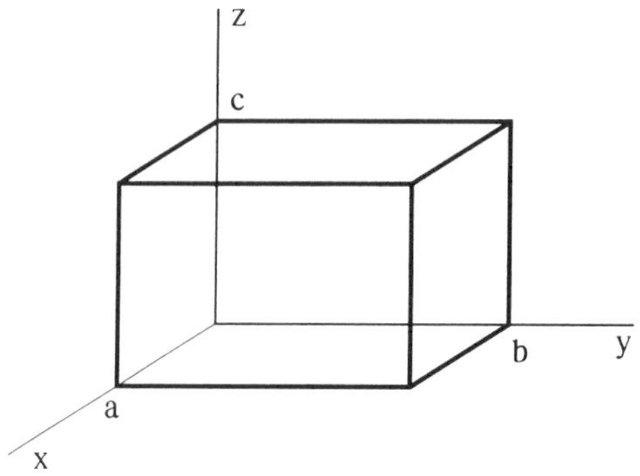

Figure 1.3. Particle in a box

1.3.1 MOTION IN ONE DIMENSION

We consider the motion parallel to the x axis. The same things happen in the other two directions. If E_x is the kinetic energy of the particle, its velocity is $+v_x$ or $-v_x$, where $v_x = (2E_x/m)^{1/2}$. It is $+v_x$ when the particle moves away from the wall at the origin of coordinates (the plane $x = 0$) towards the wall $x = a$, and it is $-v_x$ when it moves in the opposite direction. We can easily see that, according to classical mechanics, the probability of finding the particle between x and $x + \mathrm{d}x$ $(0 < x < a)$, at a moment t chosen at random, is:

$$P_{\text{cl}}(x)\,\mathrm{d}x = \frac{2\,\mathrm{d}t}{T} = \frac{\mathrm{d}x}{a} \tag{1.11}$$

where $\mathrm{d}t = \mathrm{d}x/v_x$ is the time interval spent by the particle between x and $x + \mathrm{d}x$ on its way from the wall $x = 0$ to the wall $x = a$, and in turn on its way from $x = a$ to $x = 0$, and $T = 2a/v_x$ is the period of the particle's motion, the time it takes for the particle to move from $x = 0$ to $x = a$ and back to $x = 0$. Equation (1.11) tells us that the probability of finding the particle, at a moment t chosen randomly, between x and $x + \mathrm{d}x$ does not depend on x and does not depend on the energy of the particle either. The broken line of Fig. 1.4 shows this classical result.

In the real world described by quantum mechanics, things are different. We find that the energy of the particle can have one of the following values (and no other)

$$E_{xn} = \frac{\hbar^2\pi^2 n^2}{2ma^2}, \quad n = 1, 2, 3, \ldots \tag{1.12}$$

We call the values (1.12) the energy-eigenvalues of the particle in the box; and we speak of a discrete set, or spectrum, of such eigenvalues, to

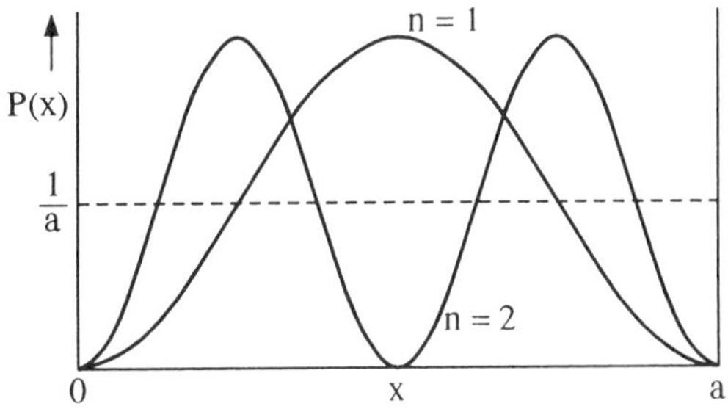

Figure 1.4. $P(x)\,\mathrm{d}x$ is the probability of finding the particle, at a moment t chosen at random, between x and $x + \mathrm{d}x$. The broken line gives the classical probability (1.11). The curves $n = 1$ and $n = 2$ are quantum-mechanical probabilities corresponding, respectively, to the ground and the first excited eigenstates of the particle

distinguish it from the corresponding continuous energy spectrum of the classical particle (the latter, as we have already mentioned, can have any energy from zero to infinity). Each eigenvalue E_{xn} corresponds to an energy-eigenstate of the particle, described by the eigenfunction

$$u_n(x, t) = u_n(x) \exp(-\mathrm{i}E_{xn}t/\hbar) \tag{1.13a}$$

$$u_n(x) = A_n \sin(n\pi x/a) \tag{1.13b}$$

where $A_n = (2/a)^{1/2}$. The wavefunction (1.13) contains all the information, that is to be had, within the framework of quantum mechanics, about the particle in the box when its energy equals E_{xn}. We see, at this stage, the similarity with the classical description of the vibrating string. A harmonic motion of the string is possible only at certain frequencies: the eigenfrequencies (1.6) of the string. The quantum-mechanical analogue of the harmonic motion of the string is a motion of the particle in the box with definite (without any uncertainty) energy, and this is possible only for certain values of the energy of the particle: the eigenvalues (1.12). The discrete spectrum of the eigenfrequencies of the string goes over to the discrete spectrum of the energy-eigenvalues of the particle in the box. The corresponding eigenfunctions (1.13) describe (indirectly) the motion of the particle, like the wavefunctions (1.5) describe the motion of the string. The analogy between the two is further established, when we discover that the most general state (any motion) of the particle in the box is described by a wavefunction of the form

$$\psi(x, t) = \sum_{n=1}^{\infty} c_n u_n(x, t) \tag{1.14}$$

where c_n are coefficients determined by the initial (at $t = 0$) state of the particle[7]. Equation (1.14) implies that, in the state described by it, the particle does not have a precise energy; its energy may be any of the allowed values (1.12) with certain probability (see Appendix B.1). We shall not be concerned, at this stage, with (1.14); we simply note the analogy between it and (1.8) which describes an arbitrary state of the vibrating string.

We noted that, we have encoded in a wavefunction all that is needed for a complete description of the state of a particle. At this stage we shall concentrate on one important property: if $\psi(\boldsymbol{r}, t)$ is the wavefunction which describes the state of a particle, then the probability of finding the particle, at time t, within a volume $\mathrm{d}V = \mathrm{d}x\,\mathrm{d}y\,\mathrm{d}z$ about the point $\boldsymbol{r}$ is given by

$$P(\boldsymbol{r}, t)\,\mathrm{d}V = |\psi(\boldsymbol{r}, t)|^2\,\mathrm{d}V \tag{1.15}$$

where $|\psi(\boldsymbol{r}, t)|$ is the absolute value of ψ given by (1.3). $P(\boldsymbol{r}, t)$ is known as the probability-density. We note (see also Appendix B.1) that, when the state of the particle is an energy-eigenstate, as in (1.13), ψ has the form:

$$\psi(\boldsymbol{r}, t) = \psi(\boldsymbol{r})\exp(-\mathrm{i}Et/\hbar) \tag{1.16}$$

and, therefore,

$$P = |\psi(\boldsymbol{r}, t)|^2 = |\psi(\boldsymbol{r})|^2 \tag{1.17}$$

The probability of finding the particle, at time t, within volume dV about $\boldsymbol{r}$ is in this case independent of the time. We note, by the way, that the same is true for all physical quantities (the momentum of the particle, etc.): When the state of the particle is an eigenstate of the energy, a physical quantity A may have, with certain probability, a series of values: $A_1, A_2, \ldots A_i, \ldots$; however, the probability $P(A_i)$ of having the value A_i does not change with the time. For this reason the eigenstates of the energy of a particle are known as the stationary states of the particle. The reader will easily convince himself that, when the state of the particle is described by a linear sum of eigenstates of the energy, as in equation (1.14), the probability density function (1.15) depends on the time; and so do other physical quantities.

Let us, however, return to the consideration of a particle in a box, and let us assume that the particle is found in one of the eigenstates (1.13). Then the probability of finding the particle between x and $x + \mathrm{d}x$ is, at any time, given by

$$P_n(x)\,\mathrm{d}x = \left(\frac{2}{a}\right)\sin^2\left(\frac{n\pi x}{a}\right)\mathrm{d}x \tag{1.18}$$

which we obtain by substituting (1.13) into (1.15), noting that in the present case $\boldsymbol{r} \to x$ and $\mathrm{d}V \to \mathrm{d}x$. In Fig. 1.4 we compare $P_1(x)$ and $P_2(x)$, corresponding to the $u_1(x, t)$ and $u_2(x, t)$ energy-eigenstates of the particle, with the corresponding classical quantity. We see that, unlike the classical probability-density, the quantum-mechanical probability-density is a function of the position and also, again unlike the classical probability, the quantum one depends on the state (1.13) of the particle. It appears that the quantum-mechanical probability-density functions of Fig. 1.4 have more in common with the wavefunctions of Fig. 1.2 which describe the displacement of a vibrating string! It is worth noting that the wavefunction (1.13) has exactly the same mathematical form as the expression in the curly brackets of (1.9) which gives the complex amplitude of the vibrating string. This is not accidental; it derives from the dual nature of a 'particle' of the quantum-mechanical world. The particle aspect of its nature allows us to speak about the probability of finding the particle (and not part of the particle) at a point $\boldsymbol{r}$ in space. Its wave nature is reflected in the fact that the complex function $\psi(\boldsymbol{r}, t)$, which carries all the information that is to be had about it within the quantum-mechanical framework, behaves like a wave. The time development of $\psi(\boldsymbol{r}, t)$ is determined by a differential equation, the equation of Schrödinger, which is similar in many respects to those equations of classical physics which describe the time development of elastic waves, electromagnetic waves, etc. In the main part of this volume we shall not be concerned with the mathematical foundation of quantum mechanics or with the mathematical analysis of the equation of Schrödinger, which leads to the determination of the energy-eigenstates and other properties of the physical systems we shall consider. We shall be content with a description of these properties and an elaboration of their physical significance. The interested reader will find a discussion of Schrödinger's equation and a brief introduction to some of the basic formulae of quantum mechanics in Appendix B.

Let us now return to our discussion of (1.13) and (1.18) which describe a particle in a box. We observe that, unlike the displacement of the string, which is given by the real part of the complex amplitude of the vibration (see (1.9)) and may therefore be positive or negative, as shown in Fig. 1.2, the probability-density function $P_n(x)$ is given by the square of the absolute value of the wavefunction and is therefore always positive, as shown in Fig. 1.4. And this is how it should be, because otherwise we could not have interpreted $P_n(x)$ as a probability density. And by an extension of the same argument, we demand that

$$\int_0^a P_n(x)\,\mathrm{d}x = 1 \tag{1.19}$$

The above integral gives the probability of finding the particle, at time t, anywhere between $x = 0$ and $x = a$. We put this probability, by convention, equal to unity. We could have put it equal to 100. We would be saying, then, that the probability of finding the particle, at time t, between x and $x + \mathrm{d}x$ is 'so much per cent'. It is obvious that the normalisation of the wavefunction according to (1.19) is responsible for the absolute value of the (normalisation) constant A_n appearing in (1.13b). It is also clear that the sign, or more generally the argument of A_n (assuming it could be a complex number), is not determined by (1.19). It turns out that replacing A_n by $|A_n|\,\mathrm{e}^{\mathrm{i}\theta}$, where θ is any real number, does not essentially change the wavefunction. The probability-density function, and the values of all other physical (measurable) quantities evaluated from the wavefunction according to the rules of quantum mechanics do not depend on the argument θ of the normalisation constant (see also Appendix B.5). Finally, we note that the same normalisation condition is satisfied by the wavefunction (1.14) which describes the most general state of the particle. In this case $P(x, t) = |\psi(x, t)|^2$ is a function of the time, but at any time ($t > 0$) we have

$$\int_0^a P(x, t)\,\mathrm{d}x = 1 \tag{1.20}$$

provided (1.20) is satisfied at $t = 0$. The above equation expresses the obvious fact that, at any time, the particle is somewhere between $x = 0$ and $x = a$.

We have already observed the discrete spectrum (1.12) of the allowed energy levels (eigenvalues) of the particle in the box. It is worth noting that zero is not one of these levels. The lowest eigenvalue of the energy is

$$E_{x1} = \frac{\hbar^2\pi^2}{2ma^2} \tag{1.21}$$

We call it the ground state energy, and the corresponding eigenstate $u_1(x, t)$ is referred to as the ground state of the particle (in the given box). Obviously, the classical picture of a motionless particle sitting in the box does not exist in the quantum world. Certainly not, when the box is of atomic dimensions. For example, putting m = mass of the electron, and $a = 2$ Å, we find $E_{x1} \simeq 10$ eV. The above observation can be understood in terms of the uncertainty principle (1.1) which, as we have seen, lies at the heart of quantum mechanics.

In the case under consideration, it is evident that the uncertainty Δx as to the position of the particle cannot be greater than the length a of the box and, therefore, using (1.1) we write

$$\Delta p_x \geqslant \frac{\hbar}{2\Delta x} > \frac{\hbar}{2a} \tag{1.22}$$

It is also evident that p_x must be of the same magnitude, if not larger, as Δp_x. We put

$$p_x \simeq \hbar/(2a) \tag{1.23}$$

which means that

$$E_{x1} = \frac{mv_x^2}{2} = \frac{p_x^2}{2m} > \frac{\hbar^2}{8ma^2} \tag{1.24}$$

in accordance with (1.21).

In general, we find that: if a particle is confined in space (the way an electron is in an atom or a molecule) then it cannot be motionless. It conducts itself in a way which is qualitatively the same with that of a particle confined in a box.

The formulae we have given for the energy eigenvalues (1.12) and for the corresponding eigenstates (1.13) are, of course, valid whatever the length a of the box, and naturally they remain valid for a large (macroscopic) box. For example, the ground state energy of an electron in a box of length $a = 1$ mm is $E_{x1} \sim 10^{-13}$ eV. And when the particle is free to move in all space ($a \to \infty$), its ground state energy tends to zero. We note also that the difference between two successive energy levels

$$E_{xn} - E_{x,n-1} = \frac{\pi^2\hbar^2}{2ma^2}(2n - 1) \tag{1.25}$$

tends to zero when $a \to \infty$. It appears that the discreteness of the energy levels is a result of the particle's confinement in space (an observation which is valid generally as we shall see in section 1.4.2). In the macroscopic box the spectrum of the energy eigenvalues becomes almost continuous or, in other words, the particle can have, practically, any energy. Finally, we note that in the case of a macroscopic box, when the spectrum of the allowed energies is practically continuous, even small energies of the particle correspond to large values of the quantum number n. In Fig. 1.5 we compare the probability-density function $P_n(x)$, when $n = 10$, with the corresponding classical quantity $P_{cl}(x)$ given by (1.11). We see that when $n \gg 1$ the classical quantity is a good approximation to the real (quantum)

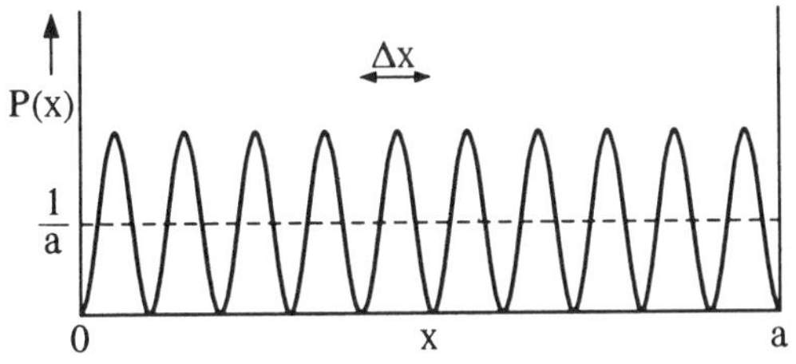

Figure 1.5. For large values of the quantum number ($n \gg 1$) the classical probability-density $P_{cl}(x)$(- - -) is a good approximation to the quantum-mechanical probability-density $P_n(x)$(—)

probability-density in the following sense: the interval Δx over which we have to average the quantum-mechanical probability-density to obtain the classical one becomes smaller as n increases. This suggests that the motion of the particle in the macroscopic box can be described, to a good approximation, by classical mechanics.

1.3.2 MOTION IN THREE DIMENSIONS

We know (see section 1.3) that the component-motions of a particle in the box of Fig. 1.3 parallel to the x, y and z directions go on independently of each other. It is also obvious that what we have said about the x-motion (parallel to the x-axis) applies to the y and z motions as well. In relation to the x-motion we have found that: the eigenvalues of the energy are given by (1.12) and the corresponding eigenstates by (1.13). The corresponding quantities for the y and z motions are obtained in the same fashion. We have

$$x:\ E_{xn} = \frac{\pi^2\hbar^2}{2m}\frac{n^2}{a^2} \qquad u_n(x, t) = \sqrt{\frac{2}{a}}\sin\frac{n\pi x}{a}\exp(-\mathrm{i}E_{xn}t/\hbar) \tag{1.26a}$$

$n = 1, 2, 3, \ldots$

$$y:\ E_{yp} = \frac{\pi^2\hbar^2}{2m}\frac{p^2}{b^2} \qquad u_p(y, t) = \sqrt{\frac{2}{b}}\sin\frac{p\pi y}{b}\exp(-\mathrm{i}E_{yp}t/\hbar) \tag{1.26b}$$

$p = 1, 2, 3, \ldots$

$$z:\ E_{zl} = \frac{\pi^2\hbar^2}{2m}\frac{l^2}{c^2} \qquad u_l(z, t) = \sqrt{\frac{2}{c}}\sin\frac{l\pi z}{c}\exp(-\mathrm{i}E_{zl}t/\hbar) \tag{1.26c}$$

$l = 1, 2, 3, \ldots$

where b and c are the dimensions of the box in the y and z directions respectively (Fig. 1.3). It is obvious that the eigenvalues of the (total) energy of the particle are

$$E_{npl} = E_{xn} + E_{yp} + E_{zl} = \frac{\pi^2\hbar^2}{2m}\left(\frac{n^2}{a^2} + \frac{p^2}{b^2} + \frac{l^2}{c^2}\right) \tag{1.27}$$

where n, p, l are any three positive integers (not necessarily different from each other). Equation (1.27) follows from the fact that the motion in one direction is independent from the motions parallel to the other two directions. We write the eigenfunctions which describe the corresponding energy-eigenstates of the particle as follows:

$$u_{npl}(\mathbf{r}, t) = u_{npl}(\mathbf{r})\exp(-\mathrm{i}E_{npl}t/\hbar) \tag{1.28a}$$

$$u_{npl}(\boldsymbol{r}) = \sqrt{\frac{8}{abc}} \sin\frac{n\pi x}{a} \sin\frac{p\pi y}{b} \sin\frac{l\pi z}{c} \tag{1.28b}$$

which we can justify by the following argument. Assuming that (1.28) is valid, the probability of finding the particle, at time t, within a volume $\mathrm{d}V = \mathrm{d}x\,\mathrm{d}y\,\mathrm{d}z$ about the point x, y, z is, according to (1.15), given by

$$P_{npl}(x, y, z)\,\mathrm{d}V = |u_{npl}(x, y, z)|^2\,\mathrm{d}V = (P_n(x)\,\mathrm{d}x)(P_p(y)\,\mathrm{d}y)(P_l(z)\,\mathrm{d}z) \tag{1.29}$$

$P_n(x)$ is obtained from (1.18), and $P_p(y)$ and $P_l(z)$ from the corresponding equations for the y and z directions. $P_n(x)\,\mathrm{d}x$ gives the probability of finding the particle between x and $x + \mathrm{d}x$ and anywhere as far as y and z are concerned, $P_p(y)\,\mathrm{d}y$ gives the probability of finding the particle between y and $y + \mathrm{d}y$ and anywhere as far as x and z are concerned, and $P_l(z)\,\mathrm{d}z$ gives the probability of finding the particle between z and $z + \mathrm{d}z$ and anywhere as far as the x and y are concerned. Because these probabilities are not correlated, the probability of finding the particle in the volume $\mathrm{d}x\,\mathrm{d}y\,\mathrm{d}z$ is, of course, given by their product, i.e. by (1.29). There is no form of wavefunction, other than (1.28), which leads to this naturally expected result. We are, therefore, convinced that (1.28) are the correct eigenfunctions. We note that (1.29) satisfies the condition

$$\int_V P_{npl}(\boldsymbol{r})\,\mathrm{d}V = 1 \tag{1.30}$$

where V is the volume of the box. Equation (1.30) is the three-dimensional version of (1.19); it tells us that the particle is to be found, at any time, somewhere in the box.

Finally we note that any (the most general) state of the particle in the box is described by a wavefunction of the form (a three-dimensional version of (1.14)):

$$\sum_{n,p,l} c_{npl} u_{npl}(\boldsymbol{r}, t) \tag{1.31}$$

where c_{npl} are coefficients to be determined by the state of the particle at $t = 0$.

1.3.3 DEGENERACY OF ENERGY-EIGENVALUES—THE ROLE OF SYMMETRY

We have emphasised the fact that the state of a physical system (all that can be known about it) is encoded in a wavefunction which describes the given state. Of particular interest to us are the stationary states of the system which are, as we have seen, the energy-eigenstates of the system. It is important to understand that an energy-eigenstate (or what amounts to the same thing, the eigenfunction which describes it) is not always uniquely determined by the corresponding eigenvalue of the energy. Often to a given energy-eigenvalue correspond two, three or more eigenstates of the physical system.[8] When this is the case, we say that the energy-eigenvalue is doubly, triply, and so on, degenerate. We shall demonstrate degeneracy by reference to the energy levels of a particle in a box whose dimension b along the y-axis (see Fig. 1.3) equals its dimension a along the x-axis. We assume that $a = b \neq c$. In this case we put $b = a$ in the equations (1.27) and (1.28) which determine the energy-eigenvalues and the corresponding eigenstates of the particle. We observe that the eigenstates $u_{npl}(\boldsymbol{r}, t)$ and $u_{pnl}(\boldsymbol{r}, t)$, with $n \neq p$, correspond to the same energy level

$$E_{npl} = E_{pnl} = \frac{\hbar^2\pi^2}{2m}\left(\frac{p^2 + n^2}{a^2} + \frac{l^2}{c^2}\right)$$

We convince ourselves that the eigenstate $|n, p, l\rangle$ ($|n, p, l\rangle$ stands for $u_{npl}(\boldsymbol{r}, t)$) is indeed different from the eigenstate $|p, n, l\rangle$ when $n \neq p$, even though the (total) energy of the particle is the same in the two states. We have (see (1.28)):

$$\begin{aligned} u_{npl}(\boldsymbol{r}) &\equiv \sqrt{\frac{8}{a^2c}}\sin\frac{n\pi x}{a}\sin\frac{p\pi y}{a}\sin\frac{l\pi z}{c} \\ &\neq u_{pnl} \equiv \sqrt{\frac{8}{a^2c}}\sin\frac{p\pi x}{a}\sin\frac{n\pi y}{a}\sin\frac{l\pi z}{c} \end{aligned}$$

when $n \neq p$; and therefore the corresponding probability densities $P_{pnl}(x, y, z)$ and $P_{npl}(x, y, z)$, obtained from (1.29), are different. It is also easy to see (from (1.26)) that if $n > p$, in the eigenstate $|n, p, l\rangle$ the particle 'moves faster' parallel to the x-axis than parallel to the y-axis, while the opposite is true when the particle is in the state $|p, n, l\rangle$.

We should note, however, that in the given box ($a = b \neq c$) not all energy levels are degenerate. It is obvious, for example, that to the energy level E_{lln}, where l and n are any two positive integers, corresponds only one eigenstate described by

$$u_{lln}(\boldsymbol{r}, t) = \sqrt{\frac{8}{a^2c}}\sin\frac{l\pi x}{a}\sin\frac{l\pi y}{a}\sin\frac{n\pi z}{c}\exp\left(-\mathrm{i}E_{lln}t/\hbar\right)$$

Let us now see what happens in a more symmetric box, i.e. when $a = b = c$. One can easily show that in this case the energy levels E_{lll} are not degenerate; on the other hand, the eigenstates u_{nll}, u_{lnl}, u_{lln}, which are given by (1.28) with $a = b = c$ and are of course different from each other when $l \neq n$, correspond to the same energy level $E_{nll} = E_{lnl} = E_{lln}$ which acquires, therefore, a three-fold degeneracy; finally we have energy levels with a six-fold degeneracy: the eigenstates u_{npl}, u_{nlp}, u_{pnl}, u_{pln}, u_{lnp}, u_{lpn} which are different from each other when $n \neq p \neq l \neq n$ correspond to the same energy level $E_{npl} = E_{nlp} = E_{pnl} = E_{pln} = E_{lnp} = E_{lpn}$, which acquires in this way a six-fold degeneracy.

It is known that the degeneracy of the energy levels of a particle moving in a potential field[9] is intimately connected with the symmetry of the potential field, and that it increases when the potential field becomes more symmetrical in space. We have just seen that in the more symmetric box ($a = b = c$) the maximum degeneracy (six-fold) of the energy levels is higher than the maximum degeneracy (two-fold) in the less symmetric box ($a = b \neq c$), and this is in turn higher than that of the asymmetric box ($a \neq b \neq c$) where none of the energy levels is degenerate. We should remember, however, that not every level is degenerate in the more symmetric field; a fact well demonstrated by the examples we have considered. It follows from the above discussion that in a spherical box, and by the same token in a potential field of spherical symmetry (such as the one seen by the electron in the hydrogen atom), there will be energy levels of the particle which are not degenerate and energy levels of multiple degeneracy which could be much larger than the six-fold degeneracy observed in the cubic box. We shall see in the next chapter that this is indeed the case.

We conclude this section with a brief discussion of symmetry-breaking and of the effect that this has on the energy spectrum of a physical system. One deals, to begin with, with a system of high symmetry, which means that some of the energy levels of the system are degenerate. Then, one introduces a *weak* perturbation to the potential energy of the system which reduces the initial symmetry and leads, therefore, to lesser degeneracy: a degenerate unperturbed level is split into a number of energy levels of lesser degeneracy. Provided the perturbation is weak, the perturbed levels remain in the vicinity of the unperturbed one and the total number of states corresponding to them equals n, if the unperturbed level was n-fold degenerate. We shall demonstrate this effect using our model of a particle in a box as an example. We begin with a cubic box ($a = b = c$). We know that certain of the energy levels of the particle in this box are six-fold degenerate: $E_{npl} = E_{nlp} = E_{pnl} = E_{pln} = E_{lnp} = E_{lpn}$, when $n \neq p \neq l \neq n$. We now perturb the system: we increase the dimension of the box parallel to the z-axis (see Fig. 1.3) by a *small* amount. The dimensions of the perturbed box are $a = b < c = a + \delta a$, $\delta a \ll a$. We can calculate the energy levels of the particle in the perturbed box from (1.27) with the same ease as for the unperturbed (cubic) box. We find that the above mentioned six-fold degenerate levels split into three two-fold degenerate levels. This is demonstrated schematically in Fig. 1.6 for the case of $(n, p, l) = (1, 2, 3)$. The unperturbed level (on the left of the diagram) is

$$E_{123} = \frac{\hbar^2\pi^2}{2m^2}\left(\frac{1^2 + 2^2 + 3^2}{a^2}\right)$$

and to it correspond six (unperturbed) states:

$$|n, p, l\rangle = |1, 2, 3\rangle, |2, 1, 3\rangle, |1, 3, 2\rangle, |3, 1, 2\rangle, |3, 2, 1\rangle, |2, 3, 1\rangle$$

given by (1.28) with $c = b = a$. The perturbed energy levels (on the right of the diagram) are:

$$E'_{123} = \frac{\hbar^2\pi^2}{2m}\left(\frac{1^2 + 2^2}{a^2} + \frac{3^2}{(a + \delta a)^2}\right), \quad E'_{132} = \frac{\hbar^2\pi^2}{2m}\left(\frac{1^2 + 3^2}{a^2} + \frac{2^2}{(a + \delta a)^2}\right),$$
$$E'_{321} = \frac{\hbar^2\pi^2}{2m}\left(\frac{3^2 + 2^2}{a^2} + \frac{1^2}{(a + \delta a)^2}\right)$$

To the lowest (E'_{123}) of these levels correspond the two states $|n, p, l\rangle' = |1, 2, 3\rangle', |2, 1, 3\rangle'$, to the middle one ($E'_{132}$) correspond the two states $|1, 3, 2\rangle', |3, 1, 2\rangle'$ and to the top level (E'_{321}) the two states $|3, 2, 1\rangle'$, $|2, 3, 1\rangle'$. The states $|n, p, l\rangle'$ of the electron in the perturbed system are obtained from (1.28) with $b = a$ and $c = a + \delta a$.

In the above example we could calculate the energy levels and corresponding eigenstates of the perturbed system as easily as those of the unperturbed system. This is not always possible. It is often the case that one can calculate the energy levels and eigenstates of the unperturbed system with relative ease, but not so those of the perturbed system. However, when the perturbation is weak one is able to obtain approximate expressions for the perturbed quantities in terms of the known (unperturbed quantities). In every case when the symmetry of the potential field is reduced by the

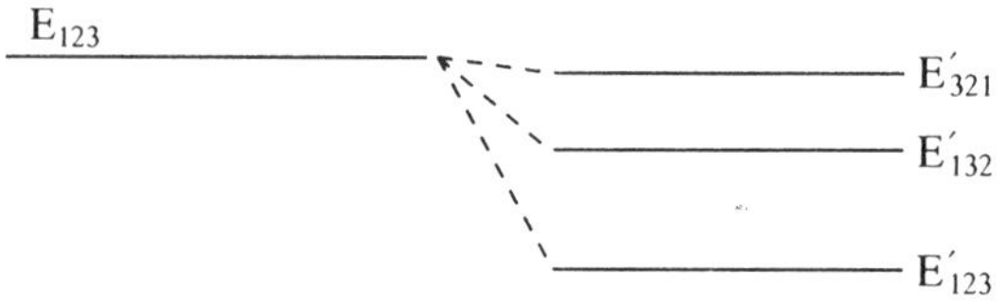

Figure 1.6. The six-fold degenerate level $(E)_{123}$ is split in the less symmetric box into three two-fold degenerate levels: $(E)'_{321}$, $(E)'_{132}$, $(E)'_{123}$

perturbation, a reduction in the degeneracy of the energy levels is to be expected, like in the above example (see Appendix B.10).

1.3.4 DENSITY OF STATES

We know that the energy levels of a particle in a box of volume V tend to a nearly continuous spectrum (see Fig. 1.7) as the volume of the box increases. This follows from (1.25) and its obvious three-dimensional analogue. A one-by-one listing of all the eigenstates corresponding to a nearly continuous spectrum of energy levels becomes tedious without being particularly useful. It turns out that for many applications of the particle-in-a-box model we need only know how many eigenstates of the particle there are with energy between E and $E + \mathrm{d}E$. We find that in a (cubic) box of volume V, when V is large enough, this is given by

$$V\rho(E)\,\mathrm{d}E$$

$$\rho(E) = \frac{4\pi}{h^3}(2m^3)^{1/2}E^{1/2} \tag{1.32}$$

By definition, $\rho(E)$, the density of states of the particle in the box, gives the number of eigenstates of the particle per unit energy and per unit volume of the box. We shall see (in Chapter 4) that the density of states of an electron in a crystal ($V \gtrsim 10^9$ Å^3) is a very useful quantity in the understanding of the properties of the crystal. It is of course different in different crystals and very rarely has the simple form of (1.32) (shown schematically in Fig. 1.7). However, in many instances (1.32), or simple variations of it, describe reasonably well the density of states of an electron in a metal or semiconductor over limited regions of energy of critical importance and, in this respect, the above equation turns out to be very useful.

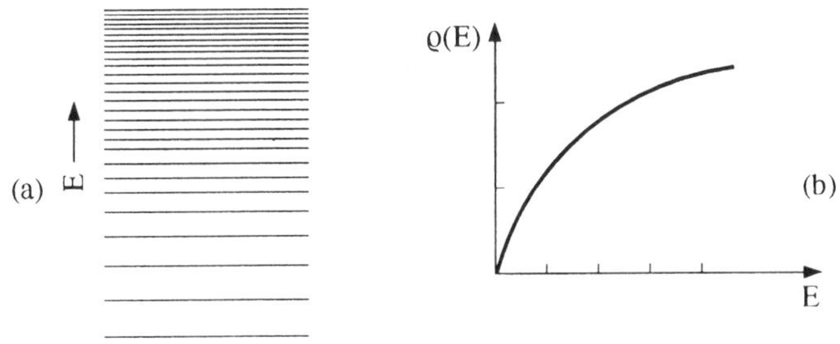

Figure 1.7. (*a*): Nearly continuous spectrum of the energy levels of a particle in a box of relatively large dimensions; (*b*): the density of states (1.32) of the particle

1.4 Potential fields

1.4.1 CLASSICAL PICTURE OF A PARTICLE IN A POTENTIAL FIELD

We say that a particle is found or moves in a potential field if the force acting on the particle is completely determined by the position $\boldsymbol{r} = (x, y, z)$ of the particle.

The potential energy $U(x, y, z)$ of a particle in a given potential field describes this field as effectively as the functions $F_x(x, y, z)$, $F_y(x, y, z)$, $F_z(x, y, z)$ which give the x, y and z-components of the force experienced by the particle when it is at x, y, z. From the potential energy function we obtain the force acting on the particle at the point x, y, z using the following relations

$$U(x, y, z) - U(x + \mathrm{d}x, y, z) = F_x(x, y, z)\,\mathrm{d}x \tag{1.33a}$$

$$U(x, y, z) - U(x, y + \mathrm{d}y, z) = F_y(x, y, z)\,\mathrm{d}y \tag{1.33b}$$

$$U(x, y, z) - U(x, y, z + \mathrm{d}z) = F_z(x, y, z)\,\mathrm{d}z \tag{1.33c}$$

where $\mathrm{d}x$, $\mathrm{d}y$, $\mathrm{d}z$ are small (infinitessimal) displacements parallel to the axes x, y, z respectively.

The potential energy as a physical quantity acquires special significance from the following fact. Within the framework of classical mechanics, one can show that when the force acting on a particle is given by (1.33), the total energy of the particle, which by definition is the sum of its kinetic energy $T(x, y, z)$ and its potential energy $U(x, y, z)$, is a constant quantity E independent of time and the position of the particle. We have

$$U(x, y, z) + T(x, y, z) = E \tag{1.34}$$

$$T(x, y, z) \equiv \frac{m}{2}(v_x^2 + v_y^2 + v_z^2) = \frac{m}{2}v^2 \tag{1.35}$$

where $\boldsymbol{v} = (v_x, v_y, v_z)$ is the velocity of the particle at the position (x, y, z). We make the following observations. First observation: Adding a constant to the potential energy does not change the force experienced by the particle, it only changes the total energy (1.34) by the said constant; which implies no more than a different choice of the zero of energy which is, in any case, chosen by some convention or other. Second observation: By definition, the kinetic energy of the particle is a positive quantity and, therefore, from (1.34) we obtain

$$E - U(x, y, z) \geqslant 0 \tag{1.36}$$

which tells us that the particle cannot be there (a region of space) where its potential energy exceeds its total energy. The region of space defined by the relation $U(x, y, z) > E$ is, according to classical mechanics, rigorously forbidden to the particle. We shall see in what follows that this rule of classical mechanics does not hold in the world of quantum mechanics. Equation (1.34) has no meaning within the framework of quantum mechanics, because the phrase 'the kinetic energy of the particle is so much at the point x, y, z' has no meaning in quantum mechanics. It implies that we can know the velocity and simultaneously the position of the particle, and this (the uncertainty principle tells us) is not possible. We shall establish the quantum-mechanical version of (1.34) in Section 1.6. But let us stay for the moment with classical mechanics, and let us follow the motion of a (classical) particle in the one-dimensional potential field of Fig. 1.8. We have

$$\begin{aligned} U(x) &= 0, \quad x < -a' \\ &< 0, \quad -a' < x < a \\ &= 0, \quad x > a \end{aligned} \tag{1.37}$$

The potential energy of the particle is negative in the region $-a' < x < +a$ and zero outside this region. Therefore, for negative values of the (total) energy E of the particle, its motion takes place in the region $-x'_E \leqslant x \leqslant x_E$, where $E - U(x) > 0$. The so-called classical turning points $+x_E$ and $-x'_E$ are defined by the relation $U(-x'_E) = U(x_E) = E$ (see Fig. 1.8). Let us follow the motion of a particle of negative energy $(-U_O < E < 0)$ assuming that initially (at $t = 0$) it is found at the turning point x_E. The kinetic energy of the particle, initially zero, increases as the particle moves from right to left, acquires its maximum value at the point $(x = 0)$ of minimum potential energy, and then decreases as the particle advances towards the left turning point $-x'_E$. There the kinetic energy of the

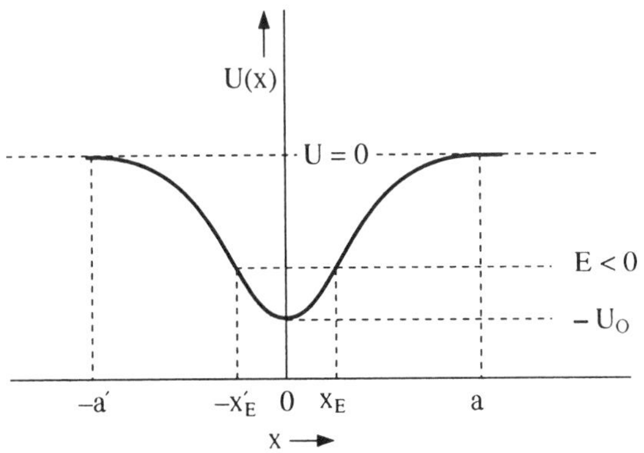

Figure 1.8. Potential field of the form (1.37)

particle goes to zero, momentarily, and the particle begins to move in the opposite direction, its kinetic energy increasing to become maximum once more at $x = 0$, then decreases again to become zero when it reaches the right turning point x_E at the end of the first period of its motion which is also the beginning of its second period of motion. Because the energy is conserved, one period will follow another in perpetuity.

If at some moment of time we cause (by some external means) an increase in the energy of the particle from its initial (negative) value E to a positive value E', the particle will move up to the top of the potential well and then away from it towards $x = +\infty$ or $x = -\infty$ with a speed $v = (2E'/m)^{1/2}$, according to (1.34).

We note that the phenomenon we have just described, which is qualitatively the same with, say, the ionisation of the hydrogen atom (detachment of the electron from the proton following irradiation), is not possible in the system of the particle in a box (Section 1.3); whatever the energy of the particle, this is reflected by the walls of the box and stays in it. We can say the same thing in a different way: the particle in the box sees (in the one dimension) the following potential field

$$\begin{aligned} U(x) &= 0, \quad 0 < x < a \\ &= \infty, \quad x < 0 \text{ or } x > a \end{aligned} \tag{1.38}$$

It is obvious that the region $x < 0$ and the region $x > a$ are forbidden to the particle whatever its energy.

1.4.2 QUANTUM-MECHANICAL DESCRIPTION OF A PARTICLE IN A POTENTIAL FIELD

We consider a particle of mass m in the potential field of Fig. 1.8. We assume that the width of the potential well has an atomic dimension $a + a' \sim 1$ Å, and a certain depth: $U(x = 0) \equiv -U_O < 0$. The quantum-mechanical description of the above system reduces to the following (see also Appendix B.1): we need to know the allowed energy levels (energy eigenvalues) of the particle in the given field and the corresponding eigenstates. We find a limited number of energy levels (E_1, E_2, . . ., E_N) between the bottom of the well and the zero of the energy (constant potential energy outside the well):

$$-U_O < E_1 < E_2 < \cdots < E_N < 0$$

The wavefunctions which describe the corresponding eigenstates decay exponentially to the left of $-x'_{E_n}$ and to the right of x_{E_n}, where $-x'_{E_n}$ and x_{E_n} are, respectively, the left and right classical turning points determined by $U(x'_{E_n}) = U(-x'_{E_n}) = E_n$ (see Fig. 1.8). These are the so-called bound states of the particle: the particle remains in the region of the well; the probability of finding the particle to the left of the left turning point or to the right of the right turning point is not zero as in classical mechanics (we shall return to this remarkable phenomenon in the next section) but, as we have already noted, goes to zero exponentially away from these turning points.

We note that the bound eigenstates of a particle correspond, always, to a discrete spectrum of energy eigenvalues. This is true whether the motion is one-dimensional (as in our example of Fig. 1.8), two-dimensional or three-dimensional (see also Fig. 1.9). This is a fundamental result of quantum mechanics and does not depend on particular features of the potential field. Of course, the number of bound eigenstates of a particle in a given potential field (we may have none, a few, or infinitely many such states), the values of the energy levels and their degeneracy, and the shape of the corresponding eigenfunctions, do depend on particular features of the potential field.

A remarkable result of quantum mechanics tells us that in an one-dimensional potential well there is, always, at least one bound state of the particle. For example, the potential well of Fig. 1.8 will hold a bound state, even if it is very shallow (as long as $-U_O < 0$) and of very limited range (as long as $a + a' > 0$). One finds, also, that in one dimension the energy levels of the discrete spectrum are never degenerate. To each one of these levels corresponds a single bound state. The situation is very different in three dimensions: there are no bound states in a very shallow three-dimensional

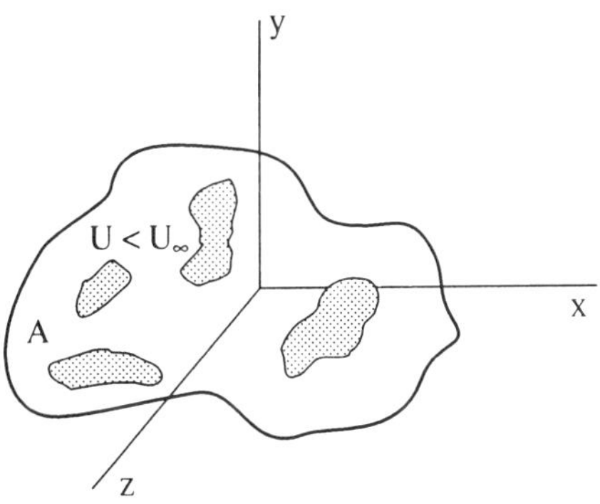

Figure 1.9. The existence of bound states of a particle implies: (i) outside a limited region A, or asymptotically, as $r \to \infty$, the potential energy $U(r)$ of the particle acquires a constant value U_∞. (ii) There is a region of space (e.g., the non-shaded area of A) where $U(r) < U_\infty$. The bound states of the particle (the corresponding eigenfunctions decay exponentially as $r \to \infty$) correspond to a discrete set (spectrum) of energy levels which extends below U_∞. Usually, when U_∞ is a finite quantity, we set $U_\infty = 0$, in which case the energy levels of the above mentioned discrete spectrum are negative quantities

potential well; and we know already that in a three-dimensional system some of the energy levels may be degenerate.

1.5 Particle in a rectangular potential well

A one-dimensional potential well which appears in most books of quantum mechanics, because it is relatively easy to obtain the eigenstates of a particle in this well (see Appendix B.2.1), is the following

$$\begin{aligned} U &= 0, \qquad x < -a \\ &= -U_O, \quad -a < x < a \\ &= 0, \qquad x > a \end{aligned} \tag{1.39}$$

The rules of quantum mechanics tell us that a particle, of mass m, in the above well has N negative (non-degenerate) energy levels, where

$$N = \text{least integer} \geqslant 4a(2mU_O)^{1/2}/h$$

where h is Planck's constant. For example, for a particle of mass m in a rectangular well whose depth $U_O = 49h^2/(128ma^2)$, we find the following four energy levels

$$E_1 = -0.94U_O,\ E_2 = -0.77U_O,\ E_3 = -0.50U_O,\ E_4 = -0.16U_O \tag{1.40}$$

(see Fig. 1.10). The corresponding bound eigenstates of the particle $u_n(x, t) = u_n(x)\exp(-\mathrm{i}\,E_n t/\hbar)$, $n = 1, 2, 3, 4$, are shown in Fig. 1.11 for $t = 0$. The eigenfunctions $u_n(x)$ (solid lines in the figure) are written as follows

$$u_n(x) = \left\{\begin{matrix} A_n\,\mathrm{e}^{\beta_n x}, & x < -a \\ C_n \cos\alpha_n x, & -a < x < a \\ A_n\,\mathrm{e}^{-\beta_n x}, & x > a \end{matrix}\right\} \tag{1.41a}$$

for $n = 1, 3$; and have the following form

$$u_n(x) = \left\{\begin{matrix} A_n\,\mathrm{e}^{\beta_n x}, & x < -a \\ C_n \sin\alpha_n x, & -a < x < a \\ -A_n\,\mathrm{e}^{-\beta_n x}, & x > a \end{matrix}\right\} \tag{1.41b}$$

for $n = 2, 4$. In every case

$$\alpha_n = [2m(E_n + U_O)]^{1/2}/\hbar \tag{1.42a}$$

$$\beta_n = [2m|E_n|]^{1/2}/\hbar \tag{1.42b}$$

and A_n and C_n are certain constants, so determined (see problem 1.3) that: (i) $u_n(x)$ and its derivative $\mathrm{d}u_n/\mathrm{d}x$ are continuous[10] (change smoothly) everywhere (including the points $x = -a$ and $x = +a$) as seen in Fig. 1.11;

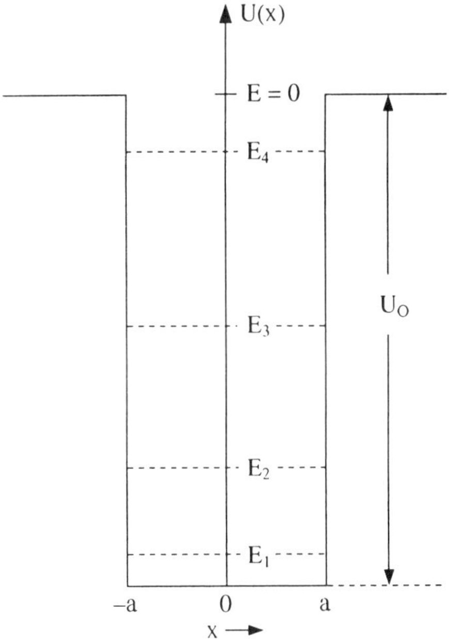

Figure 1.10. Rectangular potential well (1.39) with $U_O = 49h^2/(128ma^2)$. The negative energy levels are: $E_1 = -0.94U_O$, $E_2 = -0.77U_O$, $E_3 = -0.50U_O$, $E_4 = -0.16U_O$

and (ii) $u_n(x)$ is normalised, which means that the integral of the probability-density function $P_n(x) \equiv |u_n(x)|^2$ over all x (in principle from $x = -\infty$ to $x = +\infty$, but, as one can see from Fig. 1.11, $P_n(x)$ practically vanishes at some distance from the well) equals unity.

We observe that the eigenfunctions (1.41) have the following property

$$u_n(-x) = u_n(x) \quad \text{for } n = 1, 3 \tag{1.43a}$$

$$u_n(-x) = -u_n(x) \quad \text{for } n = 2, 4 \tag{1.43b}$$

The above equations express a symmetry of the wavefunctions, evident in Fig. 1.11, which is not at all accidental. Here is a theorem of quantum mechanics: An eigenfunction of the energy of a particle moving in a (one-dimensional) potential field possessed of the symmetry

$$U(-x) = U(x) \tag{1.44}$$

satisfies (1.43a) or (1.43b). More generally, one can show that an eigenfunction of the energy of a particle in a potential field possessed of the symmetry

$$U(-x, -y, -z) = U(x, y, z) \tag{1.45}$$

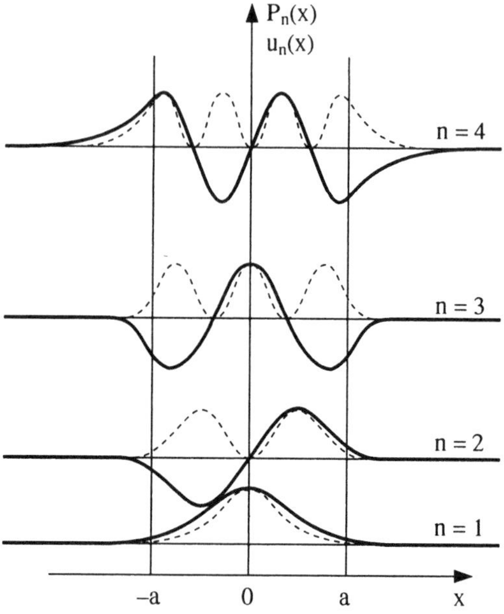

Figure 1.11. The solid lines represent the $u_n(x)$, $n = 1, 2, 3, 4$ which describe the bound eigenstates of a particle of mass m in the potential well of Fig. 1.10. The broken lines represent the corresponding probability-density functions: $P_n(x)$

satisfies the relation

$$u(-x, -y, -z) = u(x, y, z) \tag{1.46a}$$

in which case we say that the eigenfunction has even parity, or the relation

$$u(-x, -y, -z) = -u(x, y, z) \tag{1.46b}$$

in which case we say that it has odd parity.

In some cases the parity of an eigenfunction may not be obvious because the system of coordinates has not been chosen appropriately. The reader can easily show that in the case of the particle-in-a-box (Section 1.3.2) the eigenstates of the particle have a definite parity (the corresponding eigenfunctions satisfy (1.46a) or (1.46b)) provided we take the origin of coordinates at the centre of the box.

Equations (1.43) and (1.46) constitute another example of the important role played by the symmetry of the potential field in the making of the eigenstates of a pariticle found in it[11].

Let us see what else we can learn from equations (1.40)–(1.42) and Fig. 1.11. The most impressive result (after the discreteness of the energy spectrum which we have already emphasised) is the fact that $P_n(x)$, the probability of finding the particle at x, does not vanish outside the well. The fact that the particle can be found with a probability, however small, in a classically forbidden region[12] is, of course, a purely quantum-mechanical result. This phenomenon, which is common in systems of the microscopic world and of primary importance in the interpretation of phenomena which would be otherwise unexplained, must not be taken to mean that the kinetic energy of the particle can be negative. That can never happen. As we have already said (Section 1.4.1), the phrase 'when the particle is at the position x, y, z its kinetic energy is ...' has no meaning in the framework of quantum mechanics, because we cannot know the velocity (and therefore the kinetic energy) of the particle and, simultaneously, its position. What we can say is that the particle is in a certain state, which in the example we are considering may be one of the bound states of Fig. 1.11. Knowing the wavefunction which describes the state of the system we can (using well defined rules of quantum mechanics) answer the questions: What is the probability of finding the particle (at a given moment) with a certain (positive) value of kinetic energy? What is the probability of finding the particle between x and $x + \mathrm{d}x$? etc.

1.6 Mean values of physical quantities and measurements

We prepare a large number (say 1000) of identical physical systems: a particle of mass m, in the field (1.39), in the same eigenstate $u_n(x)\exp(-\mathrm{i}E_nt/\hbar)$. At a given moment we measure the position of the particle in every one of the 1000 systems. The particle is found between x_1 and $x_1 + \mathrm{d}x$ in that many (per cent) of them, between x_2 and $x_2 + \mathrm{d}x$ in so many (per cent) of them, and so on. The result of this imagined experiment[13] is the probability-density function, $P_n(x)$, shown in Fig. 1.11. By definition, the mean value of the position of a particle in the state $u_n(x)\exp(-\mathrm{i}E_nt/\hbar)$ is given by[14]

$$\langle x\rangle_n = \int_{-\infty}^{\infty} xP_n(x)\,\mathrm{d}x \tag{1.47}$$

From Fig. 1.11 it is obvious that $P_n(x) = P_n(-x)$; the particle is found at some distance to the right of the centre ($x = 0$) of the well with the same probability it is found at the same distance to the left of the centre of the well. Therefore, the mean position of the particle (1.47) equals $\langle x\rangle_n = 0$. We hasten to note that the uncertainty[15] as to the position of the particle (the spread of the 'measured' values about the mean value) is, in this instance, of the same order of magnitude as the width of the well, given that $P_n(x)$ extends over the entire well and to some degree outside it.

The potential energy $U(x)$ of the particle is a function of its position only, and this allows us to calculate the mean value of U, when the particle is in the state $u_n(x)\exp(-\mathrm{i}E_nt/\hbar)$, from the following formula

$$\langle U\rangle_n = \int_{-\infty}^{\infty} U(x)P_n(x)\,\mathrm{d}x \tag{1.48}$$

As far as the kinetic energy of the particle is concerned, we imagine again the ensemble of the one thousand identical systems. Each system: a particle in the eigenstate $u_n(x)\exp(-\mathrm{i}E_nt/\hbar)$. At some moment we 'measure' the kinetic energy of the particle in every one of the 1000 systems[16]. We find in every case a positive value of the kinetic energy; in so many (per cent) of the systems this lies between T and $T + \mathrm{d}T$, in that many (per cent) between T' and $T' + \mathrm{d}T$, and so on[17]. From these measurements we obtain the mean value $\langle T\rangle_n$ of the kinetic energy in the usual way. It turns out that

$$\langle T\rangle_n = E_n - \langle U\rangle_n \tag{1.49}$$

Using (1.48), we can rewrite (1.49) in the following form

$$\langle T\rangle_n = \int_{-\infty}^{\infty} (E_n - U(x))P_n(x)\,\mathrm{d}x \tag{1.50}$$

We observe that $(E_n - U(x))$ is the local kinetic energy of classical mechanics. It is a positive quantity inside the well $(-a < x < +a)$ and negative outside the well ($x < -a$ or $x > +a$). Given that $\langle T\rangle_n$ is always a positive quantity, the region inside the well which contributes positively to the integral (1.50) must win over the region (outside the well) which contributes negatively to (1.50). This means that $P_n(x)$, a positive quantity, must decay rapidly outside the well, which is indeed the case (see Fig. 1.11). We note also that $P_n(x)$ decays faster outside the well the lower the energy E_n of the particle.[18]

It is worth stating once more that (1.50) does not imply that we can, actually, find the particle at a pont x outside the well with negative kinetic energy. This is never the case as we have already explained.

In three dimensions, when the particle is moving in a potential field $U(\boldsymbol{r})$, we obtain, instead of (1.49),

$$\langle U\rangle_n + \langle T\rangle_n = E_n \tag{1.51}$$

E_n is the (total) energy of the particle in the eigenstate $u_n(\boldsymbol{r})\exp(-\mathrm{i}E_nt/\hbar)$, where n stands for a set of quantum-numbers; and $\langle T\rangle_n$ and $\langle U\rangle_n$ are the mean values of the kinetic and the potential energy, respectively, of the particle in the same state. The latter is given by

$$\langle U\rangle_n = \int U(\boldsymbol{r})P_n(\boldsymbol{r})\,\mathrm{d}V$$

assuming that the wavefunction has been normalised ($\int P_n(\boldsymbol{r})\,\mathrm{d}V = 1$). The above integral extends, in principle, over the entire space but in practice

$P_n(\boldsymbol{r})$ decays exponentially outside the classically allowed region and effectively vanishes at some relatively small distance from the boundary of this region.

Finally, in the general case, when the state of the particle is described by a linear sum of eigenstates, (1.51) is replaced by the following equation

$$\langle T \rangle + \langle U \rangle = \langle E \rangle \tag{1.52}$$

where $\langle \ldots \rangle$ denotes the mean of the corresponding quantity in the given state. $\langle E \rangle$ is a constant of the motion[19] (independent of time) but $\langle T \rangle$ and $\langle U \rangle$ vary with the time in the general case. We remember that $P(\boldsymbol{r}, t)$ is a function of the time in the general case (see, e.g., Section 1.3.1) and this means that $\langle U \rangle$ varies with the time and, therefore, so does $\langle T \rangle$. Equations (1.51) and (1.52) are the quantum-mechanical version of the classical equation (1.34).

We conclude this section with a remark concerning 'exactly known quantities'. We shall see in forthcoming sections of this book that, depending on the potential field, an eigenstate (of the energy) of a particle can be, also, an eigenstate of some other quantity (e.g., the momentum or the angular momentum of the particle as the case may be). Denoting this quantity by A, we can say that to the above mentioned eigenstate, $u_n(\boldsymbol{r}) \exp(-\mathrm{i}E_n t/\hbar)$, corresponds a definite value of the energy (the eigenvalue E_n) and a definite value of A (an eigenvalue A_n of A). Measurement of A, in the manner we have introduced in this section, over an ensemble of identical systems (in the same eigenstate u_n) will give, in every case, the eigenvalue A_n, in the same way that measurement of the (total) energy will given E_n in every case.

Finally we should point out that the physical process of measurement changes, as a rule, the state of the measured system; its state after the measurement will not be what it would have been if the measurement had not been performed, but a new state determined by the process of measurement. For example, to know the ground-state energy of the hydrogen atom we have to ionise the atom; to know what fraction of an electron-beam has been reflected by a target we have to collect the reflected electrons and this inevitably interrupts their state of free motion. In this respect measurement in relation to microscopic systems is very different from measurement in the macroscopic world, where it is usually possible to measure one or the other properties of a system without affecting its motion.

1.7 Non-bound states

We consider a particle in the ground-state ($E_1 = -0.94U_O$) of the rectangular potential well of Fig. 1.10. It is evident that, if the particle receives, from an incident radiation, energy greater than $0.94U_O$, it will go over into a *final* state of energy $E > 0$, which allows the particle to move away from the well towards $+\infty$ or $-\infty$ along the x axis. Since $E - U(x)$ is a positive quantity along the entire axis, nothing hinders its escape towards infinity. The same would apply if the *initial* state of the particle was one of the excited states ($n = 2, 3, 4$) described in Fig. 1.11, except that the minimum energy required to liberate the particle would be correspondingly less. The above imply that the energy E of the final, non-bound, state of the particle can take any positive value. One can, indeed, show that the non-bound states of a particle, in any given potential field[20], correspond to a continuous spectrum of energy eigenvalues. That this is the case should not surprise us; it agrees with what we have already observed in relation to the energy spectrum of a particle in a box of macroscopic ($V \to \infty$) dimensions (see 1.3.1).

1.7.1 EIGENSTATES OF A FREE PARTICLE

By definition, a particle is free if its potential energy is constant, the same over the entire space. In this case (see (1.33)) no force is acting on the particle wherever it may be. We usually set the constant value of the potential energy equal to zero; any other choice would simply add to the value of the total energy (potential + kinetic) of the particle the above mentioned constant. The wavefunction which describes a particle moving with constant velocity $\boldsymbol{v}$ (and, therefore, with constant momentum $\boldsymbol{p} = m\boldsymbol{v}$) is known as the de Broglie wave of the particle, and is, historically, one of the first important quantum-mechanical concepts. It is written as follows

$$u_k(\boldsymbol{r}, t) = C\exp(\mathrm{i}\boldsymbol{k}\cdot\boldsymbol{r} - \mathrm{i}E_k t/\hbar) = C\exp[\mathrm{i}(k_x x + k_y y + k_z z) - \mathrm{i}E_k t/\hbar] \tag{1.53}$$

where

$$\boldsymbol{k} = (k_x, k_y, k_z) = (p_x/\hbar, p_y/\hbar, p_z/\hbar) = \boldsymbol{p}/\hbar \tag{1.54}$$

$$E_k = \frac{p^2}{2m} = \frac{\hbar^2 k^2}{2m} = \frac{\hbar^2}{2m}(k_x^2 + k_y^2 + k_z^2) \tag{1.55}$$

The wavefunction (1.53) is a plane wave; which means that, at a given time t, its value is the same at all points x, y, z which lie on the same normal to the given $\boldsymbol{k}$, plane. The wavevector $\boldsymbol{k}$ (it is called so) can take any value: $-\infty < k_x, k_y, k_z < +\infty$. By specifying the value of $\boldsymbol{k}$, we specify the momentum $\boldsymbol{p}$ of the particle (1.54), the (kinetic) energy of the particle (1.55), and the corresponding eigenfunction (1.53). The normalisation constant C in (1.53) is of secondary importance and is determined by some convention or other. We note that in the present case the value of the energy *and* the value of the momentum of the particle are known with zero uncertainty. The plane wave (1.53) is an eigenstate of both the energy *and* the momentum of the particle (see also Appendix B.3).

Given that

$$E_k = \frac{\hbar^2 k_x^2}{2m} + \frac{\hbar^2 k_y^2}{2m} + \frac{\hbar^2 k_z^2}{2m} \equiv E_{k_x} + E_{k_y} + E_{k_z} \tag{1.56}$$

where E_{k_x}, E_{k_y}, E_{k_z} are, respectively, the kinetic energies of the particle along the x, y, z directions, we rewrite (1.53) as follows

$$u_{\mathbf{k}}(\mathbf{r}, t) = [A \exp(\mathrm{i}k_x x - \mathrm{i}E_{k_x} t/\hbar)][A \exp(\mathrm{i}k_y y - \mathrm{i}E_{k_y} t/\hbar)] \times [A \exp(\mathrm{i}k_z z - \mathrm{i}E_{k_z} t/\hbar)] \tag{1.57}$$

where $A = C^{1/3}$. It follows from (1.57), and what we said in Section 1.3.2 about independent (separable) motions of a particle, that $u_{\mathbf{k}}(\mathbf{r}, t)$ describes three independent motions of the particle: one parallel to the x-axis with momentum $\hbar k_x$, a second one parallel to the y-axis with momentum $\hbar k_y$, and a third one parallel to the z-axis with momentum $\hbar k_z$. It is sufficient, then, to consider free motion in one dimension. When the particle moves parallel to the x-axis, (1.53) simplifies as follows

$$u_k(x, t) = A \exp(\mathrm{i}kx - \mathrm{i}E_k t/\hbar), \; -\infty < k < +\infty \tag{1.58}$$

In this case $p = \hbar k$ is the momentum of the particle along the x direction and $E_k = \hbar^2 k^2/2m$ is the corresponding energy.

We note that the eigenstate (1.58) and the eigenstate

$$u_{-k}(x, t) = A \exp(-\mathrm{i}kx - \mathrm{i}E_k t/\hbar) \tag{1.59}$$

have the same energy. The energy eigenvalues of a free particle (in one dimension) are doubly degenerate. The two eigenstates of the same energy correspond to different values of the momentum of the particle. So, if $k > 0$, (1.58) describes a particle travelling to the right, and (1.59) describes the same particle when it travels to the left with the same energy[21].

The probability of finding the free particle described by (1.58) between x and $x + \mathrm{d}x$ is given, according to (1.15), by

$$P_k(x)\,\mathrm{d}x = |u_k(x, t)|^2\,\mathrm{d}x = |A|^2\,\mathrm{d}x \tag{1.60}$$

which tells us that the particle can be found with equal probability at any point of the x-axis. This means that the uncertainty Δx in the position of the particle is infinitely large. Indeed, since the momentum of the particle is exactly known ($p = \hbar k$), we have $\Delta p = 0$, and from the uncertainty principle we obtain $\Delta x = \hbar/\Delta p = \infty$.

We note that, in reality, the momentum of the particle is never known exactly; the uncertainty Δp may be very small, but it is never zero. This leads (see next section) to a localisation of the particle within a relatively small region of space around a mean value $\langle x \rangle$ which is of course a function of the time.

1.7.2 WAVEPACKETS OF A FREE PARTICLE

We have noted (in Section 1.3.1) that the state of a particle in a box is, in general, given by a linear sum of the eigenstates of the particle in the potential field defined by the box. The same applies in every case. The most general state of a particle in a given potential field is described by a linear sum of the eigenstates of the particle in that field. A free particle is often described by a sum of plane waves, a wavepacket, as follows:

$$\psi(x, t) = \sum_{k'} A(k') \exp(\mathrm{i}k'x - \mathrm{i}E_k t/\hbar) \tag{1.61}$$

where $A(k')$ are coefficients to be determined. For the sake of simplicity we view (1.61) as a sum of discrete terms:

$$k' = 2\pi n/L, \quad n = \ldots, -4, -3, -2, -1, 0, 1, 2, 3, 4, \ldots$$

where L is a macroscopic length $(L \to \infty)$[22].

We shall consider the special case of (1.61), when

$$A(k') = \frac{(2\alpha)^{1/2}\pi^{1/4}}{L} \exp[-\mathrm{i}(k' - k)x_o] \exp[-\tfrac{1}{2}\alpha^2(k' - k)^2] \tag{1.62}$$

The wave packet $\psi_k(x, t)$, obtained by substituting (1.62) into (1.61), consists mainly of plane waves with wavevectors k' within a relatively small region about k. In Fig. 1.12 we show the contribution to the wavepacket of the various plane waves, as determined by the values (1.62) of the $A(k')$

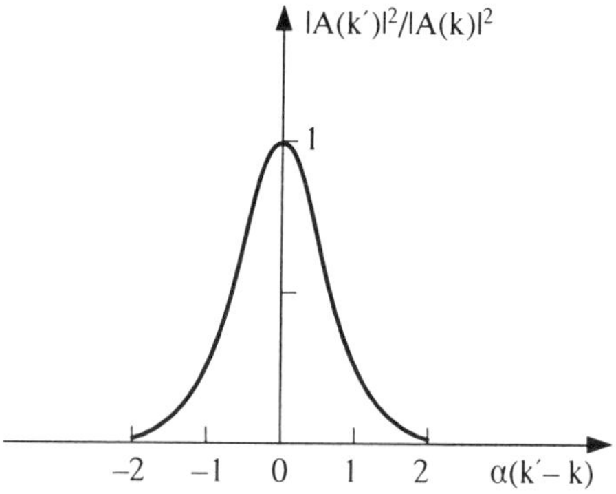

Figure 1.12. Contribution of different plane waves to the wavepacket defined by equations (1.61) and (1.62)

coefficients. We see that outside the region $(k - (1/\alpha)) \leqslant k' \leqslant (k + (1/\alpha))$ these coefficients practically vanish. We may conclude that the mean value of the momentum of the particle in the state (1.61 and 1.62) is $p \equiv \hbar k$, and that the uncertainty about this mean value is $\Delta p \simeq \hbar/\alpha$.

In Fig. 1.13 we show the real part of the wavepacket $\psi_k(x, t)$ as a function of x, at four successive moments of time $t = 0,\ t_o,\ 2t_o,\ 3t_o$; and we have taken, following Leighton[23], $t_o = m\alpha^2/\hbar$ and $\alpha = \pi/k$. The broken lines in the same figure show $|\psi_k(x, t)|$. The corresponding probability-density function $P_k(x, t) = |\psi_k(x, t)|^2$ is given by

$$P_k(x, t) = \pi^{-1/2}\left(\alpha^2 + \frac{\hbar^2 t^2}{m^2\alpha^2}\right)^{-1/2} \exp\left\{-\frac{(x - x_o - pt/m)^2}{(\alpha^2 + \hbar^2 t^2/m^2\alpha^2)}\right\} \quad (1.63)$$

We observe that, at time t, the particle is to be found with certain probability about a mean value

$$\langle x \rangle_t = x_o + pt/m \quad (1.64)$$

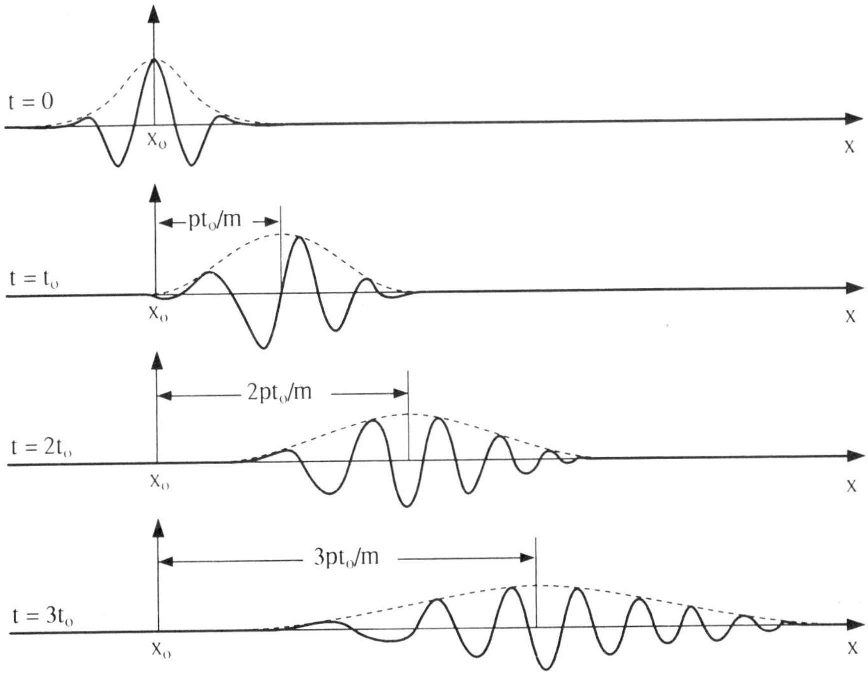

Figure 1.13. Schematic description of the wavepacket $\psi_k(x, t)$: equations (1.61–1.62) with $\alpha = \pi/k$. The solid lines represent the real part of $\psi_k(x, t)$ for $t = 0$, t_o, $2t_o$, $3t_o$, with $t_o = m\alpha^2/\hbar$. The broken lines show $|\psi_k(x, t)|$. (Reproduced with permission from R. B. Leighton, *Principles of Modern Physics* (New York: McGraw-Hill) 1959

which is the same with the position where a classical particle would be (without any uncertainty), if it was moving with constant velocity $v = p/m$, and if it was initially (when $t = 0$) at $x = x_o$. In the real world (of quantum mechanics), the particle is found with non-negligible probability within a region about the mean value $\langle x \rangle_t$ and vanishes rapidly outside this region. The uncertainty $(\Delta x)_t$ in the position of the particle, which is determined by the extension of this region, is given, according to (1.63), by

$$(\Delta x)_t \simeq (\alpha^2 + \hbar^2 t^2/m^2\alpha^2)^{1/2} \tag{1.65}$$

We see that the uncertainty in the position of the particle, initially equal to $\alpha \simeq \hbar/\Delta p$, increases with the time, and eventually, when $(t \gg m\alpha^2/\hbar)$, we obtain $(\Delta x)_t \simeq (\Delta p/m)t$. The uncertainty Δx, initially inversely proportional to Δp, becomes, after sufficient time has elapsed, proportional to Δp. We can understand this as follows: the various plane waves that contribute to the wavepacket under consideration propagate with different velocities $(v' = \hbar k'/m)$, and this leads, as time goes on, to the dispersion of the probability-density function seen in Fig. 1.13. And naturally, this dispersion is greater when the spread of the velocities in the wavepacket is larger, which means in turn a proportionally larger value of Δp. Finally, we note that (1.63) satisfies, at any time, the equation:

$$\int_{-\infty}^{\infty} P_k(x, t)\,\mathrm{d}x = 1$$

In scattering experiments, e.g., when a particle is incident on and scattered by a target, the instruments which we use to prepare the initial (incident) state of the particle, and to determine its final (after the scattering) state, allow us to prepare and measure wavepackets with very small values of Δp without any hindrance from the corresponding uncertainty in the position of the particle. The scatterer (whether it is an atom or a molecule or an area of a solid surface) sees an extended-in-space wave which differs little, as far as the coupling between the incident particle and the scatterer is concerned, from a plane wave, and the (macroscopic) measuring apparatus sees a nearly classical particle travelling in space.

1.8 Scattering by a potential barrier and the phenomenon of tunnelling

Let us assume a potential field which has the form, shown in Fig. 1.14, of a rectangular barrier

$$\begin{aligned} U(x) &= U_O, \quad -a < x < +a \\ &= 0, \quad x < -a \text{ or } x > a \end{aligned} \tag{1.66}$$

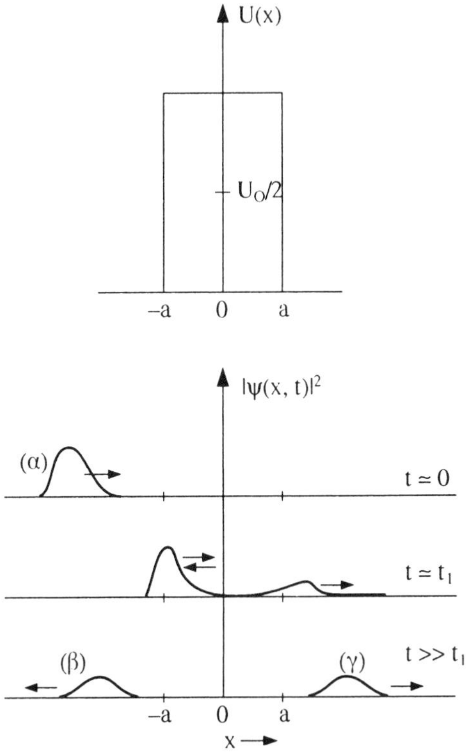

Figure 1.14. Scattering by a rectangular potential barrier. (α): incident wave, (β): reflected wave, (γ): transmitted wave

A free particle described by the wavepacket $\psi_k(x, t)$ of (1.61 and 1.62), with mean values of energy and momentum E and $p = \hbar k = (2mE)^{1/2}$ respectively,[24] is incident on the barrier from the left. At time $t = 0$, the wavepacket is found at some distance to the left of the barrier travelling to the right with velocity $v = p/m$ (curve (α) in Fig. 1.14). At some time $t \simeq t_1$ the wavepacket reaches the barrier and is affected by it (which means that the state of the particle in the barrier region is not described by (1.61) and (1.62)). After the scattering, when $t \gg t_1$, the wavefunction of the particle is made up of the two wavepackets denoted by (β) and (γ) in Fig. 1.14. The reflected wavepacket (β) travels to the left (towards $x = -\infty$) with velocity $-v$ and the transmitted wavepacket (γ) travels to the right with velocity v. The particle is found travelling towards $x = -\infty$ with probability $|B|^2$ (equal to the area below curve (β)) and travelling towards $x = \infty$ with probability $|C|^2$ (equal to the area below curve (γ)). We have, always,

$$|B|^2 + |C|^2 = |A|^2 = 1 \tag{1.67}$$

where $|A|^2$ equals the area (normalised to unity) below the curve (α) which describes the incident wavepacket.

In a real experiment a beam[25] of particles is incident on the scatterer and a fraction of them is scattered by it. The flux of the incident beam equals Nv, where N denotes the number of particles per unit volume of the beam and v the velocity of the particles. The flux of the reflected beam, when the scatterer is the potential barrier (1.66), equals $-N|B|^2v$, and the flux of the transmitted beam equals $N|C|^2v$, where $|B|^2$ and $|C|^2$ are the quantities defined above.

We introduce probability-current densities corresponding to the incident, reflected and transmitted waves of Fig. 1.14, as follows

$$\begin{aligned} &\text{Incident wave:} \quad j_a = |A|^2 v = v \\ &\text{Reflected wave:} \quad j_b = -|B|^2 v \\ &\text{Transmitted wave:} \quad j_c = |C|^2 v \end{aligned} \tag{1.68}$$

We can, then, say that the fluxes of the incident, reflected and transmitted beams are determined, respectively, by the probability-current densities (1.68) of the incident, reflected and transmitted waves. We now introduce the coefficients of reflection (R) and transmission (T) of the particle, as follows:

$$R = |j_b|/|j_a| = |B|^2/|A|^2 = |B|^2 \tag{1.69a}$$

$$T = |j_c|/|j_a| = |C|^2/|A|^2 = |C|^2 \tag{1.69b}$$

We note (see (1.67)) that

$$T + R = 1 \tag{1.70}$$

which tells us that the sum of the reflected and transmitted fluxes equals the incident flux.

The calculation of the above coefficients is made easier by the following observation (a theorem of quantum mechanics which applies generally, whatever the shape of the potential barrier). When the plane waves which make up the incident wavepacket have wavevectors within a very narrow region around a certain k, as in the wavepacket of (1.61 and 1.62), the coefficients of reflection and transmission, as defined above, are obtained from an eigenfunction of the energy of the particle, corresponding to $E = \hbar^2k^2/2m$, which has the following asymptotic form: Away from the scatterer, it is a linear sum of an incident plane wave (with momentum $\hbar k$) and plane waves (of the same energy), with momenta pointing away from the scatterer, which represent the scattered probability flux. In the example under consideration, of a particle incident on the rectangular barrier (1.66) from the left with energy $E = \hbar^2 k^2/2m$, this eigenfunction takes the form

$$\psi = \exp(\mathrm{i}kx - \mathrm{i}Et/\hbar) + B\exp(-\mathrm{i}kx - \mathrm{i}Et/\hbar) \quad \text{for } x < -a \tag{1.71a}$$

and

$$\begin{aligned}\psi &= \exp(\mathrm{i}kx - \mathrm{i}Et/\hbar) + (C-1)\exp(\mathrm{i}kx - \mathrm{i}Et/\hbar) \\ &= C\exp(\mathrm{i}kx - \mathrm{i}Et/\hbar) \quad \text{for } x > a\end{aligned} \tag{1.71b}$$

The first term of (1.71a) represents the incident wave, the second term of (1.71a) represents the reflected wave, and (1.71b) represents the transmitted wave. The coefficients B and C in (1.71) are identified with the corresponding coefficients in (1.68). The values of these coefficients are determined from the wavefunction in the region of the scatterer ($-a < x < +a$). We distinguish between two cases:

(i) $E > U_O$. In this case, in the region $-a < x < +a$ the wavefunction is written as the sum of two plane waves as follows

$$\psi = (D\,\mathrm{e}^{\mathrm{i}qx} + F\,\mathrm{e}^{-\mathrm{i}qx})\,\mathrm{e}^{-\mathrm{i}Et/\hbar}, \quad -a < x < +a \tag{1.72a}$$

$$q = [2m(E - U_O)]^{1/2}/\hbar \tag{1.72b}$$

We note that in a region of constant potential energy, the state of the particle will be described by a linear sum of plane waves corresponding to the kinetic energy of the particle, which equals $E - U_O$ in the present case. The first term in (1.72a) describes a particle travelling to the right and the second term a particle travelling to the left. There are no other possibilities in one-dimensional motion.

The wavefunction ψ and its derivative $(\mathrm{d}\psi/\mathrm{d}x)$ must be everywhere continuous functions of x (see Fig. 1.15). And therefore the value of ψ, and that of $\mathrm{d}\psi/\mathrm{d}x$, must not change as we move from a point $-a - (\delta x \to 0)$ to the left of $-a$ to a point $-a + (\delta x \to 0)$ to the right of $-a$. And the same must be true when we cross the point $x = +a$. [The continuity of the wavefunction and its derivative is necessary in order to secure the continuity of the probability-density function and that of the probability-current density. This guarantees that when the probability of finding the particle in a region of space diminishes, the particle is found with increased probability in the surrounding region; the particle does not disappear and does not appear out of nothing.] By demanding that the wavefunction and its derivative are continuous at the points $x = -a$ and $x = +a$, one determines uniquely the values of the coefficients B, C, D, F in (1.71) and (1.72) and then, by using (1.69), one obtains the reflection and transmission coefficients of the particle as functions of the incident energy (see also Appendix B.2.1).

(ii) $E < U_O$. In this case, in the region $-a < x < +a$ the wavefunction is written as a sum of two 'plane waves' which decay exponentially, the one as we move from $x = -a$ to the right and the other as we move from $x = +a$ to

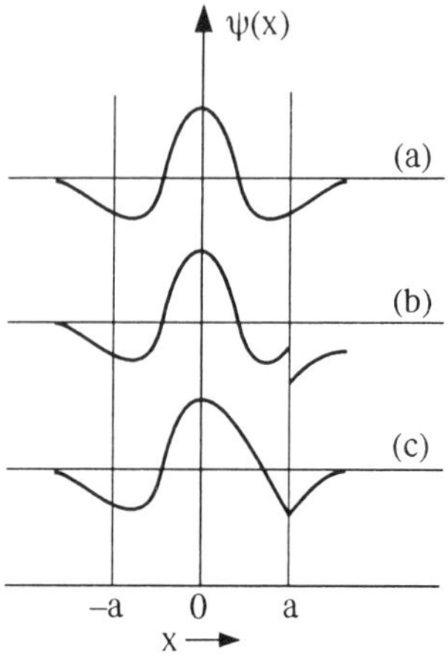

Figure 1.15. The real and imaginary parts of the wavefunction must be continuous functions everywhere and with continuous derivatives as, for example, in (*a*). The function (*b*) is not acceptable because of its discontinuity at $x = a$, and the function (*c*) is not acceptable because of the discontinuity of its derivative at $x = a$

the left, in accordance with what has been said earlier (see, e.g., Section 1.5) about the penetration of the wavefunction of a particle into classically forbidden regions of space. We write

$$\psi = (D\,\mathrm{e}^{-q'x} + F\,\mathrm{e}^{q'x})\,\mathrm{e}^{-\mathrm{i}Et/\hbar}, \quad -a < x < a$$
$$q' = \left[\frac{2m}{\hbar^2}|E - U_O|\right]^{1/2} \tag{1.73}$$

From the continuity of the wavefunction and its derivative at the points $x = -a$ and $x = +a$ one determines uniquely, as in the previous case, the coefficients B, C and D, F in (1.71) and (1.73) respectively, and obtains, through (1.69), the coefficients of reflection and transmission, which satisfy in every case (1.70).

At the end of the above calculation (see problem 1.11) one finds the following formula for the transmission coefficient[26]

$$T(E) = \left(1 + \frac{U_O^2 \sin^2([8m(E - U_O)]^{1/2}a/\hbar)}{4E(E - U_O)}\right)^{-1} \tag{1.74}$$

where E may be smaller or larger than U_O. In Fig. 1.16 we show $T(E)$ when the height of the rectangular barrier equals $U_O = 2\hbar^2/(ma^2)$. We note that the transmission coefficient does not vanish when the energy of the incident particle is smaller than the height of the barrier, and it does not equal unity, except when $E \gg U_O$, when its energy lies above the barrier height, as

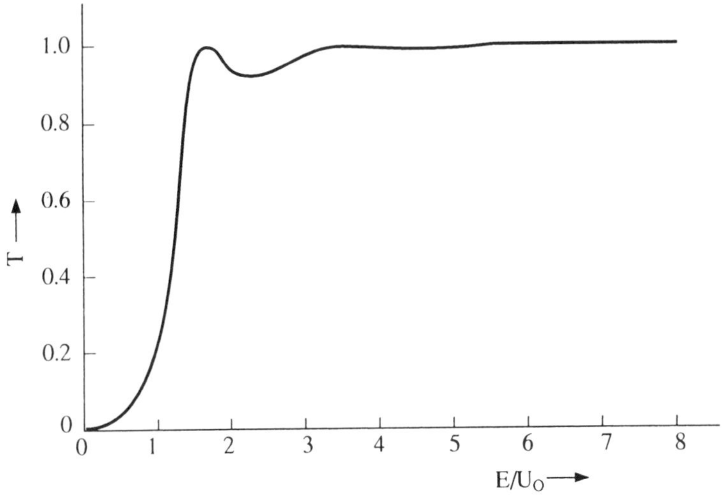

Figure 1.16. The transmission coefficient for the rectangular barrier (1.66) with $U_O = 2\hbar^2/(ma^2)$. (L. I. Schiff, *Quantum Mechanics*, 2nd edn (New York: McGraw-Hill) 1955)

would be the case in classical mechanics. The transmission of a particle through a potential barrier when its energy lies below the top of the barrier, known as the tunnelling phenomenon, should not surprise us, after what we have already said, about the extension of the wavefunction of the particle into classically forbidden regions of space. Of course, the phenomenon of tunnelling is not restricted to the penetration of the rectangular barrier we have discussed here. We note, in passing, that the transmission coefficient of a particle incident on a potential barrier which varies slowly with the position, such as the one shown in Fig. 1.17, with energy E below the

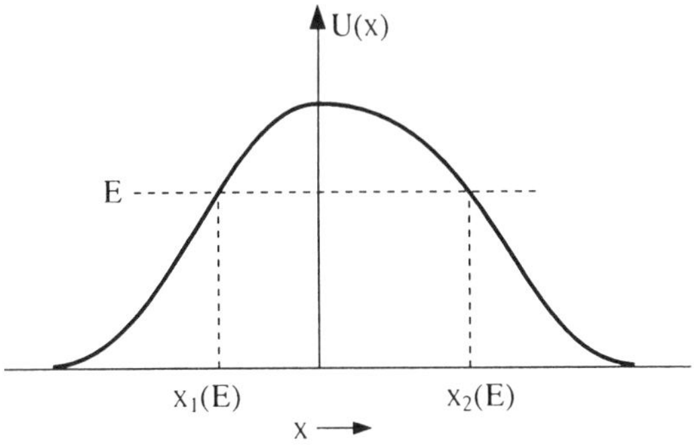

Figure 1.17. A potential barrier which varies slowly with the position

barrier maximum, is obtained, to a good approximation, from the following formula

$$T(E) = [1 + \exp(2\bar{q}L)]^{-1} \tag{1.75}$$

$L \equiv x_2(E) - x_1(E)$, where x_2 and x_1 are the classical turning points for the given energy (see Fig. 1.17), and $\bar{q}$ is the mean value of the quantity $[2m(U(x) - E)]^{1/2}/\hbar$ between x_1 and x_2.

1.9 Bound states as superpositions of plane waves

It is interesting to see how a superposition of plane waves of the same energy leads to the eigenstates of a particle in the one-dimensional box of (1.38). We demand that the particle, which is free between the walls of the box, has a definite energy E and, therefore, its momentum can be $+\hbar k$ or $-\hbar k$, where $k = (2mE)^{1/2}/\hbar$. The corresponding plane waves are

$$u_{+k} = A\exp(\mathrm{i}kx - \mathrm{i}Et/\hbar) \tag{1.76a}$$

$$u_{-k} = A\exp(-\mathrm{i}kx - \mathrm{i}Et/\hbar) \tag{1.76b}$$

Reflection of the particle at one of the walls changes its momentum from $-\hbar k$ to $\hbar k$ and the opposite happens at the other wall. We, therefore, write the wavefunction which describes the motion of the particle between the walls of the box as a linear sum of the waves (1.76a) and (1.76b), as follows

$$u_E(x, t) = (A_+ \mathrm{e}^{\mathrm{i}kx} + A_- \mathrm{e}^{-\mathrm{i}kx})\,\mathrm{e}^{-\mathrm{i}Et/\hbar} \tag{1.77}$$

where A_+ and A_- are constants to be determined. We know that the probability of finding the particle beyond the walls of the box is zero (see footnote 18), and therefore, because $u_E(x)$ must be a continuous function of x (the derivative of the wavefunction need not be continuous where the potential energy becomes infinite), we have

$$u_E(x = 0) = 0 \tag{1.78a}$$

$$u_E(x = a) = 0 \tag{1.78b}$$

Substituting (1.77) into (1.78a) we obtain

$$A_- = -A_+ \tag{1.79}$$

so that u_E becomes

$$\begin{aligned} u_E(x, t) &= A_+(\mathrm{e}^{\mathrm{i}kx} - \mathrm{e}^{-\mathrm{i}kx})\,\mathrm{e}^{-\mathrm{i}Et/\hbar} \\ &= B\sin kx\,\mathrm{e}^{-\mathrm{i}Et/\hbar} \end{aligned} \tag{1.80}$$

where $B \equiv 2\mathrm{i}A_+$ is a constant to be determined. u_E must also satisfy (1.78b), which means that

$$\sin ka = 0 \tag{1.81}$$

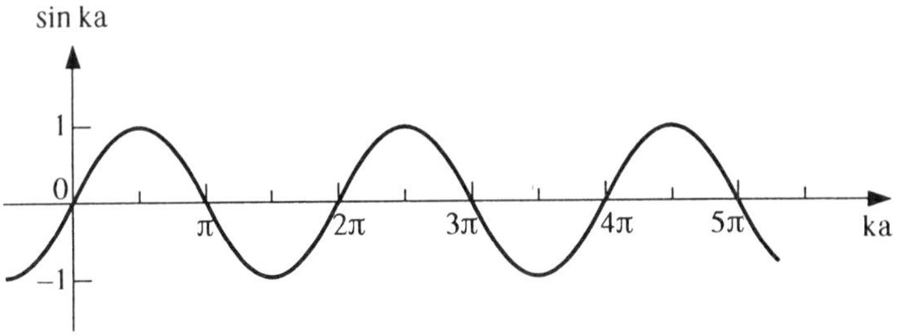

Figure 1.18. Sin ka as a function of ka

In Fig. 1.18 we show $\sin ka$ as a function of ka, and we see from this figure that (1.81) is satisfied only when[27]

$$ka = n\pi, \quad n = 1, 2, 3, \ldots \tag{1.82}$$

which means that the energy $E = \hbar^2 k^2/2m$ of the particle can have one of the following values (eigenvalues of the energy) and none other

$$E_n = \frac{\hbar^2 \pi^2 n^2}{2ma^2}, \quad n = 1, 2, 3, \ldots \tag{1.83}$$

and that the eigenstate corresponding to the eigenvalue E_n is

$$u_n(x, t) = B \sin \frac{n\pi x}{a} \mathrm{e}^{-\mathrm{i}E_n t/\hbar} \tag{1.84}$$

The constant B is determined by normalisation: $|u_n|^2 \equiv P_n(x)$ satisfies (1.19) and this leads to $B = (2/a)^{1/2}$. Equations (1.83) and (1.84) confirm the results of section 1.3.1, as expected.

Problems

1.1 The state of a particle in the potential field (1.38) at $t = 0$ is described by

$$u(x) = \frac{1}{\sqrt{2a}} \sin \frac{\pi x}{a} + \sqrt{\frac{3}{2a}} \left[\sin \frac{2\pi x}{a} \cos \frac{\pi x}{a} + \cos \frac{2\pi x}{a} \sin \frac{\pi x}{a} \right]$$

Find the wavefunction $u(x, t)$ which describes the state of the particle for $t > 0$.

***1.2** Show that the eigenfunctions $u_n(x)$, $n = 1, 2, \ldots$ defined by (1.13b) are an orthogonal set of functions, i.e., they satisfy equation (B54) of Appendix B.4.

1.3 (a) Use the continuity of $u_n(x)$ at $x = -a, +a$ and normalisation to determine A_n and C_n in equations (1.41). [The continuity of $\mathrm{d}u_n/\mathrm{d}x$ is secured when the energy of the particle is an energy-eigenvalue (see Appendix B.2.1).]

(b) Find the probability that the particle will be found in the classically forbidden region when in a bound state (1.41) of the rectangular potential well defined in Fig. 1.10.

(c) The mean kinetic energy of the particle in a bound state (1.41) can be calculated using (1.50) or the more general formula (B59) of Appendix B.5, which in the present case becomes

$$\langle T \rangle_n = -\frac{\hbar^2}{2m}\int_{-\infty}^{+\infty} u_n(x)\frac{\mathrm{d}^2 u_n}{\mathrm{d}x^2}\,\mathrm{d}x$$

Show that the two formulae give, indeed, the same result.

***1.4** A weak perturbation $\Delta U(x) = Fx$, where F is very small, is added to the potential field of Fig. 1.10. Show, using (B200) of Appendix B.10, that the energy levels (1.40) do not change to a first order approximation.

***1.5** Show that a particle of mass m in the potential field (1.39) has N discrete energy levels, where (see text following (1.39))

$$N = \text{least integer} \geqslant 4a(2mU_0)^{1/2}/h$$

1.6 Calculate the first five (in order of increasing energy) energy-eigenvalues of an electron in a box, write down the corresponding eigenfunctions, and note the degeneracy of each energy level, when the dimensions of the box are given by (i) and (ii) below

(i) $a = 1$ Å, $b = 1.5$ Å, $c = 2$ Å
(ii) $a = b = c = 1.5$ Å

1.7 Write down the energy-eigenfunctions of a particle in a box with respect to a system of coordinates the origin of which coincides with the centre of the box.

(a) Determine the parity of the eigenfunctions.

(b) Determine the average values $\langle x \rangle$, $\langle y \rangle$ and $\langle z \rangle$ when the particle is in an energy-eigenstate.

1.8 (a) Prove equation (1.32).

(b) Assuming that the energy-eigenstates of a particle in a box can be described by plane waves

$$\psi_k = \frac{1}{\sqrt{V}}\exp(\mathrm{i}\mathbf{k}\cdot\mathbf{r} - \mathrm{i}E_k t/\hbar)$$

which satisfy the periodic boundary conditions: $\psi_k(x + L, y, z) = \psi_k(x, y + L, z) = \psi_k(x, y, z + L) = \psi_k(x, y, z)$, where $L^3 = V$ is the volume of the box, show that, provided V is sufficiently large, one obtains the same density of states (1.32). What can be the physical interpretation of the above result?

***1.9** (a) Show that the wavefunction $\psi_k(x, t)$ defined by (1.61) and (1.62) has the following form:

$$\psi_k(x, t) = \pi^{-1/4}\left(\alpha + \frac{i\hbar t}{m\alpha}\right)^{-1/2} \exp\left[i(kx - Et/\hbar)\right]$$
$$\times \exp\left(-\frac{(x - x_o - pt/m)^2(1 - i\hbar t/m\alpha^2)}{2(\alpha^2 + \hbar^2 t^2/m^2\alpha^2)}\right)$$

where $p = \hbar k$ and $E = p^2/2m$.

(b) Using equation (B59) of section B.5 calculate the mean values of the position $\langle x \rangle$, the momentum $\langle p_x \rangle$, and the kinetic energy $\langle T \rangle$ of the particle at $t = 0$.

(c) Evaluate $\langle p_x \rangle$ and $\langle T \rangle$, at $t = 0$, once more by using (B60) and compare with the results of (b).

(d) Using (B61), or (B62) where appropriate, calculate, for $t = 0$, the uncertainties of the position Δx and of the momentum Δp_x of the particle, and show that

$$\Delta x \Delta p_x = \hbar/2$$

***1.10** Calculate the transmission coefficient $T(E)$ and the reflection coefficient $R(E)$ of a particle (of mass m) incident from the left on the rectangular well (1.39) with energy $0 < E < \infty$. Show that the same coefficients of reflection and transmission are obtained when the particle is incident on the well from the right (see Appendix B.2.1).

***1.11** (a) Verify that ψ described by equations (1.71) and (1.72) satisfies the equation

$$-\frac{\hbar^2}{2m}\frac{d^2\psi}{dx^2} + U(x)\psi = E\psi$$

where $U(x)$ is given by (1.66) and $E > U_O$. Determine the coefficients B, C, D, F so that ψ and its derivative are continuous at the points $x = -a$ and $x = +a$.

(b) Verify that ψ described by equations (1.71) and (1.73) satisfies the above equation when $E < U_O$; and again determine the coefficients B, C, D, F so that ψ and its derivative are continuous at the points $x = -a$ and $x = +a$.

(c) Verify equation (1.74).

***1.12** Consider the following potential field

$$U(z) = 0, \quad z < 0$$
$$= U_O, \quad z > 0$$

(a) A particle (of mass m) is incident on the above barrier from the left with energy $E < U_O$. Verify that the wavefunction which describes the state of the particle has the form $u(z, t) = u(z) \exp(-iEt/\hbar)$, where

$$u(z) = e^{ikz} + A\, e^{-ikz}, \quad z < 0$$
$$= B\, e^{-qz}, \quad z > 0$$

Determine the values of k, q, A and B, and obtain the reflection coefficient.

(b) Obtain the wavefunction which describes the state of the particle when it is incident on the barrier from the left (with energy $E > U_O$), and obtain the reflection and transmission coefficients, R and T, as functions of the energy. Verify that $T + R = 1$.

(c) Do the same as in (b) for the case when the particle is incident from the right, and compare the results with those of (b).

***1.13** (a) Write the complete wavefunction for a particle (of mass m) incident from the left, with energy $-U_O < E < 0$, on the potential barrier

$$U(x) = -U_O, x < a$$
$$= 0, x > a$$

(Put the origin of coordinates at $x = 0$).

(b) Write the complete wavefunction for a particle (of mass m) incident from the right, with energy $-U_O < E < 0$, on the potential barrier

$$U(x) = -U_O, x > -a$$
$$= 0, x < -a$$

(Put the origin of coordinates at $x = 0$.)

(c) By combining the results of (a) and (b) above, derive the conditions (B24) and (B26) of Section B.2.1 for the existence of bound states of the particle in the rectangular potential well (1.39).

1.14 Assume that the potential energy barrier of Fig. 1.17 can be described approximately by

$$U(x) = 10 - 0.2x^2 (x \text{ in Å}, U \text{ in eV})$$

Use (1.75) to evaluate approximately the transmission coefficient $T(E)$ of an electron incident on this barrier with $E < 10$ eV. Plot T versus E for $5 < E < 10$ eV.

Footnotes of Chapter 1

1. The first revolutionary steps away from classical physics came from Max Planck, Albert Einstein and Niels Bohr at the beginning of the century. The mathematical foundation of quantum mechanics as we know it today was completed in the

decade of 1920. A number of scientists, besides Bohr, contributed to it. Pioneering contributions were made by Louis de Broglie, Werner Heisenberg, Erwin Schrödinger and Paul Dirac.

2. Rutherford was led to this model of the atom from his analysis of the scattering of α-particles (nuclei of helium atoms) by the atoms of thin metal films.
3. In the sense that the best we can say about the value of the coordinate at the given moment is: it lies between $x - \Delta x$ and $x + \Delta x$. For a formal definition of the uncertainty about the mean value of a given quantity see Appendix B.5.
4. The momentum of the particle (of mass m) is defined by the relation $\boldsymbol{p} = m\boldsymbol{v} = (mv_x, mv_y, mv_z)$. Many rules of mechanics are formulated better in terms of the momentum rather than the velocity of the particle.
5. The real and imaginary parts of a complex function are, often, noted as follows: $\psi_R \equiv \mathrm{Re}\{\psi\}$, $\psi_I \equiv \mathrm{Im}\{\psi\}$.
6. A linear sum of a set of functions $\varphi_i(q)$, $i = 1, 2, \ldots, N$, where q denotes a set of variables and N may be finite or go to infinity, is by definition a sum $c_1\varphi_1 + c_2\varphi_2 + \ldots\ c_i\varphi_i + \ldots + c_N\varphi_N$, where c_i are constants, at least one of which is non zero. The functions $\varphi_i(q)$, $i = 1, 2, \ldots, N$ are linearly independent in the sense that none of them can be expressed as a linear sum of the rest.
7. The infinite sum is replaced, in practice, by a sum of a finite number of terms.
8. Here and throughout the book when we refer to the eigenstates of the system we shall mean the energy-eigenstates of the system.
9. In a potential field the force acting on the particle is determined by the position coordinates (x, y, z) of the particle (see Section 1.4). In the present case the particle experiences a force when it collides with one of the walls of the box and, therefore, the symmetry of the potential field is that of the box.
10. The continuity of the wavefunction and its derivative is an essential property of every quantum-mechanical wavefunction. See, also, Appendix B.2.1.
11. When an energy level is degenerate, we may have eigenstates of different parity corresponding to it. Evidently, a linear combination of such states produces an eigenstate of no definite parity.
12. Outside the well the potential energy of the particle is greater than its total energy E_n (when $E_n < 0$) and therefore this region is, classically, forbidden to the particle (see Section 1.4.1).
13. The above experiment is not possible in practice for technical reasons and not because it violates some basic principle. The probability-density function (the probabilty of finding the particle within $\mathrm{d}V$ about $\boldsymbol{r}$) is a very useful quantity in spite of the fact that it can not be measured as such. Its correctness (and that of the underlying wavefunction) can be checked experimentally, in an indirect way, by a variety of experiments.
14. The formula is the same with the one we use to estimate the mean value of a quantity (e.g., the length of a rod) after a series of measurements.
15. The uncertainty $(\Delta x)_n$ as to the position of the particle is obtained in the same way that one obtains the 'error' (deviation about the mean) in a series of measurements

$$(\Delta x)_n = \left[\int_{-\infty}^{\infty} (x - \langle x \rangle_n)^2 P_n(x)\,\mathrm{d}x\right]^{1/2}$$

16. Again we refer to an experiment possible in principle but not feasible in practice.
17. The percentage of systems with kinetic energy between T and $T + \mathrm{d}T$ can be obtained from the eigenfunction $u_n(x, t)$ using standard rules (see Appendix B.5).

18. The reader can better understand at this stage why in the particle-in-a-box model, described by (1.38), $P_n = 0$ outside the box. Outside the box $(E_n - U) = -\infty$ and therefore any value of $P_n(x)$, however small, would make the integral (1.50) negative.
19. Having in mind our ensemble of many identical systems we may say that, in that many (per cent) of them the particle is to be found in one of the energy eigenstates (with the corresponding energy), in so many (per cent) of them in another eigenstate, and so on. One can show (see Appendix B.5) that the above percentages are independent of time and that, therefore, $\langle E \rangle$ is independent of time.
20. We assume that the field is such that non-bound states do exist. In some cases, e.g., in the field (1.38) such states do not exist.
21. In three dimensions, for a given energy the momentum can point to any direction: The degeneracy of the energy eigenvalues tends to infinity.
22. When L is large enough the results do not depend on the value of L.
23. R. B. Leighton, *Principles of Modern Physics* New York: McGraw-Hill, 1959.
24. In our discussion of scattering phenomena we assume that the uncertainty, the spread about the mean values, of these quantities is very small, in accordance with what has been said in Section 1.7.2.
25. The particles of the beam are sufficiently away from each other, so that no interaction exists between them. All have the same velocity v, and every one is scattered by the scatterer with the same probability.
26. Note that when $x > 0$, $\sin(\sqrt{-x}) = \sin(\mathrm{i}\sqrt{x}) = (\exp(-\sqrt{x}) - \exp(\sqrt{x}))/(2\mathrm{i})$.
27. The solutions $ka = n\pi$, $n = -1, -2, \ldots$, do not lead to linearly independent eigenstates (substituting $-n$ for n in (1.84) changes the sign of the normalisation constant, which is a trivial change; see Section 1.3.1). The solution $n = 0$ $(k = 0)$ nullifies the wavefunction and is, therefore, of no physical significance.

A thing merely appears to have colour, it merely appears to be sweet or bitter. Only atoms and empty space have a real existence.

Attributed to DEMOCRITUS

2 ATOMS

2.1 The one-electron atom—introduction

We have already noted that, because the mass of the atomic nucleus is much larger than that of the electron, the centre of mass of the atom coincides, to a very good approximation, with the nucleus of the atom which, for our purposes, can be thought of as a point particle (see Section 1.1). The motion of the atom consists of a free translational motion (the kinetic energy of this motion equals $MV_{CM}^2/2$, where M is the mass of the atom and V_{CM} the velocity of its centre of mass) and of an independent internal motion of the electron about the nucleus. We are interested at this stage in the internal motion of the atom and, therefore, we disregard its translational motion. The picture we have in mind is that of an electron moving about a point-like mucleus stationed at the origin of coordinates. We choose to write the position of the electron in terms of the spherical coordinates (r, θ, φ) defined in Fig. 2.1. In this representation the potential energy of the negatively charged electron in the field of the positively charged nucleus is given by[1]

$$U(r) = -\frac{Ze^2}{4\pi\epsilon_O r} \tag{2.1}$$

Ze denotes the positive charge of the nucleus ($Z = 1$ for the hydrogen atom), $-e$ denotes the charge of the electron, and ϵ_O is a universal constant known as the permittivity of free space (see Appendix C). Equation (2.1) derives from (1.33) and a well known law of electrostatics (Coulomb's law) which states that the force between two charged particles is directed along the line joining the particles, is proportional to the product of their charges and inversely proportional to the square of the distance between them, being attractive for charges of opposite type and repulsive for charges of the same type.

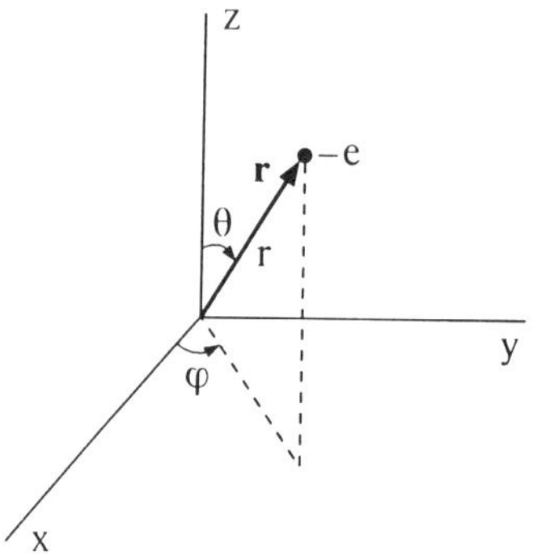

Figure 2.1. The position vector $\boldsymbol{r}$ of the electron is determined by the distance ($r = |\boldsymbol{r}|$) of the electron from the origin of the coordinates ($0 \leqslant r < \infty$), the angle θ between the z axis and $\boldsymbol{r}$ ($0 \leqslant \theta \leqslant \pi$) and the angle φ between the x axis and the projection of $\boldsymbol{r}$ on the xy plane ($0 \leqslant \varphi < 2\pi$). The spherical coordinates (r, θ, φ) are related to the Cartesian coordinates (x, y, z) as follows: $x = r \sin\theta \cos\varphi$, $y = r \sin\theta \sin\varphi$, $z = r\cos\theta$

To know the electronic properties of an one-electron atom means knowing, according to our general analysis of Chapter 1, in the first instance, the eigenvalues of the energy and the corresponding eigenstates of the electron in the potential field (2.1).

The fact that the potential field (2.1) depends only on the distance of the electron from a centre (the nucleus at the origin of the coordinates) is of particular importance and simplifies greatly the description of the eigenstates of the electron in this field. We call a field which has this property, namely that the potential energy of the particle moving in it depends only on the distance of the particle from a centre, centrally (or spherically) symmetric. The field (2.1) is not the only spherically symmetric field of physical interest. We shall later see, for example, that an electron in a many-electron atom sees a so-called mean field which is, to a good approximation, spherically symmetric. For this reason, before we deal with the eigenstates of an electron in the field (2.1), we shall look at those properties of a particle moving in a spherically symmetric field which derive solely from the spherical symmetry of the field and are, therefore, more widely valid.

2.2 Angular momentum

The energy (potential + kinetic) of a particle is not the only conserved physical quantity when the particle moves in a spherically symmetric field. The angular momentum of the particle is also conserved (it does not change with time); and this is true in classical and quantum mechanics. Let us see what the conservation of angular momentum means in classical mechanics. We consider, as an example, a periodic motion of a particle about a centre of attraction, which corresponds to a quantum-mechanical bound state of the particle.[2] In this case we can visualise the position vector of the particle, $\boldsymbol{r}(t)$ in Fig. 2.2, as a rope stretched from the centre of attraction at the origin of the coordinates to the particle. The stretched rope stays in the same plane (the plane of motion), but its length may or may not be constant as the angle (θ) between it and a specified direction (x) in the plane of motion changes during the motion of the particle. In the first case (constant length) the orbit of the particle is circular, in the second case (length varies with θ) the orbit is elliptical. However, in either case the area (shaded in Fig. 2.2) swept by $\boldsymbol{r}(t)$ in Δt seconds is a constant of the motion: in Δt seconds $\boldsymbol{r}(t)$ will sweep the same amount of area. We define the angular momentum of the particle (with respect to the centre 0) as follows: to obtain its magnitude we divide the area swept by the position vector $\boldsymbol{r}(t)$ of the particle in time Δt by Δt and multiply the result with twice the mass of the particle.[3] We complete the definition of the angular momentum noting that it is a vector, with the above magnitude, normal to the plane of the motion, defined by $\boldsymbol{r}(t)$ and the velocity $\boldsymbol{v}(t)$ of the particle and pointing in the direction a right-hand screw advances when it turns in the same way as the orbiting particle. In a centrally symmetric field, both the magnitude and the direction of the angular momentum are constant quantities. They are conserved.

In the example of Fig. 2.2 the angular momentum is normal to the xy plane and points in the positive z-direction. The general case is presented, schematically, in Fig. 2.3.

Let us now see what happens in quantum mechanics. Classical orbits of the particle, such as those of Figs. 2.2 and 2.3, have no meaning in quantum mechanics. However, from the wavefunction which describes the state of the particle we should be able to tell, at any given moment, what the probability is of the particle having an angular momentum equal to that of Fig. 2.2, to that of Fig. 2.3 or any other allowed value.

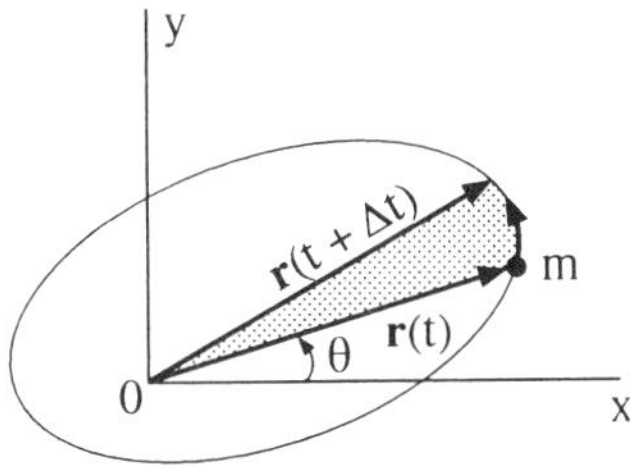

Figure 2.2. Classical orbit of a particle (m) in a centrally symmetric field with the centre of attraction (0) at the origin of the coordinates

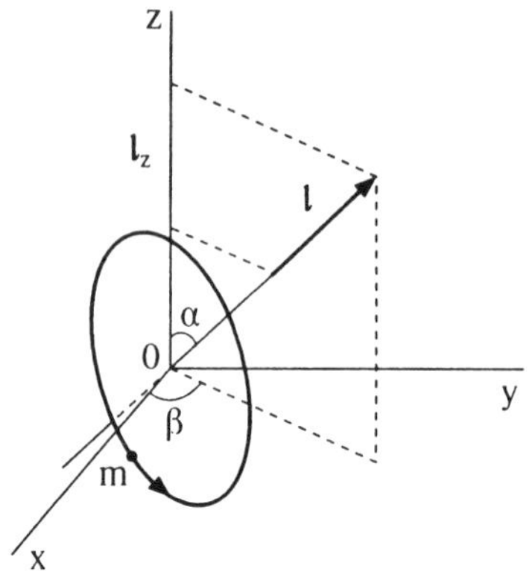

Figure 2.3. The vector $\boldsymbol{l}$ denotes the angular momentum of the particle (m) with respect to the centre (0). l_z denotes its z-component. The plane of the motion passes through the centre (0) and is normal to $\boldsymbol{l}$

One very important result of quantum mechanics tells us (see Appendix B.6.1) that the square of the magnitude of the angular momentum of a particle with respect to a centre (we assume it to be at the origin of the coordinates) can have one of the following values (and none other)

$$\boldsymbol{l}^2 = l(l+1)\hbar^2, \; l = 0, 1, 2, 3, \ldots \tag{2.2a}$$

and the projection of $\boldsymbol{l}$ on the z-axis (which we choose arbitrarily) can have, for given l, one of the following values (and none other)

$$l_z = m\hbar, \; m = -l, -l+1, \ldots, l-1, l \tag{2.2b}$$

The values (2.2a) are the eigenvalues of the square of the magnitude of the angular momentum and (2.2b) are the eigenvalues (for given l) of the z-component of the angular momentum. When the state of the particle is an eigenstate of the square of the magnitude of the angular momentum and its z-component, which means that these quantities have precise values (one of the eigenvalues (2.2a) and (2.2b) respectively), the other two components, l_x and l_y, of the angular momentum cannot be known precisely (with zero uncertainty). We find that l_x and l_y take, with non-zero probability, the values $m\hbar$ ($m = -l, -l+1, \ldots, l$) so that the mean values of these components equal zero. Referring to Fig. 2.3 we can say that knowledge with zero uncertainty of l and α, implies that the angle β obtains, with non-zero probability, every one of the allowed values between zero and 2π.

The above are complemented with the following theorem about the energy-eigenstates (stationary states) of a particle in a spherically symmetric field. We have (see Appendix B.7): the eigenstates of the energy of a particle in a spherically symmetric field are, also, eigenstates of the angular momentum[4] (with respect to the centre of the field which we take as the origin of the coordinates): An eigenstate of the particle corresponds to a certain eigenvalue of the energy E_{nl}, a certain eigenvalue $l(l+1)\hbar^2$ of the square of the magnitude of the angular momentum and a certain eigenvalue $m\hbar$ of the z-component of the angular momentum, and is described by a wavefunction of the following form

$$\psi_{nlm}(\boldsymbol{r}, t) = R_{nl}(r)Y_{lm}(\theta, \varphi)\exp(-\mathrm{i}E_{nl}t/\hbar) \tag{2.3}$$

where r, θ, φ are the spherical coordinates of the particle defined in Fig. 2.1. The functions $Y_{lm}(\theta, \varphi)$ which describe the dependence of the wavefunction on the angles θ, φ are determined by the quantum numbers l and m which specify, through (2.2), the magnitude of the angular momentum of the particle and its z-component. In Table 2.1 we present the spherical harmonics (this is how the functions $Y_{lm}(\theta, \varphi)$ are called) for $l = 0$, 1, 2 and 3; and in Figs. 2.4 and 2.5 we present in polar form the function

$$\Theta_{lm}(\theta) = |Y_{lm}(\theta, \varphi)|^2$$

Table 2.1. Spherical harmonics $Y_{lm}(\theta, \varphi)$, $l = 0, 1, 2, 3$

$Y_{00} = \left(\dfrac{1}{4\pi}\right)^{1/2}$	$Y_{3-3} = \left(\dfrac{70}{128\pi}\right)^{1/2} \sin^3\theta\, e^{-3i\varphi}$
$Y_{1-1} = \left(\dfrac{3}{8\pi}\right)^{1/2} \sin\theta\, e^{-i\varphi}$	$Y_{3-2} = \left(\dfrac{105}{32\pi}\right)^{1/2} \sin^2\theta\cos\theta\, e^{-2i\varphi}$
$Y_{10} = \left(\dfrac{6}{8\pi}\right)^{1/2} \cos\theta$	$Y_{3-1} = \left(\dfrac{42}{128\pi}\right)^{1/2} \sin\theta(5\cos^2\theta - 1)\, e^{-i\varphi}$
$Y_{11} = -\left(\dfrac{3}{8\pi}\right)^{1/2} \sin\theta\, e^{i\varphi}$	$Y_{30} = \left(\dfrac{63}{16\pi}\right)^{1/2}\left(\dfrac{5}{3}\cos^3\theta - \cos\theta\right)$
$Y_{2-2} = \left(\dfrac{15}{32\pi}\right)^{1/2} \sin^2\theta\, e^{-2i\varphi}$	$Y_{31} = -\left(\dfrac{42}{128\pi}\right)^{1/2} \sin\theta(5\cos^2\theta - 1)\, e^{-i\varphi}$
$Y_{2-1} = \left(\dfrac{15}{8\pi}\right)^{1/2} \sin\theta\cos\theta\, e^{-i\varphi}$	$Y_{32} = \left(\dfrac{105}{32\pi}\right)^{1/2} \sin^2\theta\cos\theta\, e^{2i\varphi}$
$Y_{20} = \left(\dfrac{10}{32\pi}\right)^{1/2} (3\cos^2\theta - 1)$	$Y_{33} = -\left(\dfrac{70}{128\pi}\right)^{1/2} \sin^3\theta\, e^{3i\varphi}$
$Y_{21} = -\left(\dfrac{15}{8\pi}\right)^{1/2} \sin\theta\cos\theta\, e^{i\varphi}$	
$Y_{22} = \left(\dfrac{15}{32\pi}\right)^{1/2} \sin^2\theta\, e^{2i\varphi}$	

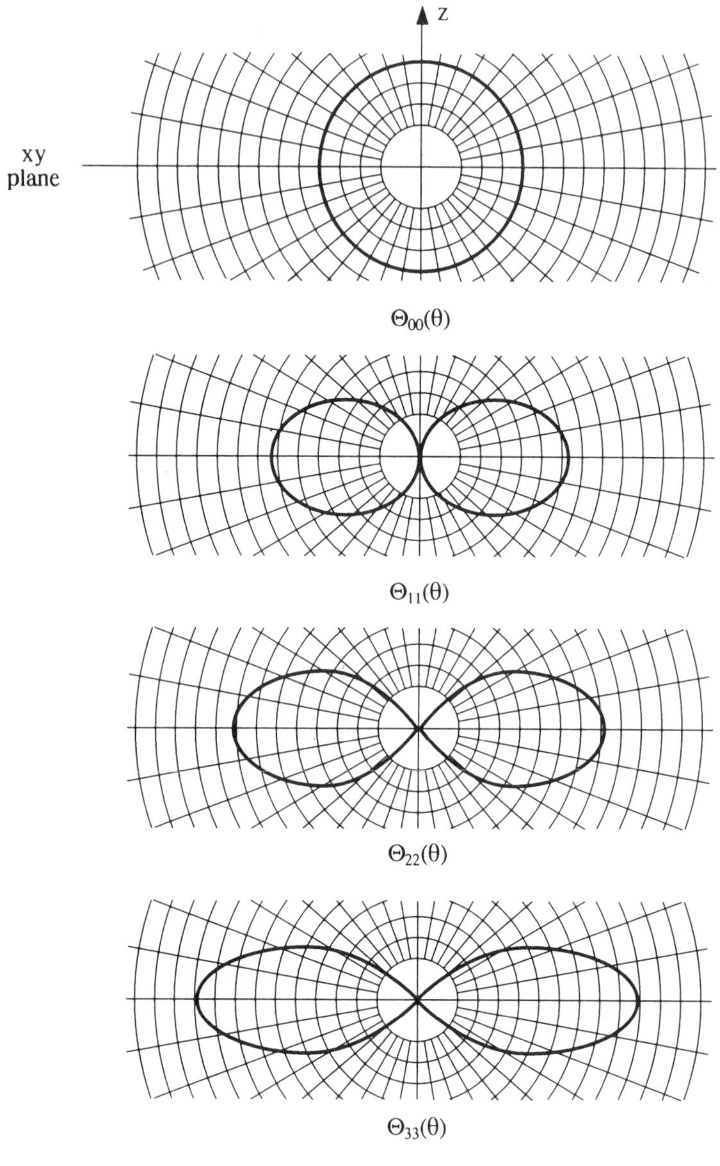

Figure 2.4. Polar representation of the functions $\Theta_{lm}(\theta) \equiv |Y_{lm}(\theta, \varphi)|^2$ for $m = \pm l$ and $l = 0, 1, 2, 3$. We note that $\Theta_{l,-m} = \Theta_{lm}$. For every point of the polar curve we read: an angle θ (this is the angle between the vertical axis (z) and the radial line crossing the curve at the point) and the corresponding value of $\Theta_{lm}(\theta)$ (given by the length of the radius of the circle passing through the point). We observe the 'concentration' of the function about the xy plane with increasing l. (Redrawn from L. Pauling and E. B. Wilson, Jr., *Introduction to Quantum Mechanics*, (New York: McGraw-Hill) 1935)

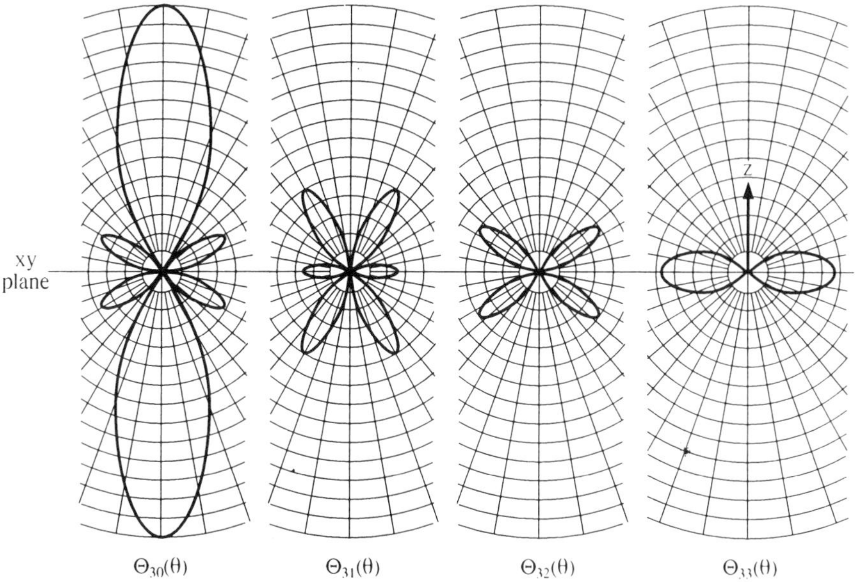

Figure 2.5. Polar representation of the functions $\Theta_{lm}(\theta) \equiv |Y_{lm}(\theta, \varphi)|^2$ for $l = 3$ and $m = 0, \pm 1, \pm 2, \pm 3$. (Redrawn from L. Pauling and E. B. Wilson, Jr., *Introduction to Quantum Mechanics*, loc. cit.)

for certain values of l and m. We note that $|Y_{lm}(\theta, \varphi)|$ does not depend on φ.

A prescription for writing down any spherical harmonic ($l = 0, 1, 2, \ldots$) and a summary of their properties can be found in Appendix B.6.1. Here we note only the following:

They are orthogonal to each other, i.e.

$$\int Y_{lm}(\Omega) Y^*_{l'm'}(\Omega)\, \mathrm{d}\Omega = \int_0^{2\pi}\int_0^{\pi} Y_{lm}(\theta, \varphi) Y^*_{l'm'}(\theta, \varphi) \sin\theta\, \mathrm{d}\theta\, \mathrm{d}\varphi,$$

$$= 1, \text{ if } l = l' \text{ and } m = m' \tag{2.4a}$$

$$= 0, \text{ otherwise}$$

where $\Omega \equiv \theta, \varphi$ and $\mathrm{d}\Omega \equiv \sin\theta\, \mathrm{d}\theta\, \mathrm{d}\varphi$ is the element of solid angle about the direction (θ, φ). [To $\mathrm{d}\Omega$ corresponds the surface element $\mathrm{d}S = r^2 \sin\theta\, \mathrm{d}\theta\, \mathrm{d}\varphi$ on the spherical surface of radius r (see Fig. 2.6).]

They satisfy the following relation

$$\sum_{m=-l}^{l} |Y_{lm}(\theta, \varphi)|^2 = \frac{2l + 1}{4\pi} \tag{2.4b}$$

Finally they are a complete set of functions, i.e. any function $f(\theta, \varphi)$,

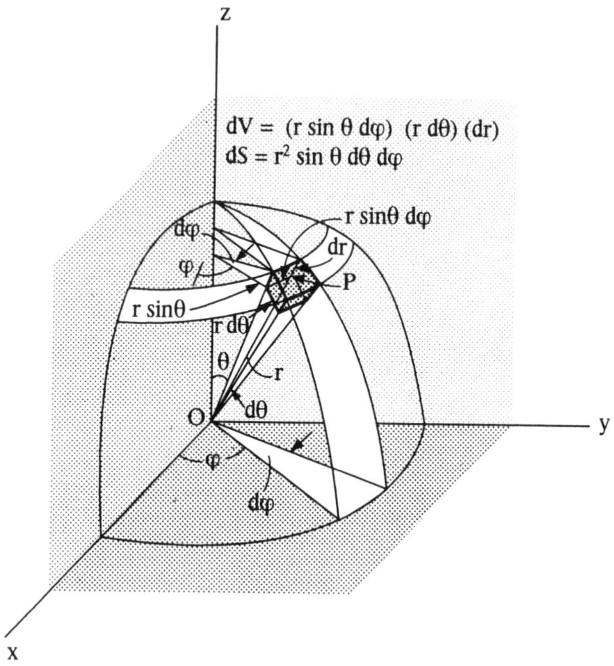

Figure 2.6. Surface element dS (on the spherical surface of radius r) and volume element dV, at the point P (with spherical coordinates r, θ, φ)

where $0 \leqslant \theta \leqslant \pi$ and $0 \leqslant \varphi < 2\pi$, can be written as a linear sum of spherical harmonics as follows

$$f(\theta, \varphi) = \sum_{l=0}^{\infty} \sum_{m=-l}^{l} c_{lm} Y_{lm}(\theta, \varphi) \tag{2.4c}$$

where c_{lm} are constants to be determined. Usually the above infinite sum reduces, in practice, to a sum of a finite number of terms.

2.3 Scattering states in a spherically symmetric field

2.3.1 THE FREE PARTICLE

A free particle moves in a constant ($U(\boldsymbol{r}) = 0$) potential field which is obviously spherically symmetric. This means that we can choose the eigenstates (of the energy) of the free particle to be, also, eigenstates of the angular momentum of the particle. In this case we obtain (see Appendix

B.7.1), instead of the plane waves of Section 1.7.1, the following spherical waves

$$\psi_{Elm}(r, \theta, \varphi; t) = j_l(kr) Y_{lm}(\theta, \varphi) \exp(-\mathrm{i}Et/\hbar) \tag{2.5}$$

where

$$k \equiv (2mE)^{1/2}/\hbar$$

$j_l(kr)$ are the so-called spherical Bessel functions (see Table 2.2). The particle (of mass m) described by (2.5) has a definite energy E, and a definite angular momentum: the square of the magnitude of the angular momentum equals $l(l+1)\hbar^2$ and its z-component equals $m\hbar$ in accordance with (2.2).

The energy spectrum of a free particle is, as we know, continuous ($0 < E < \infty$) and from (2.5) we see that to every energy level corresponds an infinite number of eigenstates: spherical waves of different angular momentum $(l, m) = (0, 0), (1, -1), (1, 0), (1, +1), (2, -2)$, etc. The situation is analogous to that of plane waves; we recall that plane waves

Table 2.2. Spherical Bessel and Hankel functions

$$j_0(x) = \frac{\sin x}{x} \qquad j_1(x) = \frac{\sin x}{x^2} - \frac{\cos x}{x}$$

$$j_2(x) = \left(\frac{3}{x^2} - \frac{1}{x}\right)\sin x - \frac{3\cos x}{x^2}, \quad j_3(x) = \left(\frac{15}{x^4} - \frac{6}{x^2}\right)\sin x - \left(\frac{15}{x^3} - \frac{1}{x}\right)\cos x$$

etc.

$$h_0^+(x) = \frac{\mathrm{e}^{\mathrm{i}x}}{\mathrm{i}x} \qquad h_1^+(x) = -\frac{\mathrm{e}^{\mathrm{i}x}}{x}\left(1 + \frac{\mathrm{i}}{x}\right)$$

$$h_2^+(x) = \frac{\mathrm{i}\,\mathrm{e}^{\mathrm{i}x}}{x}\left(1 + \frac{3\mathrm{i}}{x} - \frac{3}{x^2}\right), \qquad h_3^+(x) = \frac{\mathrm{e}^{\mathrm{i}x}}{x}\left(1 + \frac{6\mathrm{i}}{x} - \frac{15}{x^2} - \frac{15\mathrm{i}}{x^3}\right)$$

etc.[5]

Small and large argument limits:

$$x \ll 1: j_l(x) \simeq \frac{x^l}{(2l+1)!!}\left(1 - \frac{x^2}{2(2l+3)} + \cdots\right)$$

where $(2l+1)!! = (2l+1)(2l-1)(2l-3)\cdots(5)\cdot(3)\cdot(1)$

$$x \to \infty: j_l(x) \simeq \frac{1}{x}\sin\left(x - \frac{l\pi}{2}\right)$$

$$x \to \infty: h_l^+(x) \simeq (-\mathrm{i})^{l+1}\frac{\mathrm{e}^{\mathrm{i}x}}{x}$$

corresponding to momenta of different direction but of the same magnitude have the same energy.

Plane waves and spherical waves of the same energy are, of course, intimately connected. It can be shown that a plane wave can be written as a linear sum of spherical waves of the same energy as follows (we consider, for the sake of simplicity, a plane wave of energy $E = \hbar^2 k^2/2m$ propagating in the z-direction):

$$\exp(\mathrm{i}kz - \mathrm{i}Et/\hbar) = \sum_l \mathrm{i}^l [4\pi(2l+1)]^{1/2} j_l(kr) Y_{l0}(\theta) \exp(-\mathrm{i}Et/\hbar) \quad (2.6)$$

where Y_{l0} are the spherical harmonics of $m = 0$ which are functions of θ only (see Table 2.1). We could, also, write a spherical wave as a linear sum of plane waves if we wanted to. In summary, an eigenstate of the energy of a free particle can be, also, an eigenstate of the momentum (a plane wave) or an eigenstate of the angular momentum (a spherical wave) but, obviously, it cannot be an eigenstate of the momentum *and* the angular momentum.

2.3.2 SCATTERING BY A SPHERICALLY SYMMETRIC BARRIER

Equation (2.6) is particularly useful in describing the scattering of a particle by a scatterer of spherical symmetry (see Fig. 2.7). One proceeds as follows: the incident wavepacket is replaced by an incident plane wave (2.6), as we have done in treating scattering by the rectangular barrier of Section 1.8, and we seek, again in the manner of Section 1.8, to describe the scattering of the particle through a scattering wavefunction which consists, away from the scatterer, of the incident wave and outgoing waves representing the

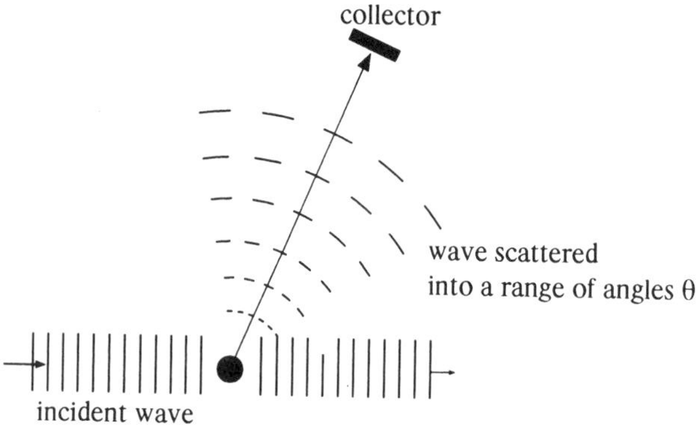

Figure 2.7. Scattering of a plane wave by a spherical scatterer centred on the origin of coordinates

scattered probability-flux. Let us assume, for example, that the scatterer is a so-called hard sphere, which means that

$$\begin{aligned} U(r) &= \infty, 0 \leqslant r \leqslant a \\ &= 0, r > a \end{aligned} \tag{2.7}$$

Away from the scatterer means, in the present case, $r > a$. There the potential equals zero and the particle is free. To obtain an expression for the scattered (outgoing part) of the wavefunction corresponding to an incident wave (2.6) we argue as follows. Because of the spherical symmetry of the scatterer, each of the spherical waves in (2.6) scatters independently of all the others. The energy and the angular momentum of the particle are both conserved during the scattering process. Now the question arises: how does one represent an outgoing (from the scatterer towards $r = \infty$) free particle with definite energy E and definite angular momentum (lm)? The answer is (see Appendix B.7.1): a spherical wave of the following form

$$\overset{+}{\psi}_{Elm}(r, \theta, \varphi; t) = h_l^+(kr) Y_{lm}(\theta, \varphi) \, \mathrm{e}^{-\mathrm{i}Et/\hbar}, \; r \geqslant a \tag{2.8a}$$

where $h_l^+(kr)$ is a spherical Hankel function (see table 2.2). Using the asymptotic expression of h_l^+ given in table 2.2 we find that asymptotically $\overset{+}{\psi}_{Elm}$ behaves as follows

$$\overset{+}{\psi}_{Elm}(r, \theta, \varphi; t) \simeq (-\mathrm{i})^{l+1} Y_{lm}(\theta, \varphi) \frac{\exp(\mathrm{i}kr - \mathrm{i}Et/\hbar)}{kr}, \; \text{as } r \to \infty \tag{2.8b}$$

which obviously represents a wave propagating radially outwards. We note (see Table 2.2) that h_l^+ diverges at the origin $(r = 0)$, but this need not worry us; we are using (2.8) in the region $r \geqslant a$ where it is appropriate.

Now, because each term in (2.6) scatters independently from all others, preserving its angular momentum, it contributes to the scattering wavefunction a corresponding term given by

$$(\mathrm{i})^l [4\pi(2l + 1)]^{1/2} j_l(kr) Y_{l0}(\theta) \exp(-\mathrm{i}Et/\hbar) + S_l h_l^+(kr) Y_{l0}(\theta) \exp(-\mathrm{i}Et/\hbar) \tag{2.9}$$

The first term in the above expression represents the incident wave (of given energy and angular momentum) and the second term represents the scattered wave (of the same energy and angular momentum). The coefficient S_l is determined by the wavefunction in the region of the scatterer, in essentially the same way as the coefficients B and C are determined in the wavefunction (1.71) for the one-dimensional rectangular barrier. In our example the potential is infinite for $r < a$, and the wavefunction vanishes inside the scatterer. Therefore, because of the continuity of the wavefunction, (2.9) must vanish at $r = a$, which means that

$$S_l = -\mathrm{i}^l [4\pi(2l + 1)]^{1/2} j_l(ka)/h_l^+(ka) \tag{2.9a}$$

The complete scattering wavefunction, denoted by ψ_E, is a sum (over all l) of the terms (2.9). We need only write down the asymptotic form of this wavefunction as $r \to \infty$; we obtain (as $r \to \infty$):

$$\psi_E(r, \theta; t) \simeq \exp(\mathrm{i}kz - \mathrm{i}Et/\hbar) + f(\theta)\frac{\exp(\mathrm{i}kr - \mathrm{i}Et/\hbar)}{r} \tag{2.10a}$$

$$f(\theta) = \sum_l [(-\mathrm{i})^{l+1} S_l/k] Y_{l0}(\theta) \tag{2.10b}$$

Noting (see Fig. 2.7) that for large values of r, a spherical wave can be regarded as a plane wave in any small space interval, we write for the probability-current density of the scattered wave the following expression[6]

$$|\psi_E(r, \theta; t)|^2 v = \left| f(\theta)\frac{\exp(\mathrm{i}kr - \mathrm{i}Et/\hbar)}{r} \right|^2 v = \frac{|f(\theta)|^2}{r^2} v$$

where $v = \hbar k/m$ is the velocity of the particle. The above expression is obtained as in the one-dimensional case of Section 1.8 (we note that j_c of (1.68) equals $|\psi(x)|^2 v$ where ψ is given by (1.71b)), by multiplying the velocity of the free particle (in the radial direction) with the probability-density at the given point. It follows from the above expression that the probability per unit time that the scattered particle will pass through a surface element $\mathrm{d}S = r^2\,\mathrm{d}\Omega$, where $\mathrm{d}\Omega = \sin\theta\,\mathrm{d}\theta\,\mathrm{d}\varphi$ is the element of solid angle in the direction (θ, φ), is

$$v|f(\theta)|^2\,\mathrm{d}\Omega$$

The ratio of the above quantity to the probability-current density v of the incident wave, known as the effective cross-section for scattering into the solid angle $\mathrm{d}\Omega$, is given by

$$\mathrm{d}\sigma = |f(\theta)|^2\,\mathrm{d}\Omega$$

We note that $\mathrm{d}\sigma$ has dimensions of area as implied by its name. Since $\mathrm{d}\sigma$ does not depend on φ we can say that

$$\mathrm{d}\sigma = 2\pi \sin\theta |f(\theta)|^2\,\mathrm{d}\theta \tag{2.11}$$

is the effective cross-section for scattering through angles in the range from θ to $\theta + \mathrm{d}\theta$.

It remains for us to evaluate $f(\theta)$ from (2.9a) and (2.10b). Provided the energy of the incident particle is not too high, only a few spherical waves corresponding to small values of the angular momentum get scattered. The reason for this can be understood from the following semiclassical argument. Let a particle with linear momentum $\hbar k$ be incident on the scatterer (as in Fig. 2.7). If there were no interaction between the particle and the scatterer, the particle would move, according to classical mechanics, along a straight line. If D is the (normal) distance from the centre of the scatterer to this straight line, the magnitude of the classical angular momentum of the particle with respect to the centre of the scatterer will be $\hbar kD$. We may assume that this quantity is equal to one of the allowed values of the angular momentum given by (2.2a). We can then say that the average value of D for a particle of angular momentum $\hbar[l(l+1)]^{1/2}$ is given, approximately, by

$$D \simeq \frac{\hbar[l(l+1)]^{1/2}}{\hbar k} \simeq \frac{l}{k}$$

If we now switch on the interaction $U(r)$ between the particle and the scatterer, we find, noting that $U = 0$ for $r > a$, that only spherical waves with $D = l/k \leqslant a$ will see the scatterer. In mathematical terms this means that

$$S_l \simeq 0 \text{ for } l > ka$$

Let us for the sake of simplicity assume that $ka \ll 1$, in which case only the term $l = 0$ contributes significantly to $f(\theta)$ of (2.10b).[7] Substituting in (2.9a) from table 2.2 we can evaluate S_0, and in turn $f(\theta)$ from (2.10b), and finally $d\sigma$ from (2.11). We find

$$d\sigma = 2\pi a^2 \sin\theta \, d\theta$$

for the effective cross-section for scattering through angles in the range from θ to $\theta + d\theta$. The effective total cross-section (for scattering into any angle), denoted by σ, we obtain by integrating the above expression over all angles $0 < \theta \leqslant \pi$. The result is

$$\sigma = 4\pi a^2$$

We conclude this section with the following note. The method we have described can be used to evaluate the differential scattering cross-section $d\sigma$ and the total scattering cross-section σ of any scatterer $U(r)$ of finite range ($U(r) = 0$ for $r > a$). The formula (2.9a) for S_l will have to be replaced in each case by a formula appropriate to the scatterer under consideration (see e.g., problem 2.2). Finally, it is worth noting the importance of scattering experiments in the investigation of the properties of various physical systems. For example, the scattering of electrons by atoms, molecules and solid surfaces tells us a lot about the structure of these systems when we are able to analyse correctly (by proper evaluation of the scattering cross-sections) the experimental data.

2.4 Bound states in a spherically symmetric field

In Fig. 2.8 we show, schematically, the discrete energy levels of a particle, of mass m, in a spherical well defined by

$$U(r) = -U_O < 0, \; r < a$$
$$= 0, \; r > a$$

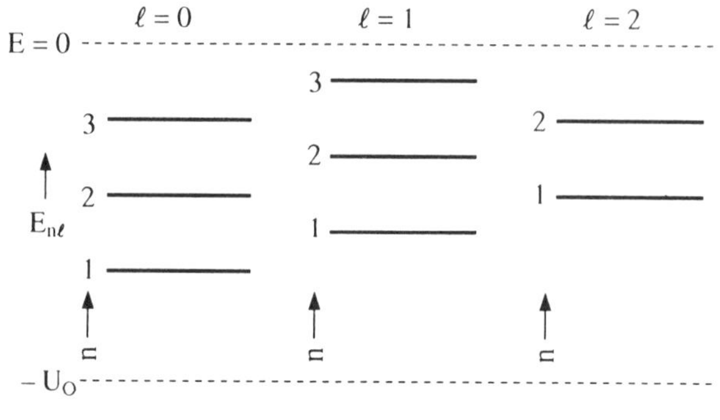

Figure 2.8. Discrete energy levels of a particle in a spherical well

The corresponding bound eigenstates of the particle are described by eigenfunctions of the form (2.3). We note that for given l (which takes the values: 0, 1, 2, ... according to (2.2a)) we have a number, N_l, of energy levels, which we denote by E_{nl}, $n = 1, 2, \ldots, N_l$ in order of increasing energy, n being the quantum number which defines besides the energy, the radial part $R_{nl}(r)$ of the corresponding eigenfunction (2.3). The number of discrete energy levels in a given well depends on the parameters, a and U_O, of the well. In the hypothetical well of our example there are three energy levels for $l = 0$, three for $l = 1$, two for $l = 2$ and none for $l > 2$. In a very shallow well there are no bound states; just above a critical depth a single bound state, of zero angular momentum ($l = 0$), is obtained. The number of bound states increases as the depth of the well increases and tends to infinity as $U_O \to \infty$ (see Section B.7.2 and problem 2.3).

We note that there are $(2l + 1)$ eigenstates: $\psi_{nlm} = R_{nl}(r)Y_{lm}(\theta, \varphi)\exp(-iE_{nl}t/\hbar)$, $m = -l, -l + 1, \ldots, l - 1, l$, corresponding to different values $(m\hbar)$ of the z-component of the angular momentum, with the same energy E_{nl}. The above degeneracy of E_{nl} follows from the spherical symmetry of the potential field: the eigenvalues of the energy of a particle in a spherically symmetric field cannot depend on the quantum number m, because a dependence on m implies a preferred z-direction which does not exist in a spherically symmetric field.

The discrete energy spectrum and the corresponding bound states of a particle in any spherically symmetric field are described in very much the same way as those of a particle in a spherical well. As a rule, the energy of a state depends on the angular momentum quantum-number l, but not on m, which means that the energy level E_{nl} is $(2l + 1)$ -fold degenerate. The corresponding eigenstates are described by wavefunctions of the form (2.3). With the angular part of the wavefunction completely determined by the angular momentum quantum-numbers (lm), one is left with the determination of the radial part $R_{nl}(r)$ of the wavefunction. For a given potential field this is determined by the magnitude of the angular momentum (quantum number l) and by the energy of the particle (quantum number n) (see Appendix B.7).

2.5 Bound states of an electron in a Coulomb field

We note that the potential energy (2.1) of an electron in the so-called Coulomb field of a nucleus is negative, increasing gradually and monotonically to zero as $r \to \infty$. We expect, therefore, the bound states of the electron in this field to correspond to a discrete set of negative energy levels. The rules of quantum mechanics lead in the present case to the following set of energy eigenvalues

$$E_n = -\frac{m_r Z^2 e^4}{32\pi^2 \epsilon_0^2 \hbar^2 n^2}, \quad n = 1, 2, 3, \ldots \tag{2.12}$$

$m_r \equiv mM/(m + M) \simeq m$, where m is the mass of the electron and M the mass of the nucleus. The approximation $m_r \simeq m$ corresponds to our putting the centre of mass of the atom at the centre of the nucleus (see Section 2.1).[8] We note that the energy levels (2.12) do not depend on the angular momentum quantum-number l, as is commonly the case with spherically symmetric fields (see Section 2.4), except in the following sense. We find that there are a number of eigenstates with the same energy E_n; they have different angular momenta: $l = 0, 1, \ldots, n - 1$; and of course for given n and l we have $(2l + 1)$ states corresponding to different values of the quantum number m: $m = -l, -l + 1, \ldots, l$. The square of the magnitude of the angular momentum of the electron is given in each case by $l(l + 1)\hbar^2$ in accordance with (2.2a) and its z-component is given by $m\hbar$ according to (2.2b). The corresponding eigenstates are described by wavefunctions of the form (2.3). In Table 2.3 we present the eigenfunctions ψ_{nlm} which correspond to the ground (the lowest) energy level, E_1, of the atom and to the two energy levels, E_2 and E_3, immediately above it. We note that E_1 is non-degenerate, E_2 is four-fold degenerate and E_3 is nine-fold degenerate. The degeneracy of the energy levels of an electron in a Coulomb field, in addition to what is expected from the spherical symmetry of this field, is a peculiarity of the Coulomb field and is known for this reason as accidental degeneracy. We can put it differently: the fact that states corresponding to different values of the quantum-number l have the same energy is an accidental phenomenon peculiar to the Coulomb field.

We verify from Table 2.3 that the radial part, $R_{nl}(r)$, of the eigenfunction depends only on the quantum-numbers n and l as expected from the general formula (2.3). A graphical representation of the R_{nl} for the eigenstates listed in Table 2.3 is given in Fig. 2.9. The following properties of R_{nl} are

Table 2.3. Eigenstates of an electron in a Coulomb field

$$\psi_{nlm} = R_{nl}(r)Y_{lm}(\theta, \varphi)\exp(-\mathrm{i}E_n t/\hbar)$$

$$\psi_{100} = 2\left(\frac{Z}{a_0'}\right)^{3/2}\exp(-Zr/a_0')Y_{00}\exp(-\mathrm{i}E_1 t/\hbar)$$

$$\psi_{200} = \frac{1}{\sqrt{8}}\left(\frac{Z}{a_0'}\right)^{3/2}\left(2-\frac{Zr}{a_0'}\right)\exp(-Zr/2a_0')Y_{00}\exp(-\mathrm{i}E_2 t/\hbar)$$

$$\psi_{210} = \frac{1}{\sqrt{24}}\left(\frac{Z}{a_0'}\right)^{3/2}\frac{Zr}{a_0'}\exp(-Zr/2a_0')Y_{10}(\theta)\exp(-\mathrm{i}E_2 t/\hbar)$$

$$\psi_{21\pm1} = \frac{1}{\sqrt{24}}\left(\frac{Z}{a_0'}\right)^{3/2}\frac{Zr}{a_0'}\exp(-Zr/2a_0')Y_{1\pm1}(\theta, \varphi)\exp(-\mathrm{i}E_2 t/\hbar)$$

$$\psi_{300} = \frac{2}{81\sqrt{3}}\left(\frac{Z}{a_0'}\right)^{3/2}\left(27-\frac{18Zr}{a_0'}+2\left(\frac{Zr}{a_0'}\right)^2\right)\exp(-Zr/3a_0')Y_{00}\exp(-\mathrm{i}E_3 t/\hbar)$$

$$\psi_{310} = \frac{4}{81\sqrt{6}}\left(\frac{Z}{a_0'}\right)^{3/2}\left(6-\frac{Zr}{a_0'}\right)\frac{Zr}{a_0'}\exp(-Zr/3a_0')Y_{10}(\theta)\exp(-\mathrm{i}E_3 t/\hbar)$$

$$\psi_{31\pm1} = \frac{4}{81\sqrt{6}}\left(\frac{Z}{a_0'}\right)^{3/2}\left(6-\frac{Zr}{a_0'}\right)\frac{Zr}{a_0'}\exp(-Zr/3a_0')Y_{1\pm1}(\theta, \varphi)\exp(-\mathrm{i}E_3 t/\hbar)$$

$$\psi_{320} = \frac{\sqrt{8}}{81\sqrt{15}}\left(\frac{Z}{a_0'}\right)^{3/2}\left(\frac{Zr}{a_0'}\right)^2\exp(-Zr/3a_0')Y_{20}(\theta)\exp(-\mathrm{i}E_3 t/\hbar)$$

$$\psi_{32\pm1} = \frac{\sqrt{8}}{81\sqrt{15}}\left(\frac{Z}{a_0'}\right)^{3/2}\left(\frac{Zr}{a_0'}\right)^2\exp(-Zr/3a_0')Y_{2\pm1}(\theta, \varphi)\exp(-\mathrm{i}E_3 t/\hbar)$$

$$\psi_{32\pm2} = \frac{\sqrt{8}}{81\sqrt{15}}\left(\frac{Z}{a_0'}\right)^{3/2}\left(\frac{Zr}{a_0'}\right)^2\exp(-Zr/3a_0')Y_{2\pm2}(\theta, \varphi)\exp(-\mathrm{i}E_3 t/\hbar)$$

$$a_0' \equiv (1 + m/M)a_0 \simeq a_0 = 4\pi\epsilon_O\hbar^2/me^2 = 0.52917\ \text{Å}$$

a_0 is known as the Bohr radius.

The eigenfunctions ψ_{nlm} are normalised, in the usual way, as follows:

$$\int |\psi_{nlm}(r, \theta, \varphi)|^2\,\mathrm{d}V = 1$$

where $\mathrm{d}V = (r^2\,\mathrm{d}r)\sin\theta\,\mathrm{d}\theta\,\mathrm{d}\varphi$ (see Fig. 2.6) and the integral extends over all space: $0 \leqslant r < \infty$, $0 \leqslant \theta \leqslant \pi$, $0 \leqslant \varphi < 2\pi$.

The eigenfunctions corresponding to the higher energy levels of the electron in the Coulomb field can be found in any book of quantum mechanics or atomic physics.

---

worth noting. For given l the number of zeros, outside the origin of coordinates, of the wavefunction, increases with n as follows. The lowest (corresponding to $n = l + 1$) has none, the next ($n = l + 2$) has one zero, the one after ($n = l + 3$) has two zeros, and so on. We observe also that, as n increases, the eigenfunction spreads further away from the nucleus; a fact which one can easily understand as follows. A larger value of n means a

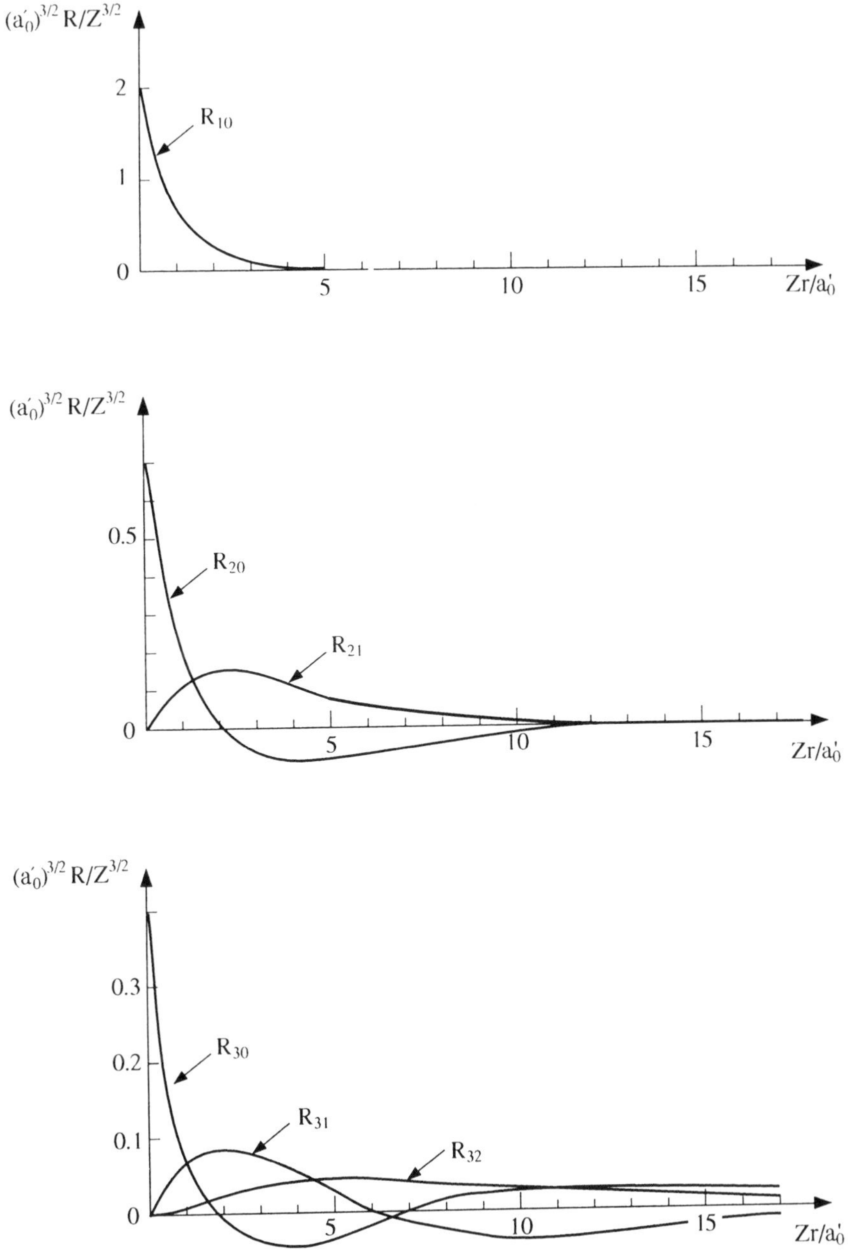

Figure 2.9. The radial part $R_{nl}(r)$, $n = 1, 2, 3$ and $l = 0, 1, 2$, of the wavefunctions listed in Table 2.3. (Redrawn from R. B. Leighton, *Principles of Modern Physics*, (New York: McGraw-Hill) 1959. Reproduced with permission)

higher value of the total energy according to (2.12) and this allows the electron to move in a larger region of space. Not only does the classically allowed region become larger, but the wavefunction penetrates further into the classically forbidden region with increasing energy (see Section 1.6). For the same reason the difference between two successive energy levels (2.12) becomes smaller as the energy increases. We remember (see Section 1.4.2) that the discreteness of the energy levels derives, ultimately, from the restriction in space of the corresponding eigenfunctions.

An electron in the eigenstate $\psi_{nlm}(\boldsymbol{r}, t)$ is found in the volume element dV about the point $\boldsymbol{r} = (r, \theta, \varphi)$ with a probability $|\psi_{nlm}|^2 dV$ which is independent of the time, as is always the case for an eigenstate of the energy, and is independent of the angle φ because $|Y_{lm}|^2$ does not depend on φ. In Fig. 2.10 we present for a number of states, 1s, 2p, etc. (the notation is explained below), a photographic representation of the corresponding probability-density function.[9] We present a sectional view of this quantity in a plane (that of the page) which contains the z-axis (vertical direction). The reader will note that the above photographic representation follows directly from the angular variation of $\Theta_{lm}(\theta)$ shown in Figs. 2.4 and 2.5 and the variation with radial distance of $R_{nl}(r)$ shown in Fig. 2.9.

It is customary, in the literature of atomic physics, to call the eigenstates of the electron with angular momentum quantum-number $l = 0$, s-states, those with $l = 1$ we call p-states, those with $l = 2$, d-states, those with $l = 3$, f-states, etc. An eigenstate of given quantum-numbers n and l is usually denoted as follows

$$1s, 2s, 3s, \ldots \text{ if } l = 0 \text{ and } n = 1, 2, 3, \ldots$$

$$2p, 3p, 4p, \ldots \text{ if } l = 1 \text{ and } n = 2, 3, 4, \ldots$$

$$3d, 4d, 5d, \ldots \text{ if } l = 2 \text{ and } n = 3, 4, 5, \ldots$$

$$4f, 5f, 6f, \ldots \text{ if } l = 3 \text{ and } n = 4, 5, 6, \ldots$$

The formulae we have given are valid for the hydrogen atom (H) when $Z = 1$, for the positive ion of helium (He^+) when $Z = 2$ (in this case an electron moves in the field of the helium nucleus which has a positive charge of $+2e$), and in general when an electron moves in the field of a bare nucleus of charge $+Ze$.

We must at this stage ask whether the above results, and in particular equation (2.12), are in agreement with the experimental data. Historically quantum mechanics was judged, at least in its early days, by its ability to explain the spectroscopic data relating to the absorption and emission of light by the hydrogen atom. It had been established, prior to the development of quantum mechanics that the hydrogen atom (similar rules apply to other atoms) absorbs (and then emits) light (electromagnetic

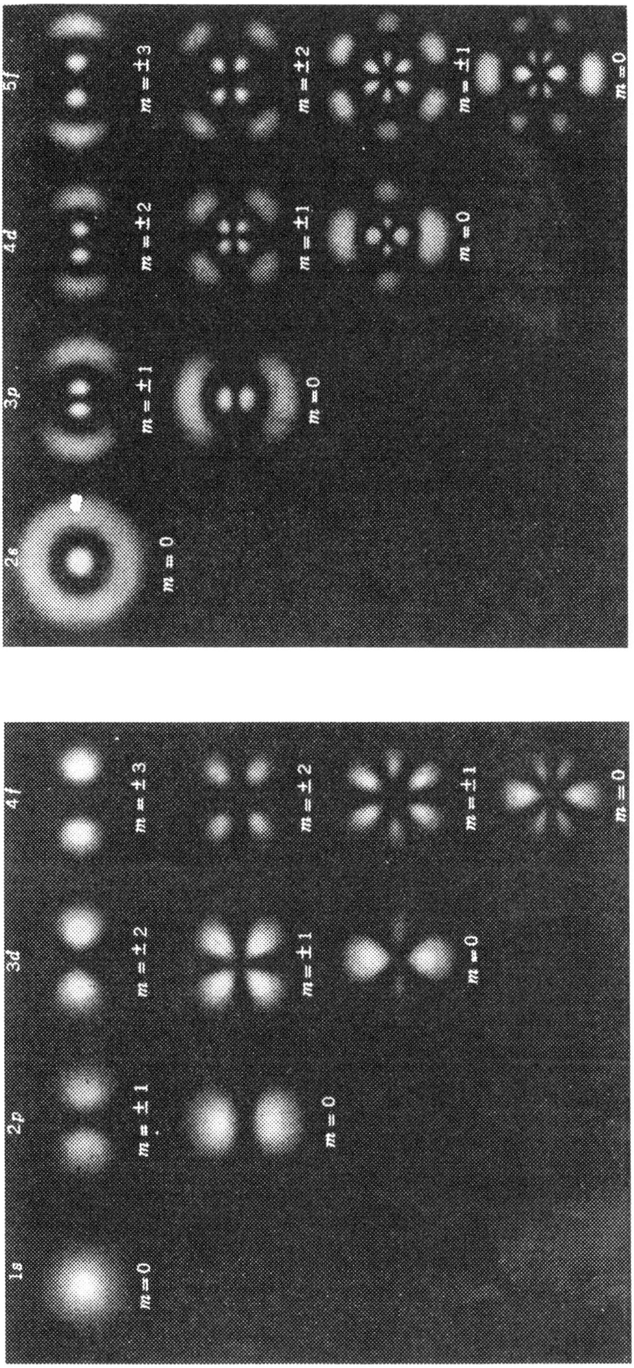

Figure 2.10. A photographic representation of the probability density function $|\psi_{nlm}|^2$, corresponding to various eigenstates, in a plane (that of the page) which contains the z-axis (vertical direction). The probability density function does not depend on the angle φ between the x-axis and the above plane. We note that the scale changes from column to column. (Reproduced with permission from R. B. Leighton, *Principles of Modern Physics* (New York: McGraw-Hill) 1959)

radiation) of certain frequencies only. The rules of classical physics (classical mechanics and classical electromagnetism) can not sustain a model of the atom which behaves in the above manner. The description of the interaction between the hydrogen atom and the electromagnetic field and the systematic calculation of the absorption and emission spectra of this atom was the first great achievement of quantum mechanics. We may add that the same quantum-mechanical rules apply (we forget for the moment the computational difficulties which increase with the complexity of the physical system under consideration) to the interaction of the electromagnetic field with matter in general (atoms, molecules, solids, etc.). We shall return to this problem in Section 2.7. At this point it is worthwhile to insert into our discussion a note concerning a collection of atoms (many millions of them) constituting a dilute gas of atoms in a given volume in a state of thermal (thermodynamic) equilibrium, which means that the gas does not lose or gain energy (of any form) from its environment.

2.6 Hydrogen gas in thermodynamic equilibrium

The atoms (or molecules) of a dilute gas move about freely except for the occasional collision between two of them or between one of them and a wall of the vessel which contains the gas. In thermodynamic equilibrium, the atoms (or molecules) are distributed as far as their translational motion is concerned in some definite way (a fraction of them having so much kinetic energy and another fraction so much and so on) which does not concern us here (see Section 2.14). We are concerned with the internal motion of the atoms (or molecules) of the gas, and in the case of atomic gases this means, simply, the motion of the electrons about the atomic nucleus. Let us concentrate on a gas of hydrogen atoms. The most general state $\psi(\boldsymbol{r}, t)$ of the hydrogen atom (we assume that the electron remains trapped in the region about the nucleus) is given according to the rules of quantum mechanics (see Appendix B.1) by a linear combination of bound states of the atom (the ψ_{nlm} of Table 2.3 with $Z = 1$); we have

$$\psi(\boldsymbol{r}, t) = \sum_{n,l,m} c_{nlm} R_{nl}(r) Y_{lm}(\theta, \varphi) \exp(-\mathrm{i}E_n t/\hbar) \tag{2.13}$$

where c_{nlm} are constant coefficients. However, we find that the behaviour of the hydrogen gas (N atoms in volume V at temperature T) is perfectly compatible with the following proposition: in thermodynamic equilibrium, every one of the atoms of the gas is to be found with certain probability in one or the other of the eigenstates $\psi_{nlm}(\boldsymbol{r}, t) = R_{nl}(r) Y_{lm}(\theta, \varphi) \exp(-\mathrm{i}E_n t/\hbar)$ of the atom and not in a (2.13)-state. Moreover we find that the number $N(nlm)$ of atoms in the eigenstate ψ_{nlm} depends only on the energy E_n of the state and the temperature:

$$N(nlm) = \text{const}\,\mathrm{e}^{-E_n/k_B T} \tag{2.14}$$

T denotes the absolute temperature in degrees Kelvin (K) of the gas and $k_B \equiv 8.628 \times 10^{-5}$ eV/K is a universal constant (Boltzmann's constant). It follows from (2.14) that in a hydrogen gas, the number N_1 of atoms in their ground states ($E_1 = -13.598$ eV) relates to the number of atoms in their ψ_{200} states ($E_2 = -3.399$ eV) as follows

$$N_2/N_1 = \exp[-(E_2 - E_1)/k_B T]$$

At room temperature, $k_B T \simeq 1/40$ eV, and therefore $N_2/N_1 \simeq \mathrm{e}^{-400}$, which means that the number of atoms with energy E_2 is negligible. The number of atoms in states of higher energy ($n > 2$) is obviously even smaller. We conclude: under normal conditions all the atoms are in their ground states. Only at very high temperatures ($k_B T > 10$ eV) are a significant number of atoms to be found in excited states.[10]

We talked about a hydrogen gas, but much the same thing applies to any gas of atoms or molecules in thermodynamic equilibrium. Assume that the eigenstates of the many-electron atom or molecule, are described by a set of quantum-numbers which we denote collectively by $\{\alpha\}$, and let the total (internal) energy of the atom or molecule in the state Ψ_α be E_α. We find that the number $N\{\alpha\}$ of atoms (molecules) in the state Ψ_α is given by

$$N(\alpha) = \text{const}\exp(-E_\alpha/k_B T) \tag{2.14a}$$

The normalisation constant in the above equation is determined from the requirement that the sum of $N(\alpha)$ over all possible states α must equal the total number of atoms (molecules) in the gas.

2.7 Interaction of atoms with the electromagnetic field

2.7.1 ATOM IN AN EXTERNAL FIELD

We know that in a hydrogen gas at not so high a temperature practically every atom will be in its ground state ψ_{100} with energy $E_1 = -13.598$ eV. Let us see what happens when the gas is irradiated with electromagnetic radiation (light) of frequency ν.

The main component of the electromagnetic radiation is, as far as we are concerned, an electric field which varies in space and time as shown in Fig. 2.11. (The magnetic field component of the electromagnetic field does not play a significant role in the phenomena we shall consider.) Because the wavelength of light is very long ($\lambda > 1000$ Å in the visible region of the electromagnetic radiation) compared to the 'diameter' of an atom (~ 1 Å), the atom sees a practically homogeneous field (the magnitude of the field is constant over a region containing the atom) which varies with the time as shown in Fig. 2.11a. This alternating electric field modifies the potential field (2.1) seen by the electron in the field of the nucleus. A new term

$$ezF_z^O \cos \omega t = eF_z^O r \cos \theta \cos \omega t \tag{2.15}$$

which describes the coupling of the atom to the electromagnetic radiation is added to (2.1). We understand (2.15) as follows: a particle of charge q in a homogeneous field F_z (in the z-direction) experiences a force qF_z in the same direction. The corresponding potential energy field is, according to (1.33), $-qF_z z = -qF_z r \cos \theta$. We obtain (2.15) by putting $q = -e$ (for the electron) and $F_z = F_z^O \cos \omega t$, where $\omega \equiv 2\pi\nu$. (Here and throughout this book we use the symbol ν for the frequency, measured in cycles/sec (Hz); and the symbol ω for the corresponding angular frequency in rad/sec. Often, when no ambiguity arises, we drop the epithet 'angular' when we refer to ω.) The contribution of the nucleus to the above coupling vanishes identically in a representation where the nucleus stays at the origin of the coordinates.

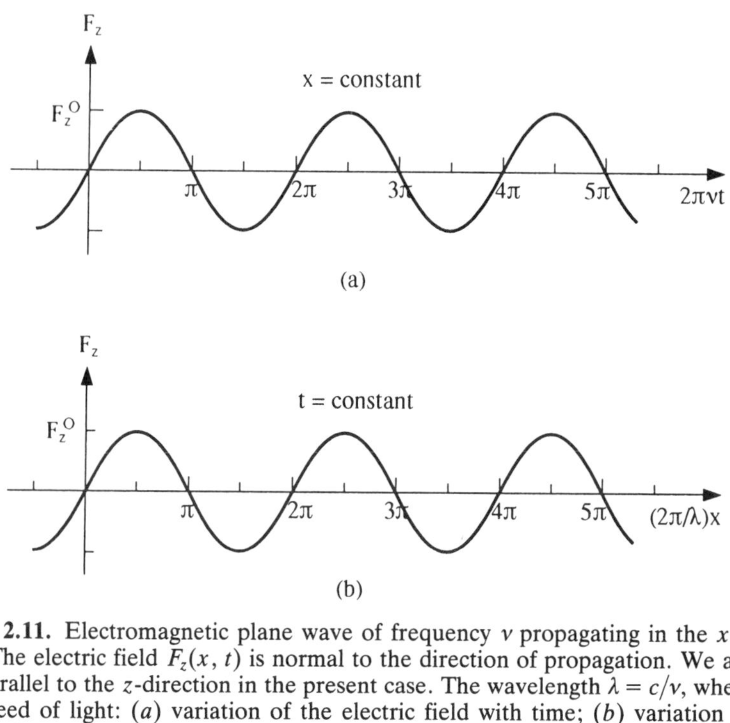

Figure 2.11. Electromagnetic plane wave of frequency ν propagating in the x-direction. The electric field $F_z(x, t)$ is normal to the direction of propagation. We assume it is parallel to the z-direction in the present case. The wavelength $\lambda = c/\nu$, where c is the speed of light: (*a*) variation of the electric field with time; (*b*) variation of the electric field in space

The state of the atom, described by ψ_{100} before the application of the external field, changes when the field is applied. We obtain

$$\psi_{100} \Rightarrow \psi(\boldsymbol{r}, t) \equiv c_{100}(t)\psi_{100} + \sum_{nlm \neq 100} c_{nlm}(t)\psi_{nlm} \tag{2.16}$$

where $\psi_{nlm} = R_{nl}(r)Y_{lm}(\theta, \varphi)\exp(-\mathrm{i}E_n t/\hbar)$ are the eigenstates of the unperturbed atom (Table 2.3). The sum (second term of (2.16)) includes all bound eigenstates of the atom except the ground state. (It is always possible to write $\psi(\boldsymbol{r}, t)$ as a linear sum of unperturbed states because the latter are a complete set of functions (see Appendix B.1); when $\hbar\omega$ is smaller than the energy required to remove the electron to infinity, only bound states contribute significantly to the above sum.) Because the interaction (2.15) is very weak compared to the interaction (2.1) between the electron and the nucleus, we expect that

$$c_{100}(t) \simeq 1$$
$$|c_{nlm}(t)| \ll 1, \; nlm \neq 100$$

The actual calculation of $\psi(\boldsymbol{r}, t)$, i.e. of the coefficients $c_{nlm}(t)$ in (2.16), will not concern us here (see Appendix B.11.1). We can summarise the results of the calculation as follows: The probability-density function $|\psi(\boldsymbol{r}, t)|^2$ obtained from (2.16) is quite different from the time-independent spherically symmetric distribution $|\psi_{100}|^2$ corresponding to the ground state of the atom (see Fig. 2.10). Under the influence of the external field, the spherically symmetric electronic cloud of the unperturbed atom is replaced by a non-symmetric distribution similar to that of an oscillating, with frequency ν, electric dipole: the electron is found, periodically, with greater probability to the left ($z < 0$) or to the right of the nucleus ($z > 0$).

The electric field induced dipole moment of the atom (nucleus + electron) is defined by the following relation (we remember that the nucleus is at the origin of coordinates and that $\boldsymbol{r}$ is the position vector of the electron with respect to this origin):

$$d_z(t) = -e\int z n(\boldsymbol{r}, t)\,\mathrm{d}V$$

which gives the z-component of the dipole moment; the other two components are zero when the external field is parallel to the z-axis; $n(\boldsymbol{r}, t)$ denotes the density of the electronic cloud at time t; it is given by $n(\boldsymbol{r}, t) = |\psi(\boldsymbol{r}, t)|^2$ where $\psi(\boldsymbol{r}, t)$ is the wavefunction (2.16). The calculation of $\psi(\boldsymbol{r}, t)$, presented in Section B.11.1, leads to the following formula for the induced dipole moment

$$d_z(t) = {\sum_j}' \frac{\omega_{j1} f_{j1}}{\omega_{j1}^2 - \omega^2} F_z^O \cos \omega t \tag{2.17}$$

where j stands for the three quantum numbers (n, l, m) which specify an eigenstate ψ_{nlm} of the atom. We note that the energy of the jth eigenstate is $E_j = E_n$ given by (2.12). The dash on the sum over j indicates that the ground state is to be excluded from the sum which, therefore, includes all states $nlm \neq 100$. By definition, $\omega_{j1} \equiv (E_j - E_1)/\hbar = (E_n - E_1)/\hbar$ and the f_{j1} are coefficients (positive or zero) which are determined by the ground state eigenfunction ψ_{100} and the corresponding (in each case) excited state $\psi_j = \psi_{nlm}$. It turns out that

$$f_{j1} = f_{nlm;100} = (2e^2/\hbar)\left|\int z\,\psi^*_{nlm}\psi_{100}\,\mathrm{d}V\right|^2 \tag{2.17a}$$

We can look at (2.17) as follows. The induced dipole moment is a sum of partial dipole moments, contributed by independent 'oscillators' of natural frequencies $\{\omega_{j1}\}$.

Finally, we note that (2.17) is valid only when $\omega \neq \omega_{j1}$, j standing for any one of the excited states of the atom. When the angular frequency ω of the incident electromagnetic wave is such that the equation[11]

$$\hbar\omega = (E_j - E_1) = (E_n - E_1) \tag{2.18}$$

is *not* satisfied whatever the value of n $(2, 3, 4, \ldots)$, the oscillating dipole does not absorb any energy from the incident radiation, and when the external field is removed the atom returns to its unperturbed state (its ground state). When the frequency of the incident wave satisfies (2.18) for some j, it is possible for the atom to absorb energy from the incident radiation, and, in this case, when the external field is removed the atom can be found (with certain probability) in excited state j (with energy $E_j = E_n = E_1 + \hbar\omega$). We find, as a rule, that a transition from a given (initial) state[12] to another (final) state is possible only when the quantum-number which defines the magnitude of the angular momentum of the final state differs from that of the initial state by unity. In other words, if the above mentioned quantum number of the initial state is l, that of the final state should be either $l + 1$ or $l - 1$. This so-called selection rule is written down as follows

$$\Delta l = \pm 1 \tag{2.19a}$$

Similarly we find that the quantum-number m, which defines the z-component of the angular momentum, obeys the selection rule

$$\Delta m = \pm 1 \text{ or } 0 \tag{2.19b}$$

We emphasise that in every case we must have

$$\nu = (E_j - E_i)/(2\pi\hbar) \tag{2.20}$$

where $E_j = E_n$ is the energy of the final state and $E_i = E_{n'}$ that of the initial state. Finally we note that the probability of an allowed transition (one that is permitted by (2.19) and (2.20)) actually happening in unit time (the transition rate) varies from transition to transition.[13]

2.7.2 DE-EXCITATION OF AN ATOM—EMISSION OF LIGHT

Let us assume that by appropriate irradiation we have excited a fraction of the atoms of a hydrogen gas. We now have in excited states (states with energy $E_n > E_1$) a larger number of atoms than ought to be there at equilibrium. And because the gas must return to equilibrium, sooner or later, the excited atoms must get rid of their excess energy by some means or other. We observe that this indeed happens in a relatively short interval of time ($\sim 10^{-8}$ s). The atom returns directly or in steps to its ground state, emitting light (electromagnetic radiation) in the process.

When the atom 'falls' from an initial (excited) state of energy E_n to a final state of lower energy $E_{n'}$ the emitted light has frequency

$$\nu_{nn'} = (E_n - E_{n'})/(2\pi\hbar) \tag{2.21}$$

We find, also, that the angular momentum quantum-number l of the final state differs from that of the initial state by plus or minus unity. The rules of equations (2.19) apply as much to the emission as to the absorption of light.

Substituting (2.12) with $Z = 1$ in (2.21) we obtain the following formula for the so-called emission lines (frequencies) of the hydrogen atom:

$$\nu_{nn'} = -\frac{m_r e^4}{64\pi^3\epsilon_0^2\hbar^3}\left(\frac{1}{n^2} - \frac{1}{n'^2}\right), \quad n > n' \tag{2.22}$$

The above frequencies or, equivalently, the corresponding wavelengths $\lambda_{nn'} = c/\nu_{nn'}$, are grouped into series as shown in Fig. 2.12. The wavelengths of a given series correspond to transitions to the same final level. In the Lyman series, the final state is the ground state ($n = 1$), in the Balmer series the final state has quantum-number $n = 2$, and in the Paschen series the final state has $n = 3$.[14]

It is obvious that the de-excitation of an atom does not always happen by a single transition to the ground state. In many instances the atom will pass through a number of states of successively lower energy before reaching the ground state.

We can now compare the above theoretical results with the experimental data, but before we do so, it is worthwhile to consider briefly in more general terms the interaction of matter (atoms, molecules, solids, etc.) with the electromagnetic field. We have described the absorption of light by the atom in terms of its interaction with the electric field of an incident electromagnetic wave, which perturbs the potential field of the free (unperturbed) atom. How is the emission of light by the atom to be

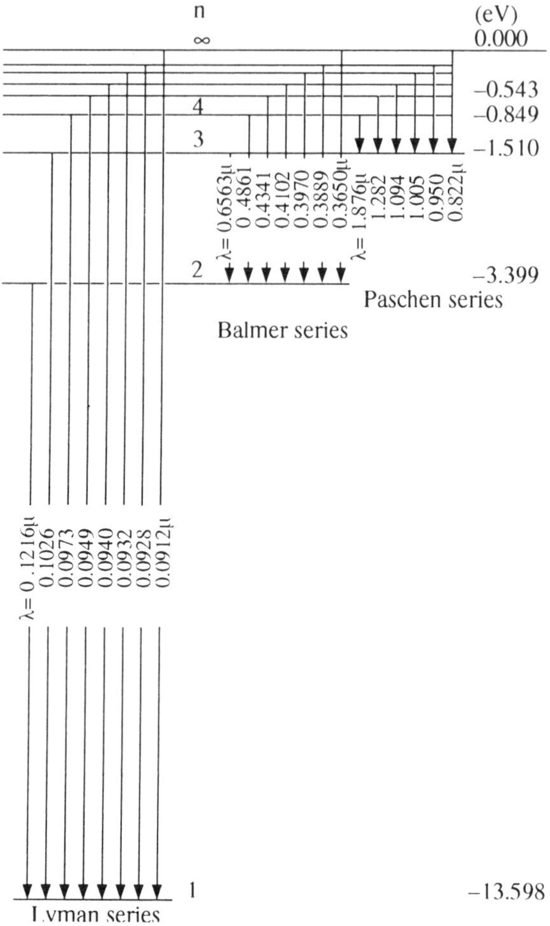

Figure 2.12. Energy-level diagram and corresponding emission spectra of the hydrogen atom according to (2.12) and (2.22)

explained? What causes the perturbation of the atom in this instance? For in the absence of perturbation the atom will stay for ever in its excited state. One must not forget that the excited eigenstates of an unperturbed atom are stationary states of the atom, as good as its ground state.

2.8 Quantum description of the electromagnetic field

A proper explanation of the emission of light by an atom requires a more systematic description of the electromagnetic field than the one we have given (Fig. 2.11). A complete description of the electromagnetic field lies beyond the scope of the present book, but we can sketch a picture of this field as follows: The electromagnetic field in an empty space of volume[15] $V = L^3$ (we disregard the interaction with atoms, molecules, etc., that may exist in the given volume) is defined by a set of independent 'oscillators'. Each oscillator is associated with a plane wave

$$\mathrm{Re}\,\{\exp(\mathrm{i}\boldsymbol{k}\cdot\boldsymbol{r} - \mathrm{i}\omega_k t)\}$$

$$\boldsymbol{k} = (k_x, k_y, k_z) = \frac{2\pi}{L}(n_x, n_y, n_z);\; n_x, n_y, n_z = 0, \pm 1, \pm 2, \ldots$$

$$\omega_k = ck = c(k_x^2 + k_y^2 + k_z^2)^{1/2} \tag{2.23}$$

where c is the speed of light, and a polarisation direction specified by a unit vector $\hat{e}$. For every wavevector $\boldsymbol{k}$, we obtain two linearly independent oscillators polarised along $\hat{e}_1$ and $\hat{e}_2$ respectively[16]: $\hat{e}_1$ and $\hat{e}_2$ are unit vectors normal to each other and to $\boldsymbol{k}$ (see Fig. 2.13).

The quantum-mechanical eigenstates (of the energy) of the oscillator $(\boldsymbol{k}, \hat{e})$ correspond to a discrete set of energy-eigenvalues given by

$$E_{\boldsymbol{k}\hat{e}}(n) = (n + \tfrac{1}{2})\hbar\omega_k,\; n = 0, 1, 2, \ldots \tag{2.24}$$

which, we note, are independent both of the direction of $\boldsymbol{k}$ and of the direction of $\hat{e}$, depending only on the magnitude of $\boldsymbol{k}$.

Classically the electric field associated with the wave (2.23) is given by $\hat{e}\,\mathrm{Re}\,\{F_k^O \exp(-\mathrm{i}\omega_k t)\exp(\mathrm{i}\boldsymbol{k}\cdot\boldsymbol{r})\}$; and there is of course a magnetic field that goes with it (see any book on electromagnetism). Quantum-mechanically one obtains, instead of $F_k(t) \equiv F_k^O \exp(-\mathrm{i}\omega_k t)$, a wavefunction (in its most general form, this consists of a linear sum of the eigenfunctions corresponding to the eigenvalues (2.24)) and a corresponding probability distribution for F_k at time t. We shall not write explicit expressions for these quantities. Let us simply note that, when the state of the $(\boldsymbol{k}, \hat{e})$-oscillator is an eigenstate of its energy, the mean value of $|F_k|^2$, is proportional to the corresponding eigenvalue given by (2.24).

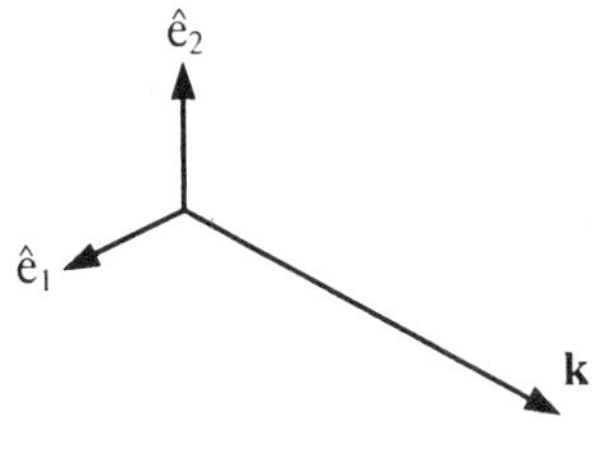

Figure 2.13. To the wavevector $\boldsymbol{k}$ correspond two 'oscillators' of different polarisation: $\hat{e}_1$ and $\hat{e}_2$ respectively

The totality of the oscillators ($-\infty < k_x,\ k_y,\ k_z < +\infty,\ \hat{e} = \hat{e}_1,\ \hat{e}_2$) constitutes the electromagnetic field in the volume $V = L^3$. At the absolute zero of temperature ($T = 0$) every one of the above oscillators is in its ground state, the state of least energy which equals $E_{k\hat{e}}(n = 0) = \hbar\omega_k/2$ according to (2.24). At higher temperatures ($T \geqslant 0$), at equilibrium, an oscillator ($\boldsymbol{k}$, $\hat{e}$) can be in any one of its eigenstates ($n = 0, 1, 2, 3, \ldots$) with certain probability given by

$$P_{k\hat{e}}(n) = [\exp(-E_{k\hat{e}}(n)/k_B T)]/Z \tag{2.25}$$

The normalisation constant Z in the above equation is determined by the requirement that the sum of (2.25) over all n must equal unity. Therefore

$$Z = \sum_n \exp(-E_{k\hat{e}}(n)/k_B T) \tag{2.25a}$$

It follows that the average energy of the oscillator ($\boldsymbol{k}$, $\hat{e}$), at temperature T, will be given by

$$\bar{E}_{k\hat{e}}(T) = \sum_n E_{k\hat{e}}(n) P_{k\hat{e}}(n) \tag{2.26}$$

Substituting (2.25) in the above equation and evaluating the sum, we obtain[17]

$$\bar{E}_{k\hat{e}}(T) = \frac{\hbar\omega_k}{2} + \frac{\hbar\omega_k}{\exp(\hbar\omega_k/k_B T) - 1} \tag{2.27}$$

The sentence 'The oscillator ($\boldsymbol{k}$, $\hat{e}$) is in its nth ($n = 1, 2, 3, \ldots$) excited state' can be stated in a different way; we can say: 'We have n *photons* of wavevector $\boldsymbol{k}$ and polarisation $\hat{e}$'. We can think of a photon (a quantum of electromagnetic excitation) as a 'particle' with a definite polarisation $\hat{e}$, definite momentum $\hbar\boldsymbol{k}$ and definite energy $\hbar\omega_k = \hbar ck$. The frequency of the photon is by definition that of the corresponding wave (2.23): $\nu_k = \omega_k/2\pi$, and its wavelength equals $\lambda_k = c/\nu_k$. We can then think of

$$n(\boldsymbol{k}, \hat{e}) = \frac{1}{\exp(\hbar\omega_k/k_B T) - 1} \tag{2.28}$$

which appears in (2.27), as the average number of ($\boldsymbol{k}$, $\hat{e}$)-photons existing at temperature T. The distribution (2.28) is a special case of the so-called Bose–Einstein distribution which we shall introduce in Section 2.14 (it is obtained from (2.72) by putting in the latter $\mu = 0$). Anticipating the classification of particles into fermions and bosons, we can say that photons are bosons.

The average number of photons with angular frequency between ω and $\omega + \mathrm{d}\omega$ in the volume V, at temperature T, is given by

$$N(\omega)\,\mathrm{d}\omega = \frac{Vg(\omega)\,\mathrm{d}\omega}{\exp(\hbar\omega/k_B T) - 1} \tag{2.29}$$

where $Vg(\omega)\,d\omega$ is the number of oscillators in the volume V with angular frequency between ω and $\omega + d\omega$ or, what amounts to the same thing, with wavevector-magnitude between $k = \omega/c$ and $k + dk = \omega/c + d\omega/c$. One can easily see that, when $V = L^3$ is sufficiently large, this equals

$$\frac{2(4\pi k^2\,dk)}{(8\pi^3/V)} = \frac{V\omega^2\,d\omega}{\pi^2 c^3} \equiv Vg(\omega)\,d\omega$$

where $4\pi k^2\,dk$ is the volume of a spherical shell in $\boldsymbol{k}$-space between k and $k + dk$, and $8\pi^3/V$ is, by the way we defined $\boldsymbol{k}$ (in (2.23)), the volume in the same space per wavevector $\boldsymbol{k}$. The factor of two takes account of the two independent polarisations of a photon of given $\boldsymbol{k}$. Substituting the above in (2.29) we obtain

$$\frac{N(\omega)\,d\omega}{V} = \frac{\omega^2\,d\omega}{(\exp(\hbar\omega/k_B T) - 1)\pi^2 c^3} \tag{2.30}$$

One usually measures the energy radiated from a photon gas; we call the radiation from a photon gas in thermal equilibrium, black-body radiation. The energy radiated per unit time, per unit surface area of the black body, with angular frequency between ω and $\omega + d\omega$ is given by

$$I(\omega, T)\,d\omega = \frac{c}{4}\hbar\omega\frac{N(\omega)\,d\omega}{V}$$

The factor (1/4), multiplying the speed of light, accounts for the fact that of the photons in V, half will propagate away from the emitting area, and those which propagate towards it will have, on average, a normal to the emitting area component of the velocity given by $c/2$. Using (2.30) and the relation $\omega = 2\pi c/\lambda$, we can rewrite the above formula as follows

$$I(\lambda, T)\,d\lambda = \frac{2\pi c^2 h}{\lambda^5[\exp(ch/\lambda k_B T) - 1]}\,d\lambda \tag{2.31}$$

Equation (2.31) gives us the energy radiated per unit time, per unit surface area of the black body in the wavelength-region between λ and $\lambda + d\lambda$. It is the first (historically) quantum-mechanical formula. It was derived by Planck in 1900 (it is known as Planck's law), who used it to determine the value of Planck's constant (h) by comparison with the, then available, experimental data. Only by introducing the quantisation of the electromagnetic excitation (described by (2.24)) could Planck reproduce the experimentally observed wavelength-distribution of the black-body radiation. This distribution, in agreement with (2.31), is shown in Fig. 2.14 for three different temperatures.

The electromagnetic field, whether in thermodynamic equilibrium or not, interacts with matter (atoms, molecules, etc.). This interaction is there even when the electromagnetic field is in its ground state (all oscillators ($\boldsymbol{k}$, $\hat{e}$) are

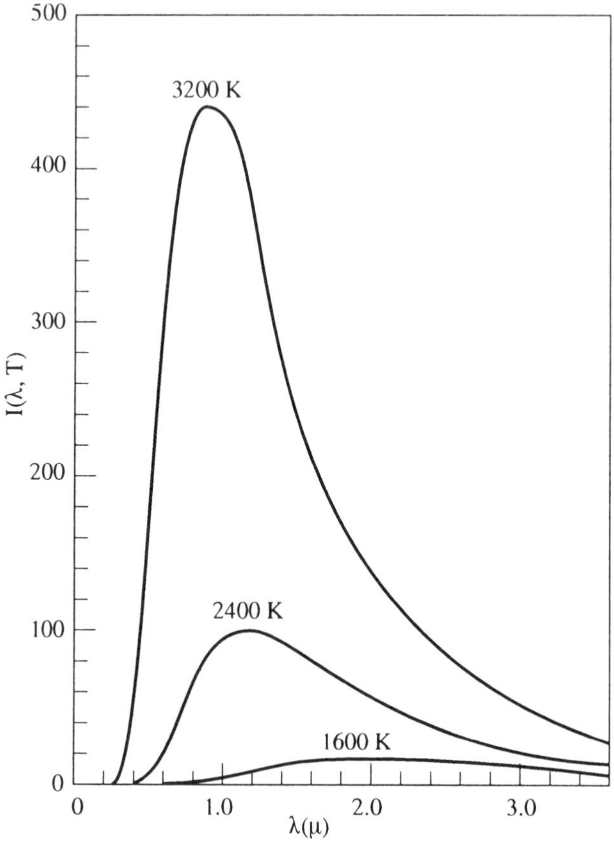

Figure 2.14. $I(\lambda, T)\,d\lambda$ is the electromagnetic energy radiated from a black body in wavelengths between λ and $\lambda + d\lambda$ per unit area per unit time, in units of watts per square centimeter per micron. We note that the visible region of the electromagnetic radiation extends from $\lambda \simeq 4000$ Å (violet) to $\simeq 8000$ Å (red). $\mu \equiv 10^4$ Å. (Redrawn from R. B. Leighton, *Principles of Modern Physics*, (New York: McGraw-Hill) 1959. Reproduced with permission)

in their ground states). It is this interaction, of matter with the electromagnetic field in its ground state (we assume that $T = 0$), which is responsible for the de-excitation of an atom: energy $\hbar\omega = 2\pi\hbar\nu_{nn'} = (E_n - E_{n'})$ is transferred from the atom to an oscillator $(\boldsymbol{k}, \hat{e})$, with $k = \omega/c$. The oscillator is thereby excited from its ground state ($n = 0$) to its first excited ($n = 1$): a photon of frequency $\nu_{nn'}$ is created. We now see that equation (2.21) is a statement of conservation of energy between atoms (matter) and the electromagnetic field.

We can describe the absorption of light by an atom in similar fashion. The incident (on the atom) electromagnetic radiation consists of photons of certain frequency, which implies oscillators $(\boldsymbol{k}, \hat{e})$ of appropriate wave-vectors in excited states. One of these oscillators interacting with the atom drops, say, from its nth state to its $(n-1)$th state, releasing energy $E_{\boldsymbol{k},\hat{e}}(n) - E_{\boldsymbol{k},\hat{e}}(n-1) = \hbar ck = 2\pi\hbar\nu$ which is taken by the atom.

It is worth noting that the quantum-number n (of the oscillator) changes by unity in both the above processes. Photons are created one at a time, and they are annihilated (absorbed by matter) one at a time.

We emphasise that the interaction between matter (atoms, molecules, etc.) and the electromagnetic field is weak and this allows us, as we have seen, to disregard it to begin with, and to speak of stationary states of atoms, molecules, etc., as if the electromagnetic field did not exist, and similarly to speak of stationary states of the electromagnetic field (or of the oscillators $(\boldsymbol{k}, \hat{e})$ which constitute this field) as if matter did not exist. This means that a hydrogen atom (analogous statements can be made about other atoms, molecules, etc.) in an excited state ψ_{nlm}, $n > 1$ will remain in this state for some time (it turns out that this is of the order of $\sim 10^{-8}$ s) before the interaction with the electromagnetic field causes it to transit to another state $\psi_{n'l'm'}$ of lower energy creating (emitting) a photon in the process. The exchange of energy (photons) between matter and the electromagnetic field goes on, of course, all the time; even when the two of them are in thermal equilibrium in themselves and with each other. Only, in equilibrium, the number of atoms or molecules (of a gas) in a given eigenstate, and similarly the number of photons of given $\boldsymbol{k}$, $\hat{e}$ do not change, on average, with time. These are given respectively, as we have seen, by equations (2.14a) and (2.28). It is also evident from these equations that at ordinary temperatures only low-frequency (invisible) photons are available for this interplay between matter and the electromagnetic field.

2.9 Fine structure of the energy levels of the hydrogen atom

Let us now return to the emission spectra of the hydrogen atom (Fig. 2.12). We must say at this point that every emission line in these spectra has a finite width (the emitted light is not rigorously monochromatic) which represents the 'uncertainty' ΔE about the energy of the excited atom, which derives from the interaction of the atom with the electromagnetic field and is related to the finite lifetime τ of the excited state as follows

$$\Delta E = \hbar/\tau \tag{2.32}$$

Putting $\tau \simeq 10^{-8}$ s, we find that $\Delta E \simeq 10^{-8}$ eV.

Spectroscopic measurements show that the theoretical results we have presented on the frequency (Fig. 2.12) of the emission lines of the hydrogen atom describe the corresponding experimental data well, but not exactly. Very accurate measurements reveal a spectrum of energy levels more complex than that of (2.12). This is shown in Fig. 2.15 (solid lines) together with that of (2.12) (broken lines). We note that the displacement of the real energy levels (solid lines) relative to the approximate levels (broken lines) has been overemphasised in the figure. The scale for the displacement is ten thousand times larger than that which defines the difference between the broken lines of the diagram. Besides the displacement of the energy levels, one observes in the diagram of Fig. 2.15 a doubling of all energy levels which correspond to states with angular momentum quantum-number $l > 0$. Both effects, the displacement, and the splitting of the levels corresponding to $l > 0$, are less pronounced the larger the values of the quantum-numbers

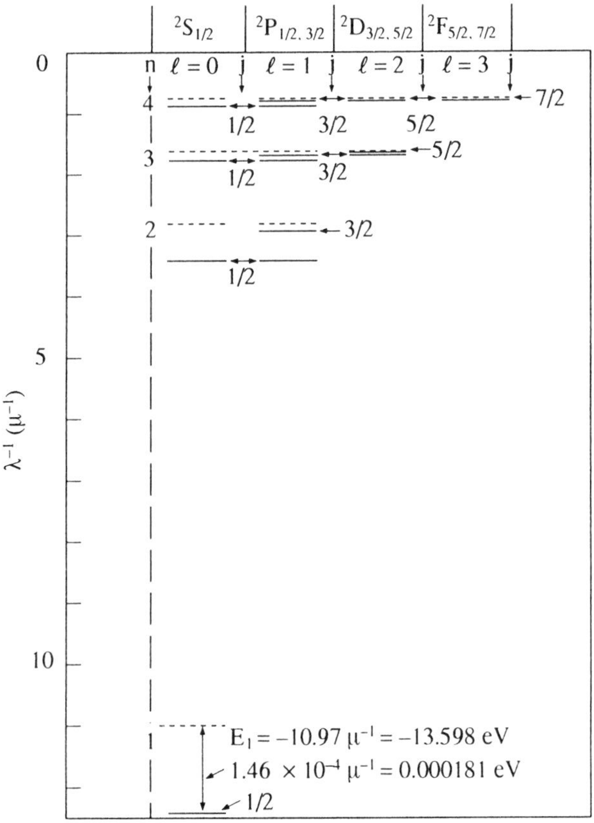

Figure 2.15. Fine structure of the energy levels of the hydrogen atom (solid lines). The broken lines correspond to the energy levels (2.12)

n and l. In the case of the levels $n, l = 4, 2$ and $4, 3$ the splitting is already too small to be discerned in the diagram of Fig. 2.15.

It is evident that the theory we have described (Schrödinger's theory) must be improved upon (extended in some way) in order to account for the fine structure of the energy spectrum of the hydrogen atom shown in Fig. 2.15. This has been achieved by Dirac.

2.10 Spin

Guided by Dirac's theory we discover that the electron possesses an 'intrinsic angular momentum' which we call the spin of the electron and denote it by $\boldsymbol{\sigma}$. The square of the magnitude of this intrinsic angular momentum has the precise value (eigenvalue of $\boldsymbol{\sigma}^2$)[18]

$$\boldsymbol{\sigma}^2 = \tfrac{1}{2}(1 + \tfrac{1}{2})\hbar^2 = \tfrac{3}{4}\hbar^2 \tag{2.33a}$$

and its component parallel to the z-axis (chosen arbitrarily) takes the following two values (eigenvalues of σ_z) and none other

$$\sigma_z = \frac{\hbar}{2} \text{ or } -\frac{\hbar}{2} \tag{2.33b}$$

We note that (2.33a) and (2.33b) are obtained from (2.2a) and (2.2b) for the orbital angular momentum if we put in these equations $l = \frac{1}{2}$. And, again as in the case of the orbital angular momentum, when σ_z has a precise value ($\hbar/2$ or $-\hbar/2$) the other two components of the spin can not be known precisely: σ_x and σ_y have, with equal probability, the values $\hbar/2$ and $-\hbar/2$, so that their mean values equal zero (see Appendix B.6.2). We should emphasise that, unlike orbital angular momentum, spin has not a classical analogue. One could be tempted to think of the spin as the angular momentum of an internal rotation, about an axis passing through the centre of mass of the electron, which is quantised with respect to its magnitude and direction. However, for a point-particle like the electron this would be identically zero. The fact that there is no classical analogue to the spin need not worry us. The experimental data, not only in relation to the energy spectrum of hydrogen but generally, confirm its existence beyond any doubt, and convince us that Dirac's theory is correct.

This theory indicates, and experiment confirms, that because of its spin the electron possesses an intrinsic magnetic moment[19]

$$\boldsymbol{\mu} = -\frac{e}{m}\boldsymbol{\sigma} \tag{2.34a}$$

where m is the mass of the electron. And we note that, according to (2.33b), the z-component of this magnetic moment can have one of the following two values and none other

$$\mu_z = e\hbar/2m \text{ or } -e\hbar/2m \tag{2.34b}$$

The quantity $\mu_B \equiv e\hbar/2m$ is known as the Bohr magneton.

We know, also, that a magnetic moment is associated with the orbital angular momentum of the electron. We have

$$\boldsymbol{M} = -\frac{e}{2m}\boldsymbol{l} \tag{2.35}$$

which means that the eigenvalues of the magnitude and of the z-component of the magnetic moment $\boldsymbol{M}$ are determined by (2.2). The magnetic moment (2.35) corresponds to the magnetic moment of an elementary electric circuit, as shown schematically in Fig. 2.16 (see problem 2.9), except that in the microscopic world this is a quantised quantity.

Now imagine a hydrogen atom: an electron orbiting around the nucleus. The magnetic moment associated with its spin will exert a torque on the magnetic moment associated with its orbital motion and vice versa, in the same way that two ordinary magnetic needles will exert a torque on each other. We refer to this interaction between the magnetic moments associated with the spin and orbital motions of the electron as the *spin–orbit* interaction. The spin–orbit interaction is partly responsible for the fine structure of the energy-spectrum of hydrogen (Fig. 2.15). Further correction is necessary, according to Dirac, to account for the fact that the kinetic energy of a particle (of the electron in the present case), when it moves very fast, is not the one assumed in Schrödinger's theory but somewhat smaller.[20] As a matter of fact, both phenomena, the spin–orbit interaction and the somewhat smaller kinetic energy, derive from the requirements of the special theory of relativity[21] which apply equally well to classical and quantum mechanics. And we may add at this point that since the spin–orbit interaction is a relativistic interaction, it, also, is larger when the speed of the electron is on the average larger. We can now understand why the

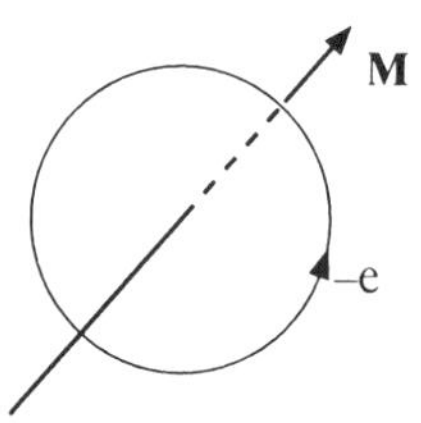

Figure 2.16. The 'orbit' of the electron corresponds to a circular electric current and gives rise to a magnetic moment

changes to the non-relativistic spectrum of the hydrogen levels (broken lines in Fig. 2.15) brought about by Dirac's relativistic theory are more pronounced for states of small n and l. In these states the electron stays nearer to the nucleus (see Fig. 2.9), where the 'local kinetic energy' $(E_n - U(r))$ is larger, which means, in turn, larger values for the mean kinetic energy and speed of the electron (see Section 1.6).[22]

We shall return to a discussion of the energy levels of the hydrogen atom and of the eigenstates to which they correspond in Section 2.13. We shall close this section with the following remark: if we were to disregard the small spin–orbit and kinetic energy corrections to the energy levels of the hydrogen atom, then the degeneracy of every level (broken line in Fig. 2.15) would be doubled. For every eigenstate ψ_{nlm}, as given in Table 2.3, we would now have two eigenstates; these two eigenstates would have the same energy E_n and the same orbital wavefunction (that of Table 2.3), but would differ with respect to the z-component of the spin, the one would be an eigenstate of σ_z with eigenvalue $\hbar/2$ (a so-called spin-up state) and the other would be an eigenstate of σ_z with eigenvalue $-\hbar/2$ (a spin-down state). We must now introduce a means of describing these states.

2.11 Description of the eigenstates of a particle with spin ($s = \frac{1}{2}$)

Let us imagine a disc, such as a coin, with two sides A and B moving (perhaps under the influence of some force) on a plane: the surface of a table. We have two ways of putting the disc on the table. In the first, which we call 'up', the visible side (VS) of the disc is A; in the second which we call 'down' the VS of the disc is B. The 'state' of the disc is completely defined if we know the position $\boldsymbol{r}_\parallel = (x, y)$ of the centre of the disc (we disregard any rotation of the disc about its centre) and its VS which can be 'up' or 'down'. In a classical framework we would know $\boldsymbol{r}$ and VS as functions of the time. In the most general motion, the VS of the disc would not be constant but would change with time, like its position. It will be 'up' for some time, then 'down' for awhile, then 'up' again, and so on. In the quantum world the state of the disc would be described by a wavefunction which would tell us, at any given time t, what the probability is of finding the disc at $\boldsymbol{r}_\parallel$ (between x and $x + dx$ and between y and $y + dy$) with VS 'up', and what the probability is of finding it at $\boldsymbol{r}_\parallel$ with VS 'down'. We expect this most general state to be a linear sum of eigenstates of the energy of the system under consideration. Let us see what the form could be of these eigenstates. We distinguish between two cases (1 and 2 below):

Case 1. In this case we know precisely (with zero uncertainty) the energy *and* the VS of the disc. The eigenstates of the energy are also eigenstates of the VS of the disc. The corresponding eigenfunctions may be written in the following form[23]

$$\Psi_{n+} = \psi_{n+}(\boldsymbol{r}_\parallel)\begin{pmatrix}1\\0\end{pmatrix}\exp(-\mathrm{i}E_{n+}t/\hbar) \tag{2.36a}$$

for VS 'up', and

$$\Psi_{n-} = \psi_{n-}(\boldsymbol{r}_\parallel)\begin{pmatrix}0\\1\end{pmatrix}\exp(-\mathrm{i}E_{n-}t/\hbar) \tag{2.36b}$$

for VS 'down'. We understand the above formulae as follows

$$\begin{pmatrix}1\\0\end{pmatrix} \text{ stands for VS 'up'}, \quad \begin{pmatrix}0\\1\end{pmatrix} \text{ stands for VS 'down'}$$

The orbital part of the eigenfunction ($\psi_{n+}(\boldsymbol{r}_\parallel)$ in (2.36a) and $\psi_{n-}(\boldsymbol{r}_\parallel)$ in (2.36b)) is, as usual, a real or complex function of x, y. It and the eigenvalue of the energy (E_{n+} in (2.36a) and E_{n-} in (2.36b)) may depend on the VS of the disc, as implied by the above formulae, but quite often it does not.

Let us assume that the state of the disc is described by (2.36a). Then, the probability of finding the disc in the area element $\mathrm{d}A = \mathrm{d}x\,\mathrm{d}y$ about $\boldsymbol{r}_\parallel$ with VS 'up' is

$$P_{n+}(\boldsymbol{r}_\parallel)\,\mathrm{d}A = |\psi_{n+}(\boldsymbol{r}_\parallel)\exp(-\mathrm{i}E_{n+}t/\hbar)|^2\,\mathrm{d}A = |\psi_{n+}(\boldsymbol{r}_\parallel)|^2\,\mathrm{d}A \tag{2.37}$$

and the probability of finding it with VS 'down' is zero. Ψ_{n+} is normalised as follows

$$\int P_{n+}(\boldsymbol{r}_\parallel)\,\mathrm{d}A = 1 \tag{2.38}$$

where, by definition, the integral extends over all the (x, y) space.

We interpret (2.36b) in the same manner. In this case, the probability of finding the disc within $\mathrm{d}A$ about $\boldsymbol{r}_\parallel$ with VS 'down' is

$$P_{n-}(\boldsymbol{r}_\parallel)\,\mathrm{d}A = |\psi_{n-}(\boldsymbol{r}_\parallel)|^2\,\mathrm{d}A \tag{2.39}$$

and the probability of finding it with VS 'up' is zero. Ψ_{n-} is normalised as follows

$$\int P_{n-}(\boldsymbol{r}_\parallel)\,\mathrm{d}A = 1 \tag{2.40}$$

Case 2. In this case the VS of the disc is not precisely known. The eigenstates of the energy of the disc are not eigenstates of the VS of the disc, and the corresponding eigenfunctions have the form

$$\Psi_n = \left\{ \varphi_{n+}(\boldsymbol{r}_\parallel)\begin{pmatrix}1\\0\end{pmatrix} + \varphi_{n-}(\boldsymbol{r}_\parallel)\begin{pmatrix}0\\1\end{pmatrix} \right\} \exp(-\mathrm{i}E_n t/\hbar) \tag{2.41}$$

which we can also write as follows

$$\Psi_n = \begin{pmatrix}\varphi_{n+}(\boldsymbol{r}_\parallel)\\ \varphi_{n-}(\boldsymbol{r}_\parallel)\end{pmatrix} \exp(-\mathrm{i}E_n t/\hbar) \tag{2.41a}$$

E_n denotes, as usual, the energy of the eigenstate, and φ_{n+} and φ_{n-} are the orbital parts of the wavefunction associated, respectively, with the VS 'up' and the VS 'down' components of the wavefunction. We interpret (2.41) as follows: the probability of finding the disc in the area element dA about $\boldsymbol{r}_\parallel$ with VS 'up' is

$$P_{n+}(\boldsymbol{r}_\parallel)\,\mathrm{d}A = |\varphi_{n+}(\boldsymbol{r}_\parallel)|^2\,\mathrm{d}A \tag{2.42a}$$

and the probability of finding it in dA about $\boldsymbol{r}_\parallel$ with VS 'down' is

$$P_{n-}(\boldsymbol{r}_\parallel)\,\mathrm{d}A = |\varphi_{n-}(\boldsymbol{r}_\parallel)|^2\,\mathrm{d}A \tag{2.42b}$$

The probability of finding the disc with VS 'up' anywhere in (x, y) space is given by

$$S_{n+} = \int P_{n+}(\boldsymbol{r}_\parallel)\,\mathrm{d}A \tag{2.43a}$$

and that of finding it with VS 'down' anywhere in (x, y) space is:

$$S_{n-} = \int P_{n-}(\boldsymbol{r}_\parallel)\,\mathrm{d}A \tag{2.43b}$$

Ψ_n is normalised as follows

$$S_{n+} + S_{n-} = 1 \tag{2.44}$$

We expect that any state, the most general, of the disc can be written, as is the case generally (see, e.g., Section 1.3.1), as a linear sum of the eigenstates of the energy; we have

$$\Psi = \sum_n c_n \begin{pmatrix}\varphi_{n+}(\boldsymbol{r}_\parallel)\\ \varphi_{n-}(\boldsymbol{r}_\parallel)\end{pmatrix} \exp(-\mathrm{i}E_n t/\hbar) \tag{2.45}$$

where c_n are coefficients to be determined by the initial ($t = 0$) state of the disc. We remember that n represents, in general, a set of quantum numbers.

The transportation of the above formulae from a disc with VS to a particle with spin ($s = \frac{1}{2}$)[24] is obtained through the following substitutions

$$\text{disc} \Rightarrow \text{particle}$$

$$\boldsymbol{r}_{\|} = (x, y) \Rightarrow \boldsymbol{r} = (x, y, z)$$

$$\mathrm{d}A \equiv \mathrm{d}x\,\mathrm{d}y \Rightarrow \mathrm{d}V \equiv \mathrm{d}x\,\mathrm{d}y\,\mathrm{d}z$$

$$\int \ldots \mathrm{d}A \Rightarrow \int \ldots \mathrm{d}V$$

$$\text{VS 'up'} \Rightarrow \text{spin 'up'}$$

$$\text{VS 'down'} \Rightarrow \text{spin 'down'}$$

The reader can now reread Section 2.11 having in mind the above substitutions to know the type of wavefunctions which describe a particle with spin $s = \frac{1}{2}$. The forms (2.36), (2.41) and (2.45) are known as spinor-wavefunctions (see also Appendix B.6.2).

2.12 Eigenstates of a particle in a spin-independent potential field

When the potential field a particle 'sees' does not depend on the state of its spin (whether it is up or down), the eigenvalues of its energy, naturally, do not depend on the spin. We obtain two eigenstates with the same energy and the same orbital motion corresponding, respectively, to a spin-up and a spin-down state. These are described by the following spinor-eigenfunctions

$$\Psi_{n+} = \psi_n(\boldsymbol{r})\begin{pmatrix}1\\0\end{pmatrix}\exp(-\mathrm{i}E_n t/\hbar) \equiv \begin{pmatrix}\psi_n(\boldsymbol{r})\exp(-\mathrm{i}E_n t/\hbar)\\0\end{pmatrix} \quad (2.46\text{a})$$

$$\Psi_{n-} = \psi_n(\boldsymbol{r})\begin{pmatrix}0\\1\end{pmatrix}\exp(-\mathrm{i}E_n t/\hbar) \equiv \begin{pmatrix}0\\\psi_n(\boldsymbol{r})\exp(-\mathrm{i}E_n t/\hbar)\end{pmatrix} \quad (2.46\text{b})$$

We remember that n can be a set of quantum-numbers, in which case we may have a number of states with different orbital parts (different $\psi_n(\boldsymbol{r})$) corresponding to the same energy. In any case whatever the degeneracy of an energy level might be without the spin, the existence of the spin doubles it.

For example: the motion of an electron in a box does not depend on its spin. Its energy eigenvalues are, therefore, given by the formulae of Section 1.3.2 (we put m in those formulae equal to the mass of the electron). The eigenstates of the electron (with spin taken into account) are described by spinor-eigenfunctions which are obtained from the given (orbital) eigenfunctions by multiplication with the appropriate spin factor in each case. For example, we know (see Section 1.3.3) that in a box of dimensions $a = b \neq c$ the following two (orbital) eigenfunctions

$$u_{npl}(\boldsymbol{r}, t) = \sqrt{\frac{8}{a^2 c}} \sin\frac{n\pi x}{a} \sin\frac{p\pi y}{a} \sin\frac{l\pi z}{c} \exp(-\mathrm{i}E_{npl}t/\hbar) \quad (2.47a)$$

and

$$u_{pnl}(\boldsymbol{r}, t) = \sqrt{\frac{8}{a^2 c}} \sin\frac{p\pi x}{a} \sin\frac{n\pi y}{a} \sin\frac{l\pi z}{c} \exp(-\mathrm{i}E_{pnl}t/\hbar) \quad (2.47b)$$

where $n \neq p$ and l are integer numbers, correspond to the same energy

$$E_{npl} = E_{pnl} = \frac{\hbar^2\pi^2}{2m}\left(\frac{p^2 + n^2}{a^2} + \frac{l^2}{c^2}\right) \quad (2.48)$$

The presence of spin makes two spinor-eigenstates out of each one of the above orbital states. We obtain

$$\Psi_{npl+} = \begin{pmatrix} u_{npl}(\boldsymbol{r}, t) \\ 0 \end{pmatrix} \quad (2.49a)$$

$$\Psi_{npl-} = \begin{pmatrix} 0 \\ u_{npl}(\boldsymbol{r}, t) \end{pmatrix} \quad (2.49b)$$

$$\Psi_{pnl+} = \begin{pmatrix} u_{pnl}(\boldsymbol{r}, t) \\ 0 \end{pmatrix} \quad (2.50a)$$

$$\Psi_{pnl-} = \begin{pmatrix} 0 \\ u_{pnl}(\boldsymbol{r}, t) \end{pmatrix} \quad (2.50b)$$

where the + (−) sign denotes a spin up (down) state. The above four spinor-eigenstates correspond to the same energy (2.48). The presence of spin has doubled the degeneracy of this level from two to four.

Finally, it is obvious that spin doubles the number of eigenstates[25] with energy between E and $E + \mathrm{d}E$ of a particle in a box of volume V and we must, therefore, modify (1.32) for the density of states accordingly. The density of states of a particle, with spin $s = \frac{1}{2}$, in a box is

$$2\rho(E) = \frac{8\pi}{h^3}(2m^3)^{1/2} E^{1/2} \quad (2.51)$$

2.13 Spinors of the hydrogen atom

If we disregard the kinetic-energy and spin-orbit corrections to the energy of a hydrogen atom, the presence of spin by itself, leads to a doubling of the degeneracy of the levels (2.12). We can describe the corresponding spinor-eigenstates in the same way as for an electron in a box. In Table 2.4

Table 2.4. Spinor-eigenstates of the hydrogen atom when the spin–orbit interaction is neglected

Energy[a]	Orbital eigenfunction[b]	Spinors[c]
E_n	$\psi_{nlm} = R_{nl}(r)Y_{lm}(\theta, \varphi)\exp(-\mathrm{i}E_n t/\hbar)$	$\Psi^0_{nlm+} = R_{nl}Y_{lm}\begin{pmatrix}1\\0\end{pmatrix}\exp(-\mathrm{i}E_n t/\hbar)$
		$\Psi^0_{nlm-} = R_{nl}Y_{lm}\begin{pmatrix}0\\1\end{pmatrix}\exp(-\mathrm{i}E_n t/\hbar)$
E_1	$R_{10}Y_{00}\exp(-\mathrm{i}E_1 t/\hbar)$	$R_{10}Y_{00}\begin{pmatrix}1\\0\end{pmatrix}\exp(-\mathrm{i}E_1 t/\hbar)$
		$R_{10}Y_{00}\begin{pmatrix}0\\1\end{pmatrix}\exp(-\mathrm{i}E_1 t/\hbar)$
E_2	$R_{20}Y_{00}\exp(-\mathrm{i}E_2 t/\hbar)$	$R_{20}Y_{00}\begin{pmatrix}1\\0\end{pmatrix}\exp(-\mathrm{i}E_2 t/\hbar)$
		$R_{20}Y_{00}\begin{pmatrix}0\\1\end{pmatrix}\exp(-\mathrm{i}E_2 t/\hbar)$
E_2	$R_{21}Y_{10}\exp(-\mathrm{i}E_2 t/\hbar)$	$R_{21}Y_{10}\begin{pmatrix}1\\0\end{pmatrix}\exp(-\mathrm{i}E_2 t/\hbar)$
		$R_{21}Y_{10}\begin{pmatrix}0\\1\end{pmatrix}\exp(-\mathrm{i}E_2 t/\hbar)$
E_2	$R_{21}Y_{11}\exp(-\mathrm{i}E_2 t/\hbar)$	$R_{21}Y_{11}\begin{pmatrix}1\\0\end{pmatrix}\exp(-\mathrm{i}E_2 t/\hbar)$
		$R_{21}Y_{11}\begin{pmatrix}0\\1\end{pmatrix}\exp(-\mathrm{i}E_2 t/\hbar)$
E_2	$R_{21}Y_{1-1}\exp(-\mathrm{i}E_2 t/\hbar)$	$R_{21}Y_{1-1}\begin{pmatrix}1\\0\end{pmatrix}\exp(-\mathrm{i}E_2 t/\hbar)$
		$R_{21}Y_{1-1}\begin{pmatrix}0\\1\end{pmatrix}\exp(-\mathrm{i}E_2 t/\hbar)$

[a] E_n is given by (2.12) (with $Z = 1$).
[b] The $R_{nl}(r)$ are given in Table 2.3 ($Z = 1$) and the spherical harmonics $Y_{lm}(\theta, \varphi)$ in Table 2.1.
[c] The zero upper-index on Ψ^0 reminds us that spin–orbit interaction has been neglected. We, often, refer to the $\Psi^0_{nlm\pm}$ as the unperturbed eigenstates of the hydrogen atom.

we list the orbital eigenfunctions and the corresponding spinors (obtained in the above manner) corresponding to the first two energy levels, E_1 and E_2, of the atom. We see that E_1 (the ground-energy level) is two-fold degenerate, and E_2 is eight-fold degenerate.

We must now consider how to describe the eigenstates corresponding to the observed energy levels of the atom, represented by the solid lines of Fig. 2.15. The spin–orbit interaction means that the spin angular momentum of the electron exerts a torque on the orbital angular momentum of the electron and, by the same token, the orbital angular momentum exerts a torque on the spin of the electron. Therefore, these quantities cannot be, rigorously, conserved precisely known quantities. However, when the interaction is weak (and it is weak in the present case), the *magnitude* of the orbital angular momentum is such to a very good approximation, and of course the magnitude of the spin, defined by (2.33a), is always the same. But the z-components of the spin and of the orbital angular momentum are not. An eigenstate of the energy is also an eigenstate of the square of the magnitude of the orbital angular momentum with eigenvalue $l(l+1)\hbar^2$ (l remains a good quantum number), but it is not an eigenstate of the z-component of the orbital angular momentum or the z-component of the spin.

We now introduce the total angular momentum, denoted by $\boldsymbol{J}$, of the electron (in the centrally symmetric field of the nucleus); it is the sum of the orbital and spin angular momenta:

$$\boldsymbol{J} \equiv \boldsymbol{l} + \boldsymbol{\sigma} \tag{2.52}$$

It turns out (see Section B.8) that the allowed values (eigenvalues) of $\boldsymbol{J}^2$ are determined from those of the orbital angular momentum ($l(l+1)\hbar^2$) and that of the spin ($\frac{1}{2}(\frac{1}{2}+1)\hbar^2$), as follows:

For $l \neq 0$:

$$J^2 = j(j+1)\hbar^2,\ j = l + \tfrac{1}{2},\ l - \tfrac{1}{2} \tag{2.53}$$

For $l = 0$:

$$J^2 = j(j+1)\hbar^2,\ j = \tfrac{1}{2}$$

For given j, the z-component of $\boldsymbol{J}$ takes the following values (eigenvalues of J_z)

$$J_z = m_j\hbar,\ m_j = -j, -j+1, \ldots, j-1, j \tag{2.54}$$

The importance of the total angular momentum lies in the fact that its magnitude and its z-component can be known precisely, together with the energy of the atom. The energy depends on j but not on m_j, which means that an energy level will be $(2j+1)$-fold degenerate.

We can now return to the energy-level diagram of the hydrogen atom shown in Fig. 2.15. Let us forget the accidental degeneracy of the Coulomb field (unperturbed levels of different $l < n$ have the same energy E_n) and let us concentrate on a particular level, say the $n = 2$, $l = 1$. The degeneracy of this unperturbed level (when we disregard states corresponding to $n = 2$, $l = 0$) is six. The six states corresponding to this level are (see Table 2.4) Ψ^0_{nlm+} and Ψ^0_{nlm-} with $n = 2$, $l = 1$ and $m = 0, 1, -1$. When the kinetic energy correction and the spin–orbit coupling are taken into account the above energy level is shifted downward and is split into two energy levels (solid lines in Fig. 2.15). We find that the higher-energy one is four-fold degenerate, the corresponding eigenstates have total angular momentum quantum numbers

$$j = l + \tfrac{1}{2} = 1 + \tfrac{1}{2} = \tfrac{3}{2} \text{ and } m_j = \tfrac{3}{2}, \tfrac{1}{2}, -\tfrac{1}{2}, -\tfrac{3}{2}$$

The lower-energy one is two-fold degenerate; the corresponding eigenstates have total angular momentum quantum numbers

$$j = l - \tfrac{1}{2} = 1 - \tfrac{1}{2} = \tfrac{1}{2} \text{ and } m_j = \tfrac{1}{2}, -\tfrac{1}{2}$$

We see that the total number (six) of the perturbed states equals the number of unperturbed states of the same n and l. (We remember that l is a good quantum number of the perturbed states as well.)

The other energy levels n, $l \neq 0$ and corresponding states can be described in the same way. Each unperturbed level is shifted downward and is split into two energy levels. One of them corresponds to states of total angular momentum quantum-number $j = l + \frac{1}{2}$ and is $[2(l + \frac{1}{2}) + 1]$-fold degenerate, and the other corresponds to states of total angular momentum quantum-number $j = l - \frac{1}{2}$ and is $[2(l - \frac{1}{2}) + 1]$-fold degenerate. We note once more that the number of states corresponding to the two levels is $2(2l + 1)$ which equals the degeneracy of the corresponding unperturbed level.[26]

The splitting of the energy levels, when $l > 0$, can be understood as follows. In the state of the larger total angular momentum $(j = l + \frac{1}{2})$ the magnetic moments associated with the spin and the orbital angular momentum are parallel to each other; in the state of the smaller total angular momentum $(j = l - \frac{1}{2})$ they are antiparallel to each other; and we know that the interaction energy between two parallel magnetic moments is different from that between two antiparallel magnetic moments. The above situation does not arise for states with $l = 0$ and for such states no splitting occurs.

Dirac's theory takes into account, as we have already noted in Section 2.10, not only the spin–orbit interaction but also the kinetic energy correction introduced in that section. When the two are taken into account one obtains the following formula for the energy levels of the one-electron atom[27]

$$E_{nj} = E_n + \frac{Z^2|E_n|\alpha^2}{4n^2}\left(3 - \frac{4n}{j + \frac{1}{2}}\right) \tag{2.55}$$

where E_n is the unperturbed energy level given by (2.12) and α is the so-called fine-structure constant, a pure number given by

$$\alpha = \frac{e^2}{4\pi\epsilon_0 \hbar c} = \frac{1}{137.0377} \tag{2.56}$$

We remember that for given n, the states of different j are obtained according to (2.53), with $l = 0, 1, \ldots, n - 1$. It is evident from (2.55) that the correction to Schrödinger's formula (2.12) due to relativistic effects (we recall that both the kinetic-energy and the spin–orbit corrections are relativistic in origin) represented by the second term in (2.55) becomes smaller as n and j get larger (see Fig. 2.15 and relevant comments in Section 2.10). It is also evident that the relativistic correction will be larger the heavier the atom (when Z gets larger).

The eigenstates of the atom corresponding to the corrected energy levels (2.55) are described, to a very good approximation, by linear sums of the unperturbed spinor-eigenfunctions of Table 2.3, as follows

$$\Psi_{nljm_j} = \sum_{m=-l}^{l} \{C^{+}_{nl}(jm_j; m)\Psi^{0}_{nlm+} + C^{-}_{nl}(jm_j; m)\Psi^{0}_{nlm-}\} \exp[-i(E_{nj} - E_n)t/\hbar] \tag{2.57}$$

where $C^{+}_{nl}(jm_j; m)$ and $C^{-}_{nl}(jm_j; m)$ are known coefficients[28]. We note that the eigenstate Ψ is defined by a set of quantum-numbers $\{n, l, j, m_j\}$ which includes the orbital angular momentum quantum number l, which reminds us that the magnitude of the orbital angular momentum is a conserved, precisely known, quantity, besides the energy specified by n and j, and besides the magnitude of the total angular momentum specified by j and its z-component specified by m_j.

Finally, it is worth noting that the transitions between atomic states which determine the absorption and emission spectra of light of an one-electron atom obey the following selection rules (which replace (2.19))

$$\begin{aligned} \Delta n &= \text{any value} \\ \Delta l &= \pm 1 \\ \Delta j &= \pm 1 \text{ or } 0 \\ \Delta m_j &= \pm 1 \text{ or } 0 \end{aligned} \tag{2.58}$$

We shall say no more on this subject except to assure the reader that the theoretical results we have described are in excellent agreement with the experimental data for the hydrogen atom and with those on one-electron ions.[29]

In many applications we are not interested in the fine structure of the energy spectrum of the system under consideration. It is sufficient for us to know how many eigenstates there are with energy between E and $E + \Delta E$, where ΔE need not always be very small. It is obvious that if the displacement and the splitting of degenerate levels and the shift of non-degenerate ones, due to some interaction or other, does not change the number of states in the interval ΔE, we can neglect the above interaction and the fine structure of the energy spectrum associated with it.

2.14 Fermions and bosons

Every particle[30] has its own intrinsic angular momentum, its own spin, the square of the magnitude of which equals

$$\boldsymbol{\sigma}^2 = s(s+1)\hbar^2 \tag{2.59}$$

where s has a well defined value, an integer or half-integer, for each particle. For given s the z-component σ_z of the spin may have one of the following values (eigenvalues of σ_z) and none other

$$\sigma_z = m_s\hbar,\ m_s = -s, -s+1, \ldots, s-1, s \tag{2.60}$$

When the value of σ_z is precisely known (the state of the particle is an eigenstate of σ_z corresponding to one of the above eigenvalues) the other two components, σ_x and σ_y, of the spin cannot be known precisely; the state of the particle can not be an eigenstate of σ_z and σ_x (or σ_y) at the same time.

If s is an integer we say that the particle is a boson. The photon with $s = 1$ and the helium atom (He^4) with $s = 0$ are examples of bosons.

If s is a half-integer ($\frac{1}{2}, \frac{3}{2}, \frac{5}{2}$, etc.) we say that the particle is a fermion. The electron with $s = \frac{1}{2}$ and the proton, also with $s = \frac{1}{2}$, are fermions.

We shall see in what follows that the properties of a system which consists of identical particles in some potential field or other (e.g., electrons in an atom or a crystal, atoms of He^4 in a gas, etc.) depend to a large degree on whether the particles are fermions or bosons.

Let us consider two independent identical particles in a box of dimensions a, b, c (Fig. 1.3). 'Independent' means that the particles do not interact with each other; no force is exerted on one by the other. In this case we expect the energy of the system (the two particles in the box) to be given by the sum of the energies of the two particles and the wavefunction of the system to be a product of two wavefunctions describing the states of the two particles respectively.[31]

Let us assume that the two particles are fermions with spin $s = \frac{1}{2}$ (e.g., non-interacting electrons)[32] and that the particle '1' is in the eigenstate

$$\Psi_{npl+}(1) = \begin{pmatrix} u_{npl}(\mathbf{r}_1, t) \\ 0 \end{pmatrix} = \begin{pmatrix} u_{npl}(\mathbf{r}_1) \\ 0 \end{pmatrix} \exp(-\mathrm{i}E_{npl}t/\hbar) \tag{2.61}$$

and the particle '2' is in the eigenstate

$$\Psi_{n'p'l'-}(2) = \begin{pmatrix} 0 \\ u_{n'p'l'}(\mathbf{r}_2, t) \end{pmatrix} = \begin{pmatrix} 0 \\ u_{n'p'l'}(\mathbf{r}_2) \end{pmatrix} \exp(-\mathrm{i}E_{n'p'l'}t/\hbar) \tag{2.62}$$

where $\mathbf{r}_1 = (x_1, y_1, z_1)$, $\mathbf{r}_2 = (x_2, y_2, z_2)$ denote the position coordinates of particles '1' and '2' respectively; and npl and $n'p'l'$ are given triads of integers which define the energies, through (1.27), and the orbital parts, through (1.28), of the corresponding one-particle eigenstates. We have assumed that particle '1' has spin up and that particle '2' has spin down. We write the wavefunction for the system of the two particles, in accordance with what has been said above about independent particles, as follows

$$\Psi_{\substack{npl+\\ n'p'l'-}}(1,2) = \begin{pmatrix} u_{npl}(\boldsymbol{r}_1) \\ 0 \end{pmatrix} \begin{pmatrix} 0 \\ u_{n'p'l'}(\boldsymbol{r}_2) \end{pmatrix} \exp\left[-\mathrm{i}(E_{npl} + E_{n'p'l'})t/\hbar\right] \tag{2.63}$$

which represents (we think) an eigenstate of the two-particle system corresponding to the energy eigenvalue $E_{npl} + E_{n'p'l'}$.

In the same spirit we consider a system of two independent (non-interacting) bosons. Let us assume that we have two such particles, of zero spin, in the box of Fig. 1.3. If particle '1' is in the eigenstate $u_{npl}(\boldsymbol{r}_1)\exp(-\mathrm{i}E_{npl}t/\hbar)$ and particle '2' in the eigenstate $u_{n'p'l'}(\boldsymbol{r}_2)\exp(-\mathrm{i}E_{n'p'l'}t/\hbar)$, we write the wavefunction for the system of the two particles as follows

$$\Psi_{\substack{npl\\ n'p'l'}}(1,2) = u_{npl}(\boldsymbol{r}_1)u_{n'p'l'}(\boldsymbol{r}_2)\exp\left[-\mathrm{i}(E_{npl} + E_{n'p'l'})t/\hbar\right] \tag{2.64}$$

which represents (we think) an eigenstate of the two-particle system corresponding to the energy eigenvalue $E_{npl} + E_{n'p'l'}$.

Let us examine more carefully the wavefunctions (2.63) and (2.64). We read (2.63) as follows. Particle '1' is in the state $|npl+\rangle$ and particle '2' is in the state $|n'p'l'-\rangle$. This sentence has meaning in physical science only if we can *distinguish* particle '1' from particle '2'. We can put the question differently, as follows. Is the eigenstate (2.63) different from

$$\Psi_{\substack{npl+\\ n'p'l'-}}(2,1) = \begin{pmatrix} u_{npl}(\boldsymbol{r}_2) \\ 0 \end{pmatrix} \begin{pmatrix} 0 \\ u_{n'p'l'}(\boldsymbol{r}_1) \end{pmatrix} \exp\left[-\mathrm{i}(E_{npl} + E_{n'p'l'})t/\hbar\right] \tag{2.65}$$

which tells us that particle '2' is in the state $|npl+\rangle$ and particle '1' in the state $|n'p'l'-\rangle$. And, similarly, we can ask: Is the eigenstate (2.64) different from

$$\Psi_{\substack{npl\\ n'p'l'}}(2,1) = u_{npl}(\boldsymbol{r}_2)u_{n'p'l'}(\boldsymbol{r}_1)\exp\left[-\mathrm{i}(E_{npl} + E_{n'p'l'})t/\hbar\right] \tag{2.66}$$

In both cases the answer must be negative, when the particles are identical. We cannot, under any circumstances, distinguish one from the other particle. That is what we mean when we say that the particles are identical. The above truth must be reflected in the wavefunction which describes a system of two identical particles. We must have

$$\Psi(2,1) = \Psi(1,2) \tag{2.67}$$

or

$$\Psi(2,1) = -\Psi(1,2) \tag{2.68}$$

where 1 and 2 denote the position and spin coordinates of the two particles.

We cannot exclude (2.68) because the change of sign does not change the physical state which the wavefunction represents (see Appendix B.5).

Bosons. Experience tells us that bosons obey (2.67). The wavefunction which describes a system of two bosons is symmetric with respect to interchange of the position and spin (if $s \neq 0$) coordinates of the two particles. (By spin coordinates we mean the possible orientations of the spin as defined by (2.60).) Therefore, the correct wavefunction for the two-particle system we have been looking at is not (2.64) or (2.66) but the symmetrised version of these two, i.e.

$$\Psi_{\substack{npl\\n'p'l'}}(1,2) = \frac{1}{\sqrt{2}}\{u_{npl}(\boldsymbol{r}_1)u_{n'p'l'}(\boldsymbol{r}_2) + u_{npl}(\boldsymbol{r}_2)u_{n'p'l'}(\boldsymbol{r}_1)\} \times \exp[-\mathrm{i}(E_{npl} + E_{n'p'l'})t/\hbar] \tag{2.69}$$

which, obviously, satisfies the condition (2.67). The factor $(2)^{-1/2}$ has been introduced in (2.69) for normalisation purposes (we assume that n, p, $l \neq n'$, p', l'; see prolem 2.11). Obviously, the eigenstate (2.69) corresponds to energy eigenvalue $E_{npl} + E_{n'p'l'}$ of the system.

At this stage we can generalise from our example of two non-interacting particles to a system of N particles (bosons) which may or may not interact with each other. We have: the wavefunction which describes a state of a system of N identical particles (bosons) is symmetric with respect to an interchange of the position and spin (if $s \neq 0$) coordinates of any two particles of the system:

$$\Psi(1, 2, 3, \ldots, i, \ldots, j, \ldots, N) = \Psi(1, 2, 3, \ldots, j, \ldots, i, \ldots, N) \tag{2.70}$$

for any pair i, j. We remember that $1, 2, \ldots, i, \ldots$ denote the position and spin (if $s \neq 0$) coordinates of particles $1, 2, \ldots, i, \ldots$ respectively.

If the particles are not interacting with each other (the N particles move independently in a given, potential field which is the same for all), the form of the N-particle wavefunction simplifies as follows. Let $\{E_\alpha\}$ be the energy eigenvalues and $\{u_\alpha\}$ the corresponding one-particle eigenstates[33] in the given potential field; and let us assume that the N particles occupy the states u_{α_i}, $i = 1, 2, \ldots N$. The corresponding wavefunction of the N-particle system which satisfies (2.67) takes the following form (when the u_{α_i} ($i = 1, \ldots, N$) are all different from each other):

$$\begin{aligned}\Psi_{\alpha_1\alpha_2\ldots\alpha_N}&(1, 2, \ldots, N)\\ &= \frac{1}{\sqrt{N!}}\{u_{\alpha_1}(1)u_{\alpha_2}(2)\ldots u_{\alpha_i}(i)\ldots u_{\alpha_j}(j)\ldots u_{\alpha_N}(N)\\ &\quad + u_{\alpha_1}(2)u_{\alpha_2}(1)\ldots u_{\alpha_i}(i)\ldots u_{\alpha_j}(j)\ldots u_{\alpha_N}(N) + \ldots\\ &\quad + u_{\alpha_1}(1)u_{\alpha_2}(2)\ldots u_{\alpha_i}(j)\ldots u_{\alpha_j}(i)\ldots u_{\alpha_N}(N) + \ldots\}\\ &\quad \times \exp(-\mathrm{i}(E_{\alpha_1} + E_{\alpha_2} + \ldots + E_{\alpha_N})t/\hbar)\end{aligned} \tag{2.71}$$

where the sum in $\{\ldots\}$ includes all the terms corresponding to all possible arrangements (sequences) of the numbers 1, 2, ..., N. We have in total $N! \equiv 1 \cdot 2 \cdot 3 \ldots \cdot (N-1) \cdot N$ such arrangements. We read (2.71) as follows. We have N particles in the (one-particle) eigenstates $u_{\alpha_1}, u_{\alpha_2}, \ldots, u_{\alpha_N}$. The energy of the system in the state (2.71) is $E_{\alpha_1} + E_{\alpha_2} + \ldots E_{\alpha_N}$.

We emphasise, however, that it is perfectly legitimate to have more than one particle in a state u_α. For example, we can put $\alpha_2 = \alpha_1$ in (2.71), meaning that we have two particles in the state u_{α_1}. In this case some terms in the curly brackets of (2.71) become identical (e.g., the second becomes identical with the first term) and of the identical terms only one should be kept. The number of different terms in the curly brackets of (2.71) is in general given by $N!\,(N_1!\,N_2!\ldots)^{-1}$, where N_i is the number of particles in the state u_{α_i} and the product within the brackets extends over all occupied states. We note that $N!\,(N_1!\,N_2!\ldots)^{-1}$ equals the number of ways we can distribute N particles so that N_1 of them are in the state u_{α_1}, N_2 of them in the state u_{α_2} and so on. The normalisation constant $(N!)^{-1/2}$ of (2.71) must, accordingly, be replaced by $((N_1!\,N_2!\ldots)/N!)^{-1/2}$ (see problem 2.11). The important thing to remember is that we can have more than one particle (bosons) in the same one-particle state. For example, in a gas of N bosons of zero spin in a box we may have two or more particles in certain eigenstates (1.28). We note, in particular, that at the absolute zero of temperature ($T = 0$) all the particles are found in the (one-particle) eigenstate of least energy, i.e. in the eigenstate $(n, p, l = 1, 1, 1)$.

At any temperature ($T \geqslant 0$) the average number $n(\alpha)$ of particles (bosons) in the eigenstate u_α is given by[34]

$$n(\alpha) = \frac{1}{\exp\left[(E_\alpha - \mu)/k_{\mathrm{B}}T\right] - 1} \tag{2.72}$$

where μ is a constant (known as the chemical potential) determined, at any given temperature, by the requirement

$$\sum_\alpha n(\alpha) = N \tag{2.73}$$

where the sum extends over all one-electron eigenstates, and N is the total number of particles in the system. In our example of N spinless particles in a box, α would be the three integers (n, p, l) specifying a state (1.28), E_α would be given by (1.27) and the sum (2.73) would include all states (1.28).

We emphasise that formula (2.72), known as the Bose–Einstein distribution, is valid for any system of N non-interacting bosons moving in the same potential field. We note that, at a given temperature, the number n of particles in a given state (α) depends, according to (2.72), only on the energy of the particle in that state. In Fig. 2.17 we show the variation of n

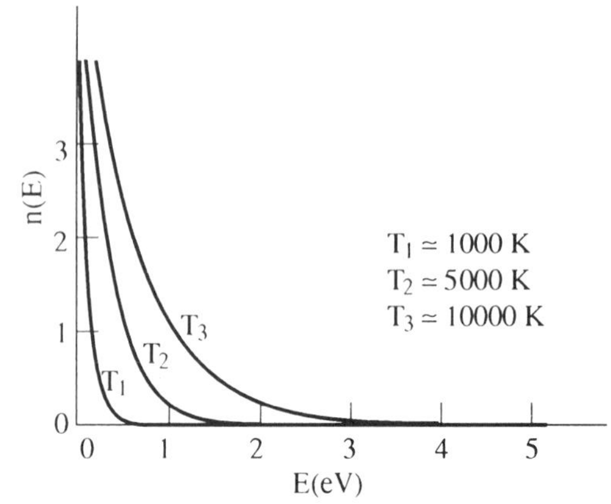

Figure 2.17. Bose–Einstein distribution; $\mu = 0$

with the energy E of the particle for three different temperatures when $\mu = 0$. It is worth noting that, when $E_\alpha - \mu \gg k_B T$,

$$n(\alpha) \simeq \text{const} \cdot \exp(-E_\alpha / k_B T) \tag{2.74}$$

For example, at sufficiently high temperatures practically all the particles of a dilute gas of non-interacting bosons will have translational energies satisfying the above condition and (2.74) will describe well the corresponding energy distribution. We note in this respect that in a gas of non-interacting bosons, $\mu \leqslant 0$ at any temperature.[35]

We note, in passing, a most interesting phenomenon concerning a gas of non-interacting bosons (N non-interacting bosons in a box of volume V; $V \to \infty$ with N/V constant). We find that the Bose–Einstein distribution (2.72) implies the following phenomenon. When $T \leqslant T_c$, where T_c is a definite temperature (determined by the atomic mass and the density N/V of the gas), a significant fraction of the particles are found in the ground state (the eigenstate $u_{111}(\boldsymbol{r}, t)$ of (1.28) with energy E_{111} which tends to zero when $V \to \infty$). In Fig. 2.18 we show this fraction as a function of the temperature. We see that when $T > T_c$ the fraction of particles which are in their ground states is practically zero. *Abruptly*, when $T = T_c$ it begins to increase as the temperature decreases to become unity (all the particles are in the ground state) at $T = 0$. The phenomenon is known as Bose–Einstein condensation. The only relevant Bose system that exists at sufficiently low temperatures is liquid He^4. The specific heat of this system which naturally depends on the energy distribution of the particles, becomes logarithmically infinite at a temperature of 2.18 K, and it is reasonable to assume that this anomaly is the manifestation of Bose–Einstein condensation modified by interatomic forces (these forces are weak but cannot be neglected in the

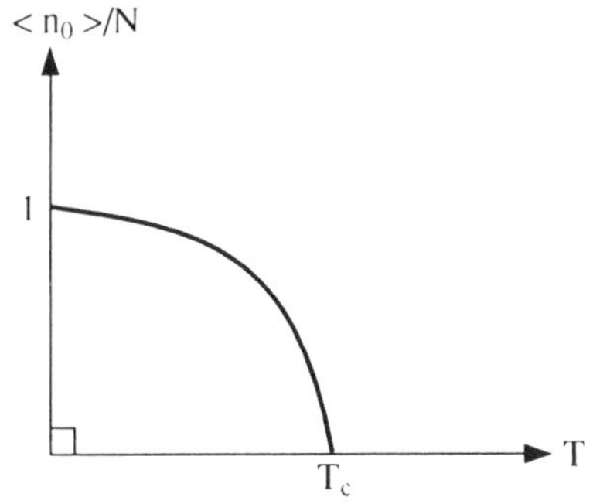

Figure 2.18. The fraction of particles (bosons) of an ideal gas in the ground state ($E = 0$) as a function of the temperature. $\langle n_0 \rangle$ denotes the mean value of the number of particles in the ground state and N is the total number of particles in the gas

liquid state of He^4). This interpretation is supported by the fact that no such anomaly is observed in the specific heat of liquid He^3 whose atoms are fermions and by the fact that the theoretical value of T_c for He^4 (obtained from (2.72) and (2.73)) is 3.14 K, which is very reasonable. We note that helium liquefies at 4.25 K.

Fermions. Experience tells us that fermions obey (2.68). The wavefunction which describes a system of two fermions is antisymmetric with respect to interchange of the position and spin coordinates of the two particles. Therefore, the correct wavefunction for the two-particle system we were considering at the beginning of our discussion is not (2.63) or (2.65) but the antisymmetrised version of these two, i.e.

$$\Psi_{\substack{npl+\\ n'p'l'-}}(1,2) = \frac{1}{\sqrt{2}}\left\{\begin{pmatrix} u_{npl}(\boldsymbol{r}_1) \\ 0 \end{pmatrix}\begin{pmatrix} 0 \\ u_{n'p'l'}(\boldsymbol{r}_2) \end{pmatrix} - \begin{pmatrix} u_{npl}(\boldsymbol{r}_2) \\ 0 \end{pmatrix}\begin{pmatrix} 0 \\ u_{n'p'l'}(\boldsymbol{r}_1) \end{pmatrix}\right\}$$
$$\times \exp\left[-\mathrm{i}(E_{npl} + E_{n'p'l'})t/\hbar\right] \quad (2.75)$$

which, obviously, satisfies the condition (2.68). And again a factor $(2)^{-1/2}$ has been introduced for normalisation purposes (see problem 2.12). The energy of the system in the above eigenstate is, of course, $E_{npl} + E_{n'p'l'}$.

We generalise (2.68) to a system of N fermions which may or may not interact with each other as follows. The wavefunction which describes the state of a system of N identical particles (fermions) is antisymmetric with respect to an interchange of the position and spin coordinates of any two particles of the system:

$$\Psi(1, 2, 3, \ldots, i, \ldots, j, \ldots, N) = -\Psi(1, 2, 3, \ldots, j, \ldots, i, \ldots, N) \quad (2.76)$$

for any pair i, j. We remember that 1, 2, . . . i . . . denote the position and spin coordinates of particles 1, 2, . . . i, . . . respectively.

If the particles are not interacting with each other (the N particles move independently in a given potential field which is the same for all, the form of the N-particle wavefunction simplifies as follows (we assume that the energy eigenvalues $\{E_\alpha\}$ and the corresponding eigenstates $\{u_\alpha\}$ of one particle in the given field are known, and that the N particles occupy the states u_{α_i} $i = 1, 2, \ldots, N$):

$$\begin{aligned}
&\Psi_{\alpha_1\alpha_2\ldots\alpha_N}(1, 2, \ldots, N) = \\
&\frac{1}{\sqrt{N!}}\{(-1)^{P(1,2,\ldots,i,\ldots,j,\ldots,N)}u_{\alpha_1}(1)u_{\alpha_2}(2)\ldots u_{\alpha_i}(i)\ldots u_{\alpha_j}(j)\ldots u_{\alpha_N}(N) \\
&+ (-1)^{P(2,1,\ldots,i,\ldots,j,\ldots,N)}u_{\alpha_1}(2)u_{\alpha_2}(1)\ldots u_{\alpha_i}(i)\ldots u_{\alpha_j}(j)\ldots u_{\alpha_N}(N) + \cdots \\
&+ (-1)^{P(1,2,\ldots,j,\ldots,i,\ldots,N)}u_{\alpha_1}(1)u_{\alpha_2}(2)\ldots u_{\alpha_i}(j)\ldots u_{\alpha_j}(i)\ldots u_{\alpha_N}(N) \\
&+ (-1)^{P(2,1,\ldots,j,\ldots,i,\ldots,N)}u_{\alpha_1}(2)u_{\alpha_2}(1)\ldots u_{\alpha_i}(j)\ldots u_{\alpha_j}(i)\ldots u_{\alpha_N}(N) + \cdots\} \\
&\times \exp[-\mathrm{i}(E_{\alpha_1} + E_{\alpha_2} + \cdots + E_{\alpha_N})t/\hbar]
\end{aligned} \tag{2.77}$$

where the sum in $\{\ldots\}$ includes all the terms corresponding to all possible sequences of the numbers 1, 2, . . ., N. Every such sequence can be obtained from the natural sequence 1, 2, 3, . . ., N by successive interchanges of pairs of particles; the sequence is called even (odd) if it is brought about by an even (odd) number of such interchanges. The even or odd character of a given sequence, determines the value of $P(1, 2, \ldots, j, \ldots, i, \ldots, N)$ associated with the sequence: $P = 1$ for an even sequence and $P = -1$ for an odd sequence. Having the above in mind one can easily show that (2.77) satisfies the condition (2.68).[36] We read (2.77) as follows. We have N particles occupying the one-particle eigenstates u_{α_i}, $i = 1, 2, \ldots, N$; the energy of the system is $E_{\alpha_1} + \cdots E_{\alpha_N}$. We observe that if $u_{\alpha_i} = u_{\alpha_j}$ for a pair i, j, the wavefunction (2.77) vanishes; in this case the interchange i, $j \Rightarrow j$, i or, what is equivalent to it, the interchange $\alpha_i, \alpha_j \Rightarrow \alpha_j, \alpha_i$, changes the sign of (2.77) without changing the wavefunction! We have

$$\Psi_{\ldots\alpha_i\ldots\alpha_j\ldots} = -\Psi_{\ldots\alpha_j\ldots\alpha_i\ldots} = -\Psi_{\ldots\alpha_i\ldots\alpha_j\ldots} \tag{2.78}$$

The first equation of (2.78) derives from the form of the wavefunction and the second from the fact that $u_{\alpha_i} = u_{\alpha_j}$. But $\Psi = -\Psi$ means that $\Psi = 0$ for only zero is equal to its opposite. We conclude, therefore, that it is not possible to have two particles (fermions) in the same one-particle state. This rule is known as Pauli's *exclusion principle*[37].

For example, if we have N independent particles (fermions with $s = \frac{1}{2}$) in a box, then at zero temperature these will occupy the N one-particle eigenstates of lower energy. In Fig. 2.19 we show the lower energy levels of

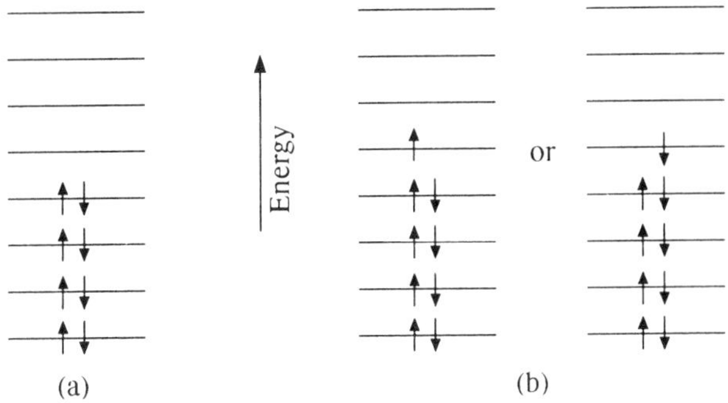

Figure 2.19. Schematic description of the ground state of a system of N independent particles (fermions with $s = \frac{1}{2}$) in a box of dimensions a > b > c. (*a*) $N = 8$ (*b*): $N = 9$. The horizontal lines denote doubly degenerate (because of the spin) levels. The arrow ↑ (↓) denotes an occupied spin up (down) state

a particle in a box of dimensions $a > b > c$. Because the particle has spin $s = \frac{1}{2}$, each level is doubly degenerate (there is no additional degeneracy in the given box): to each level correspond two eigenstates, one with spin up and the other with spin down. In Fig. 2.19*a* we describe, schematically, the ground state of a system of eight particles. When there are nine particles in the box, the 'ninth' can have spin up or spin down, as shown in Fig. 2.19*b*. It is evident that in the first case and, in general, when there is an even number of particles in the given box, the ground state of the system is non-degenerate; in the second case and, in general, when there is an odd number of particles in the given box, the ground state of the system is degenerate.

At any temperature ($T \geqslant 0$) and for any system of non-interacting fermions moving in a given potential field which is the same for all, the average number $n(\alpha)$ of particles in the eigenstate u_α is given by[34]

$$n(\alpha) = \frac{1}{\exp\left[(E_\alpha - \mu)/k_B T\right] + 1} \tag{2.79}$$

where μ (the chemical potential) is a constant determined, at any given temperature, by the requirement

$$\sum_\alpha n(\alpha) = N$$

where N is the total number of particles in the system. We note that $n(\alpha)$ depends only on the energy of the particle, for given temperature, and that $n(\alpha) \leqslant 1$ (the denominator in (2.79) is always larger or equal to unity) in accordance with Pauli's exclusion principle. Formula (2.79) is known as the Fermi–Dirac distribution. In Fig. 2.20 we show n as a function of the energy

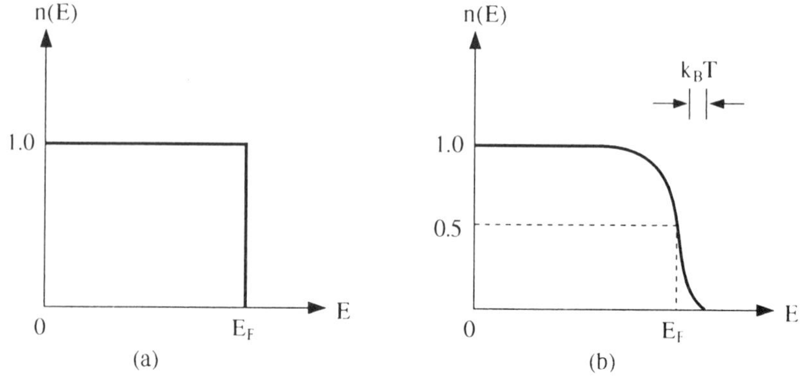

Figure 2.20. Fermi Dirac distribution. We assume that $k_B T \ll E_F$ in which case $\mu(T) \simeq \mu(0) = E_F$: (*a*) $T = 0$; (b) $T > 0$

of the particle when $T = 0$, and when $0 < k_B T \ll E_F$, where $E_F \equiv \mu(0)$ is the chemical potential at zero temperature (known, also, as the Fermi energy or the Fermi level); and we have assumed that when $k_B T \ll E_F$, $\mu(T) = E_F$. We see that, when $T = 0$ all eigenstates u_α with energy $E_\alpha < E_F$ are occupied ($n(\alpha) = 1$), and those with energy $E_\alpha > E_F$ are empty ($n(\alpha) = 0$). If there is an eigenstate with energy equal to E_F then the probability of it being occupied is 50% ($n = \frac{1}{2}$), as it happens in the example of Fig. 2.19(*b*). It is evident from Fig. 2.20 that a non-zero temperature changes the zero-temperature distribution significantly only when $k_B T$ is of the same order of magnitude or greater than the difference between the highest occupied energy level and the lowest empty one (at $T = 0$).

Finally, it is worth noting that when $E_\alpha - \mu \gg k_B T$, formula (2.79) reduces to the form (2.74). For example, in a dilute gas of non-interacting fermions at sufficiently high temperatures, most of the particles have (translational) energies satisfying the above condition and, then, one can use (2.74) instead of (2.79).

We conclude this section with a brief reference to the so-called Boltzmann (classical) statistics, which takes its mathematical expression in (2.74). We have seen that the translational energy of a dilute gas, of either fermions or bosons, at high temperatures, is distributed in the same manner, in accordance with (2.74). The reason for this can be described roughly as follows. In a dilute gas at high temperatures, the particles are away from each other in space and spread out energetically and, therefore, their behaviour is not affected by their being non-distinguishable; they behave like classical particles.[38] In a dilute gas of non-interacting classical particles (N particles in a large box of volume V, in thermal equilibrium at temperature T), the fraction of particles which have a translational velocity between (v_x, v_y, v_z) and $(v_x + dv_x, v_y + dv_y, v_z + dv_z)$ is given by

$$p(v)\,dv_x\,dv_y\,dv_z = \text{const}\exp[-E(v)/k_B T]\,dv_x\,dv_y\,dv_z \qquad (2.80)$$

where $E(v) = mv^2/2$ is the kinetic energy of the particle.[39] The normalisation constant in the above formula (which is essentially the same with (2.74)) is determined from the requirement that the integral of the above quantity over all velocities must equal unity (see problem 2.14). Finally, we note the similarity of the above formula with that ((2.14a)) which determines the distribution of the particles (atoms or molecules) with respect to their internal motion.

2.15 The *N*-electron atom

In the one-electron atom the potential energy (2.1) of the system comes form the electrostatic interaction between the nucleus and the electron of the atom. In the case of an atom, or ion, with N electrons the potential energy is constituted, by an obvious extension of (2.1), as follows

$$U(\boldsymbol{r}_1, \boldsymbol{r}_2, \ldots, \boldsymbol{r}_N) = -\sum_{i=1}^{N}\frac{Ze^2}{4\pi\epsilon_O r_i} + \sum_{i\neq j}\frac{e^2}{4\pi\epsilon_O|\boldsymbol{r}_i - \boldsymbol{r}_j|} \qquad (2.81)$$

The first term of (2.81) derives from the electrostatic interaction (Coulomb attraction) between the nucleus and the electrons of the atom. We assume as in the case of the one-electron atom, and for the same reasons, that the nucleus stays motionless at the origin of the coordinates. The charge of the nucleus equals Ze. In a neutral atom $N = Z$, in a positively charged ion $N < Z$, and in a negatively charged ion $N > Z$. In what follows we shall be concerned with neutral atoms ($N = Z$ is known as the atomic number of the atom). The position vector of the ith electron ($i = 1, 2, \ldots, N$) is denoted by $\boldsymbol{r}_i$. The second term of (2.81) comes from the electrostatic interaction (Coulomb repulsion) between the electrons; the sum includes all pairs i, j with $i \neq j$. The presence of this term (the interaction between the electrons) makes the calculation of the energy eigenvalues and corresponding N-electron eigenfunctions of the atom extremely difficult. We have to be satisfied, at least to begin with, with approximate solutions. A first approximation (and a good one as we shall see) is obtained as follows: we replace the complex system described by (2.81) with a system of N electrons moving, independently from each other in a mean potential energy field $U(\boldsymbol{r})$, which we construct as follows:

$$U(\boldsymbol{r}) = U_c(\boldsymbol{r}) + U_{ex}(\boldsymbol{r}) \qquad (2.82)$$

The first term of (2.82) gives the potential energy of an electron in the electrostatic field due to the nucleus (first term of (2.83)) and the electronic cloud of the N electrons of the atom (second term of (2.83)). We write

$$U_c(\boldsymbol{r}) = -\frac{Ze^2}{4\pi\epsilon_O r} + \frac{e^2}{4\pi\epsilon_O}\int\frac{n(\boldsymbol{r}')\,\mathrm{d}V'}{|\boldsymbol{r}-\boldsymbol{r}'|} \tag{2.83}$$

where $-en(\boldsymbol{r}')\,\mathrm{d}V'$ denotes the electronic charge in the volume element $\mathrm{d}V' = \mathrm{d}x'\,\mathrm{d}y'\,\mathrm{d}z'$ about $\boldsymbol{r}'$; and the integral extends over all space as usual. We note that $n(\boldsymbol{r})$ is the density of the electronic cloud[40] from all the electrons of the atom and includes, therefore, a contribution from the electron which by assumption moves in the potential field (2.83). Because the electron does not interact with itself, we must add to (2.83) a certain amount of (negative) potential energy to cancel the interaction of the electron with itself included in (2.83). $U_{\text{ex}}(\boldsymbol{r})$ in (2.82) represents this correction. We find an approximate expression for it as follows. To begin with, we note that the electronic cloud is due to electrons of spin up and spin down; denoting these quantities by $n_\uparrow(\boldsymbol{r})$ and $n_\downarrow(\boldsymbol{r})$, respectively, we write

$$n(\boldsymbol{r}) = n_\uparrow(\boldsymbol{r}) + n_\downarrow(\boldsymbol{r}) \tag{2.84}$$

and, if we neglect some small spin-polarisation that may exist, we obtain

$$n_\uparrow(\boldsymbol{r}) = n_\downarrow(\boldsymbol{r}) = \tfrac{1}{2}n(\boldsymbol{r}) \tag{2.85}$$

We now identify the excessive energy (interaction of an electron with its own cloud) with the potential energy of an electron at the centre of a *uniformly* charged sphere with charge density $-en(\boldsymbol{r})/2$ (i.e. equal to that of the electronic cloud of the same spin at the point $\boldsymbol{r}$ where the electron is) and radius R such that

$$\frac{4\pi}{3}R^3(\tfrac{1}{2}n(\boldsymbol{r})) = 1 \tag{2.86}$$

which leads to

$$R = [2\pi n(\boldsymbol{r})/3]^{-1/3} \tag{2.87}$$

Equation (2.86) tells us that there is a total of one electronic charge in the above sphere (volume: $4\pi R^3/3$). The potential energy of an electron at the centre of this sphere equals $3e^2/(8\pi\epsilon_O R)$ (see problem 2.15). We must subtract the above amount from (2.83), which means that

$$U_{\text{ex}}(\boldsymbol{r}) = -\frac{3e^2}{8\pi\epsilon_O R} = -\frac{3\alpha e^2}{4\pi\epsilon_O}\left[\frac{3}{8\pi}n(\boldsymbol{r})\right]^{1/3} \tag{2.88}$$

where $\alpha \equiv (4\pi/3)^{2/3}/2$. We can say that (2.88) is the potential energy of an electron at the centre of a sphere which is *positively* charged, with a total charge $+e$ uniformly distributed within the sphere. We can think of the above positively charged sphere which is known as the *Fermi hole*, as arising from the removal from the region surrounding the electron, of the electronic cloud of the same spin as the electron. We note, in this respect, the close relation between the *Fermi hole* and the Pauli exclusion principle which in the present case demands that we shall not have two electrons at the same position with the same spin[37]. The potential energy (2.88) is known as the exchange energy.

It turns out that the model we have described[41] gives better results, in comparison with more accurate methods and in comparison with the experimental data, if one treats the coefficient α in (2.88) as an adjustable parameter which varies somewhat from atom to atom. It has been established that the optimum value of α lies in every case within the region: $\frac{2}{3} < \alpha < 1$.

We note that the density $n(\boldsymbol{r})$ of the electronic cloud is not necessarily spherically symmetric. One can prove (see next section) that the contribution to $n(\boldsymbol{r})$ from so-called closed shells is spherically symmetric, but that of open shells (other than s-shells) is not. On the whole, however, $n(\boldsymbol{r})$ does not differ that much from a spherical distribution and this allows us to replace it by such a distribution, as shown schematically in Fig. 2.21. We denote the spherically symmetric distribution by $\bar{n}(r)$. The substitution $n(\boldsymbol{r}) \rightarrow \bar{n}(r)$ makes the Coulomb energy (2.83) and the exchange energy (2.88) spherically symmetric.

In conclusion: In the $X\alpha$-model of the atom, the electrons of the atom move, independently of each other, in the spherically symmetric potential field

$$U(r) = U_c(r) + U_{ex}(r) \tag{2.89a}$$

$$U_c(r) = -\frac{Ze^2}{4\pi\epsilon_O r} + \frac{e^2}{4\pi\epsilon_O}\int \frac{\bar{n}(r')\,\mathrm{d}V'}{|\boldsymbol{r} - \boldsymbol{r}'|} \tag{2.89b}$$

$$U_{ex}(r) = -\frac{3\alpha e^2}{4\pi\epsilon_O}\left[\frac{3}{8\pi}\bar{n}(r)\right]^{1/3} \tag{2.89c}$$

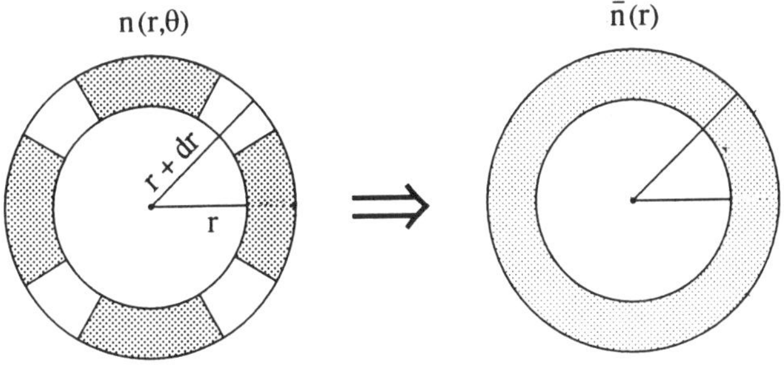

Figure 2.21. The electronic cloud in the region between r and $r + \mathrm{d}r$ is distributed uniformly within this region

We note that near the nucleus the first term of (2.89b) dominates and, therefore, $U \rightarrow -\infty$ when $r \rightarrow 0$; $U(r)$ increases monotonically with r and should tend to zero as $r \rightarrow \infty$, as follows

$$\underset{r\rightarrow\infty}{U(r)} = -\frac{e^2}{4\pi\epsilon_O r} \tag{2.90}$$

which tells us that as the electron moves towards infinity it leaves behind it a positively charged ion, to which it is attracted by a Coulomb force. Let us, then, assume that when $r = r_c$, the potential energy, as evaluated from (2.89), equals $U(r_c) = -e^2/(4\pi\epsilon_O r_c)$. Then, for $r > r_c$ we evaluate $U(r)$ from (2.90).[42]

2.15.1 SELF-CONSISTENT ELECTRONIC STRUCTURE OF MANY-ELECTRON ATOMS

Let us assume that we know the potential field $U(r)$ as defined by (2.89) and (2.90). Then we can calculate using the rules of quantum mechanics, and an electronic computer[43], the energy eigenvalues and corresponding bound eigenstates of an electron in this potential field. We know (see Sections 2.4 and 2.12) that the bound eigenstates of an electron in a spherically symmetric and spin-independent potential field are described by eigenfunctions of the following form

$$\Psi_{nlm+} = R_{nl}(r)Y_{lm}(\theta, \varphi)\begin{pmatrix}1\\0\end{pmatrix}\exp(-\mathrm{i}E_{nl}t/\hbar), \text{ for spin up}$$

and (2.91)

$$\Psi_{nlm-} = R_{nl}(r)Y_{lm}(\theta, \varphi)\begin{pmatrix}0\\1\end{pmatrix}\exp(-\mathrm{i}E_{nl}t/\hbar), \text{ for spin down}$$

where $R_{nl}(r)$ is a real function of r; and we remember that the energy level, the eigenvalue E_{nl}, of a given eigenstate depends only on the quantum numbers l ($l = 0, 1, 2, \ldots$) and n. For given l, the number n specifies the successively larger (less negative) energy levels of the electron.

For $l = 0$ (s-states) we have $n = 1, 2, 3, \ldots$ and the corresponding energy levels and eigenstates are denoted by: 1s, 2s, 3s, . . .

For $l = 1$ (p-states) we have $n = 2, 3, 4, \ldots$ and the corresponding energy levels and eigenstates are denoted by: 2p, 3p, 4p, . . .

For $l = 2$ (d-states) we have $n = 3, 4, 5, \ldots$, and the corresponding energy levels and eigenstates are denoted by: 3d, 4d, 5d, . . .

For $l = 3$ (f-states) we have $n = 4, 5, 6, \ldots$ and the corresponding energy levels and eigenstates are denoted by: 4f, 5f, 6f, . . .

etc.[44]

We recall that for given l the quantum number m takes the values

$m = -l, \ -l+1, \ \ldots, \ l-1, \ l$, and because the corresponding states, described by (2.91), all have the same energy E_{nl}, the latter has a degeneracy of $2(2l+1)$. It follows that an s-level is doubly degenerate (in this case the degeneracy derives solely from the two states, up and down, of the spin), a p-level is six-fold degenerate, and so on. The $2(2l+1)$ eigenstates of an electron corresponding to the eigenvalue E_{nl} constitute the shell (as it is called) nl.

Because of the Pauli exclusion principle (see Section 2.14), when an atom with N electrons is in its ground state (the state of least energy of the atom), the N electrons occupy N different one-electron eigenstates (2.91) beginning with those of the first shell (corresponding to the lowest of the one-electron energy levels); when the first shell is complete, electrons occupy states of the second in the energy ladder shell, then those of the third shell and so on, until all electrons have been placed. The ground state of the atom will be described by a wavefunction of the form (2.77), where u_{α_i} will be the above N one-electron eigenstates.

Every occupied one-electron state $\Psi_{nlm\pm}$ contributes to the density of the electronic cloud (to the spherically symmetric distribution $\bar{n}(r)$) the amount $[R_{nl}(r)]^2/(4\pi)$[45]. Now let q_{nl} be the number of eigenstates of the shell nl which are occupied. An empty shell has $q_{nl} = 0$; a partly occupied shell (we call it an open shell) has $0 < q_{nl} < 2(2l+1)$; and a fully occupied one (we call it a closed shell) has $q_{nl} = 2(2l+1)$. We can then write

$$\bar{n}(r) = \sum_{nl} q_{nl}[R_{nl}(r)]^2/(4\pi) \tag{2.92}$$

and we note that

$$\sum_{nl} q_{nl} = N \tag{2.93}$$

where N is the number of electrons in the atom.

We see at this stage that the density $\bar{n}(r)$ of the electronic cloud, which we obtain after some calculation (in accordance with well defined rules of quantum mechanics), is the quantity which determines (through (2.89)) the potential field which lies at the beginning of the calculation of $\bar{n}(r)$. In order to admit the result of the calculation, we must make sure that the $\bar{n}(r)$ we calculate for a given potential field $U(r)$, leads back to it through equations (2.89). The situation is described schematically in Fig. 2.22. In practice, we achieve this as follows. We begin with a more or less arbitrary potential field $U_0(r)$, we calculate for this field the one-electron eigenstates $\Psi_{nlm\pm}$, and obtain through (2.92) the first approximation $\bar{n}_1(r)$ to the density of the electronic cloud, and using it in (2.89) we obtain the first approximation $U_1(r)$ to the potential field. We repeat the calculation to obtain the second approximations, $\bar{n}_2(r)$ and $U_2(r)$, of the two quantities, and we keep

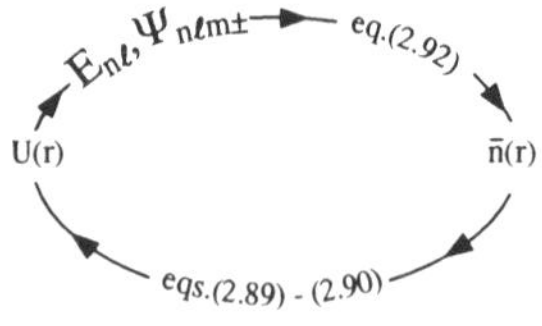

Figure 2.22. Self-consistent calculation of the electronic cloud and of the potential field an electron sees in a many-electron atom

repeating the calculation until convergence (as shown below) is obtained. We find:

$$\begin{aligned} U_0 \to \bar{n}_1(r) \to U_1(r) \to \bar{n}_2(r) \to U_2(r) \ldots \\ \to \bar{n}_i(r) \to U_i(r) \to \bar{n}_{i+1}(r) = \bar{n}_i(r) \to U_i(r) \end{aligned} \tag{2.94}$$

i.e. after i successive approximations, the output potential equals the input potential, and the same applies to the density of the electronic cloud. In this way the self-consistency of these quantities is achieved as shown, schematically, in Fig. 2.22

$$\begin{aligned} U(r) &= U_i(r) \\ \bar{n}(r) &= \bar{n}_i(r) \end{aligned} \tag{2.95}$$

The energy eigenvalues E_{nl} and the corresponding eigenstates $\Psi_{nlm\pm}$ of an electron in the self-consistent potential feld (2.95) are admitted as the appropriate solutions for the atom under consideration. This kind of calculation has been done for the known atoms and the results are tabulated in extensive tables of the relevant literature. In Table 2.5 we present the results of calculations by Herman and Skillman[46]. The atomic number Z (number of electrons of the atom) is noted in the first column and the symbol of the atom in the second column of the table. The closed and partially occupied shells of the atom are listed in the third column with the inner shells appearing in parenthesis (the separation of shells into inner and outer ones is explained in Section 2.15.3). The number of electrons in each shell appears as an exponent to that shell (when this exponent is unity it is omitted). In the remaining columns of the table we have the absolute values ($|E_{nl}| = -E_{nl}$) of the one-electron energy levels corresponding to the above shells in Hartree units (1 Hartree = 27.21 eV).

Let us see, for example, what we can learn from Table 2.5 about the nitrogen atom (symbol N). We read that it has seven electrons. Two electrons are found in the 1s shell with energy -14.8685 Hartrees (we presume that this means that to remove an electron from this shell to infinity we must supply to the atom a minimum of energy equal to 14.8685 Hartrees), and another two electrons are found in the 2s shell with energy -0.84795 Hartrees. These two shells are closed (fully occupied). The remaining three electrons of the atom are found in the 2p shell with energy -0.4225 Hartrees. This shell is obviously open, one needs six electrons to fill up a p-shell.

The theoretical results of Table 2.5 can be checked experimentally as follows. It is possible to remove an electron from an inner shell (nl) of a particular atom by radiating it with photons of energy $\hbar\omega$ greater than the ionisation energy $|E_{nl}|$ of the given shell (noted in the table). Knowing the energy of the absorbed photon and that (E') of the ejected electron (which we can measure) we determine E_{nl} from the relation: $E_{nl} = E' - \hbar\omega$. Similarly, we can ionise an atom by bombarding it with electrons, or other particles, of sufficiently high energy. The incident particle can knock off an electron from an inner shell, provided its energy exceeds a certain minimum which determines the ionisation energy of the shell.

One can show using the rules for adding angular momenta (Section B.8) that the removal of one electron from a closed shell (an inner shell is of course closed) leads to a state which is spectroscopically equivalent to that of a single electron in the same shell and, therefore, one obtains, as a result of the spin–orbit interaction, a fine structure of the energy spectrum (doubling of the p, d, f, etc., levels) similar to that of Fig. 2.15. Also, the large kinetic energy of the electrons of inner shells introduces a relativistic correction (analogous to the one discussed in Section 2.10 for the one-electron atom) to the numbers given in Table 2.5. These corrections are relatively easy to make, and one can say that the final results describe the experimental data well (see also Section 2.15.5).

The above discussion applies to the ionisation of a closed shell. Similar arguments apply to an open outer shell, when there is only one electron in the given shell, as is the case with the s-shells of Na, K, etc.. In such cases we can again compare directly the values given in Table 2.5 (after relativistic corrections have been made where necessary) with the corresponding experimental ionisation energies. In other cases when, for example, there are more than two electrons in the open shell, the situation is more complicated. We shall describe it in the next section.

At this stage we should say that in their calculations Herman and Skillman put $\alpha = 1$ in the expression (2.88) for the exchange energy, which (this was not realised at the time) is not the most appropriate value for this parameter, for it does not lead to optimum results for the total energy of the atom (see equation (2.97)). Putting $\alpha = 2/3$ turned out to be a more suitable choice for that purpose. When the one-electron energy levels (E_{nl}) are calculated from the formulae we have given with $\alpha = 2/3$, the results are not dramatically different from those of Table 2.5. They are in reasonably good agreement with the experimental data (the difference between them would not exceed ten per cent or so in typical cases) but they are not as good as those of Table 2.5. There are ways[41] to better the mean-field treatment of

Table 2.5. Absolute values of the one-electron energy levels of the lighter atoms, in Hartree units (Herman and Skillman, reproduced with permission from J. C. Slater, *Quantum Theory of Matter* (New York: McGraw Hill) 1968)

Z	Atom		1s	2s	2p	3s	3p	3d	4s	4p
1	H	1s	0.500							
2	He	$1s^2$	0.8605							
3	Li	$(1s^2)2s$	2.199	0.20195						
4	Be	$(1s^2)2s^2$	4.349	0.3006						
5	B	$(1s^2)2s^22p$	7.1865	0.46195	0.2449					
6	C	$(1s^2)2s^22p^2$	10.689	0.64475	0.33015					
7	N	$(1s^2)2s^22p^3$	14.8685	0.84795	0.42225					
8	O	$(1s^2)2s^22p^4$	19.728	1.0720	0.52045					
9	F	$(1s^2)2s^22p^5$	25.269	1.31745	0.6251					
10	Ne	$(1s^2)2s^22p^6$	31.495	1.584	0.7355					
11	Na	$(1s^22s^22p^6)3s$	39.025	2.3615	1.3345	0.18885				
12	Mg	$(1s^22s^22p^6)3s^2$	47.475	3.276	2.072	0.25255				
13	Al	$(1s^22s^22p^6)3s^23p$	56.83	4.3575	2.9735	0.36225	0.1791			
14	Si	$(1s^22s^22p^6)3s^23p^2$	67.02	5.5435	3.977	0.49875	0.2401			
15	P	$(1s^22s^22p^6)3s^23p^3$	78.055	6.842	5.090	0.6294	0.30695			
16	S	$(1s^22s^22p^6)3s^23p^4$	89.945	8.255	6.314	0.7650	0.3781			

17	Cl	$(1s^2 2s^2 2p^6) 3s^2 3p^5$	102.68	9.785	7.6515	0.9062	0.45335			
18	Ar	$(1s^2 2s^2 2p^6) 3s^2 3p^6$	116.27	11.4325	9.1035	1.0534	0.53265			
19	K	$(1s^2 2s^2 2p^6 3s^2 3p^6) 4s$	131.045	13.530	11.004	1.47605	0.8664		0.1543	
20	Ca	$(1s^2 2s^2 2p^6 3s^2 3p^6) 4s^2$	146.76	15.8135	13.090	1.9375	1.24115		0.19935	
21	Sc	$(1s^2 2s^2 2p^6 3s^2 3p^6) 3d 4s^2$	163.145	17.9865	15.065	2.2154	1.44155	0.2654	0.21545	
22	Ti	$(1s^2 2s^2 2p^6 3s^2 3p^6) 3d^2 4s^2$	180.385	20.2605	17.1395	2.49065	1.6830	0.31395	0.2289	
23	V	$(1s^2 2s^2 2p^6 3s^2 3p^6) 3d^3 4s^2$	198.475	22.6445	19.323	2.76855	1.852	0.35915	0.24095	
24	Cr	$(1s^2 2s^2 2p^6 3s^2 3p^6) 3d^5 4s$	217.225	24.9115	21.3895	2.8557	1.8455	0.23945	0.2156	
25	Mn	$(1s^2 2s^2 2p^6 3s^2 3p^6) 3d^5 4s^2$	237.235	27.7545	24.025	3.3402	2.2388	0.44295	0.26265	
26	Fe	$(1s^2 2s^2 2p^6 3s^2 3p^6) 3d^6 4s^2$	257.905	30.4785	26.542	3.63455	2.4456	0.48125	0.27255	
27	Co	$(1s^2 2s^2 2p^6 3s^2 3p^6) 3d^7 4s^2$	279.445	33.324	29.179	3.93925	2.6600	0.52075	0.28235	
28	Ni	$(1s^2 2s^2 2p^6 3s^2 3p^6) 3d^8 4s^2$	301.85	36.285	31.93	4.2500	2.878	0.55755	0.29155	
29	Cu	$(1s^2 2s^2 2p^6 3s^2 3p^6) 3d^{10} 4s$	324.85	39.075	34.51	4.317	2.8545	0.37155	0.25455	
30	Zn	$(1s^2 2s^2 2p^6 3s^2 3p^6 3d^{10}) 4s^2$	349.2	42.56	37.775	4.8965	3.3305	0.6291	0.30925	
31	Ga	$(1s^2 2s^2 2p^6 3s^2 3p^6 3d^{10}) 4s^2 4p$	374.6	46.32	41.315	5.619	3.946	1.0200	0.41885	0.18095
32	Ge	$(1s^2 2s^2 2p^6 3s^2 3p^6 3d^{10}) 4s^2 4p^2$	400.9	50.255	45.025	6.381	4.5985	1.44505	0.52855	0.23415
33	As	$(1s^2 2s^2 2p^6 3s^2 3p^6 3d^{10}) 4s^2 4p^3$	428.15	54.365	48.91	7.1875	5.2935	1.91145	0.6377	0.2913
34	Se	$(1s^2 2s^2 2p^6 3s^2 3p^6 3d^{10}) 4s^2 4p^4$	456.25	58.65	52.975	8.037	6.029	2.4170	0.74765	0.35075
35	Br	$(1s^2 2s^2 2p^6 3s^2 3p^6 3d^{10}) 4s^2 4p^5$	485.35	63.12	57.215	8.9285	6.8055	2.96225	0.85925	0.41235
36	Kr	$(1s^2 2s^2 2p^6 3s^2 3p^6 3d^{10}) 4s^2 4p^6$	515.3	67.765	61.625	9.864	7.6245	3.54875	0.9728	0.47595

ionisation energies we have presented, but their examination lies beyond the scope of the present book (see also Section 2.15.5).

2.15.2 ENERGY-LEVEL DIAGRAM OF A COMPLEX ATOM

Let us take carbon (C) as an example. One of its two outer shells, the 2p shell, holds two electrons instead of its full complement of six. This means that the two electrons can be in any two of the six 2p eigenstates $\Psi_{21m\pm}$ ($m = -1, 0, 1$) corresponding to the one-electron level $E_{2p} = -0.33015$ Hartrees. The total energy of the atom in the approximation we are considering (the electrons move independently in the mean field $U(r)$) is the same in every case and, therefore, we have 15 different states of the atom (corresponding to the 15 different ways we can choose two out of the above mentioned six 2p states) with the same energy. In other words, within the mean field approximation, the ground energy of carbon (let us denote it by ε^0) is fifteen-fold degenerate.

There is, however, a residual electrostatic interaction between the electrons, which comes from the difference between the actual potential energy field of the electrons given by (2.81) and the mean field of (2.89)[47]. And there is, also, the weaker magnetic interaction between the spins and orbital angular momenta of the electrons. When these interactions are taken into account the symmetry of the mean field approximation is destroyed and the mean field energy level ε^0 is split into five levels, as shown schematically in Fig. 2.23. The lowest of these corresponds to the actual ground state of the atom.

Let us consider the eigenstates corresponding to the above levels. It can be shown that, because space is isotropic, the magnitude of the total angular momentum of the atom and its z-component are conserved quantities that can be known exactly (with zero uncertainty) besides the energy: the (energy)-eigenstate of the atom is also an eigenstate of the total angular momentum (the square of its magnitude and its z-component).

The total angular momentum $\boldsymbol{J}$ of the atom is the sum of the total orbital angular momentum $\boldsymbol{L}$ and the total spin $\boldsymbol{S}$ of the electrons of the atom[48]. The contribution to $\boldsymbol{L}$ and $\boldsymbol{S}$ from the closed shells of the atom vanishes

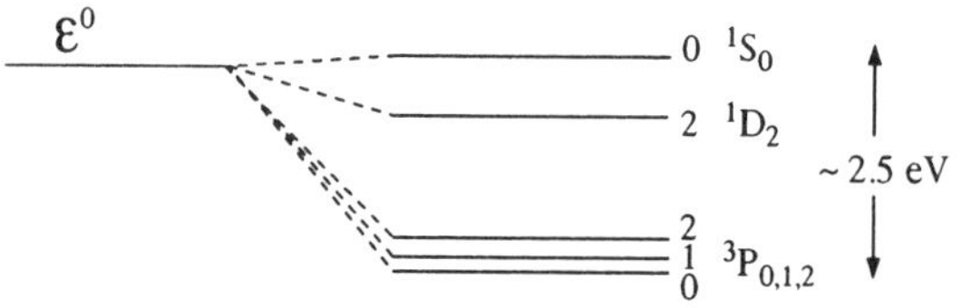

Figure 2.23. The mean-field ground-energy level ε^0 of the carbon atom splits into five energy levels, the lowest of which gives the actual ground-energy of the atom

identically and, therefore, $\boldsymbol{L}$ and $\boldsymbol{S}$ are determined by the orbital angular momenta and spins, respectively, of the electrons in the open shells of the atom (see Appendix B.8). The magnitudes of $\boldsymbol{L}$ and $\boldsymbol{S}$ (but not their z-components), when the interaction between $\boldsymbol{L}$ and $\boldsymbol{S}$ is small, are themselves conserved quantities that can be known exactly besides the energy and the total angular momentum of the atom. Therefore, an eigenstate of the atom is associated with a definite value of the energy, a definite value of the square of the magnitude of the total orbital angular momentum $L(L+1)\hbar^2$ ($L = 0$, 1 or 2 in the present case), a definite value of the square of the magnitude of the total spin $S(S+1)\hbar^2$, ($S = 0$ or 1 in the present case), a definite value of the square of the magnitude of the total angular momentum $J(J+1)\hbar^2$ (for given L and S, J can be any integer in the range: $|L - S| \leqslant J \leqslant L + S$), and a definite value of J_z, which, for given J, can take any of the $(2J+1)$ values $M_j\hbar$ ($M_j = -J, -J+1, \ldots, J$). The energy of the state depends on L, S and J but not on M_j (the energy of the atom cannot depend on the arbitrary choice of the z-axis), which means that the corresponding energy eigenvalue has a $(2J+1)$-fold degeneracy.

Let us now return to the diagram of Fig. 2.23. The five energy levels of the atom correspond to eigenstates of definite total angular momentum (a definite value of J) which is shown next to each level. The lowest (the ground level) corresponds to $J = 0$ and is, therefore, non degenerate. The second level corresponds to $J = 1$ and is, therefore, three-fold degenerate, the third level corresponds to $J = 2$ and is, therefore, five-fold degenerate, the same applies to the fourth level and, finally, the fifth level corresponds to $J = 0$ and is non degenerate. This means that we have, in all, 15 different states, which is how it should be. We remember that the unperturbed (mean field) level ε^0 is fifteen-fold degenerate, and that the number of eigenstates does not change when the symmetry of the field is broken by a relatively small perturbation of the potential field.

The eigenstates of the atom can be described (see Appendix B.10) by linear sums of the unperturbed eigenstates of the atom. In the case of the carbon atom, the unperturbed eigenstates of the atom corresponding to the energy ε^0 are six-electron wavefunctions of the form (2.77), or equivalent Slater determinants; the six one-electron states, denoted by u_{α_i} in (2.77), are chosen as follows. Four of them are the four independent states of the electron in the two closed shells (1s and 2s), the remaining two one-electron states can be, as we have already stated, any two of the six one-electron states of the 2p shell, which results in 15 different (linearly independent) Slater determinants corresponding to the 15 unperturbed states of the atom with energy ε^0. We can construct 15 different combinations (linear sums) of these determinants which are eigenstates of the total orbital momentum (its magnitude), the total spin (its magnitude) and the total angular momentum (its magnitude and its z component). These combinations describe approximately the actual states of the atom corresponding to the energy levels shown in Fig. 2.23. The symbols next to the energy levels of Fig. 2.23 have the following meaning. The letter tells us the magnitude, $\hbar\sqrt{L(L+1)}$ of the total orbital angular momentum of the atom. We denote a state of the atom of total orbital angular momentum $L = 0, 1, 2, 3, \ldots$ by S, P, D, F ... respectively. The upper index on the left of the letter is determined by the magnitude $\hbar\sqrt{S(S+1)}$ of the total spin of the atom; the index equals $(2S+1)$. Finally the lower index on the right of the letter equals J meaning that the magnitude of the total angular momentum of the atom equals $\hbar\sqrt{J(J+1)}$. We read, for example, that the ground state of the carbon atom is a $^3P_0$ state, which means that the magnitude of the total orbital angular momentum equals $\hbar\sqrt{1(1+1)} = \sqrt{2}\hbar$, the magnitude of the total spin equals $\hbar\sqrt{1(1+1)} = \sqrt{2}\hbar$, and that of the total angular momentum is zero.

It can be recognised at this stage that, for an accurate evaluation of the energy required to remove an electron from the 2p shell of carbon, one must calculate not only the energy corresponding to the above mentioned $^3P_0$ state, but also the ground state energy of the positive ion of carbon, and take the difference between the two. We shall not discuss this any further. We simply note that the experimental value of this quantity is 0.414 Hartrees, which is larger than the mean-field value of this quantity (0.330 Hartrees, according to Table 2.5) as, indeed, we expect it to be.

The energy-level diagram of Fig. 2.23 exhibits a number of features which are typical of atomic spectra and are worth noting. One finds (Hund's rule) that for a given configuration of the electrons (by configuration we mean the distribution of the electrons between different shells) the states of the atom with the higher value of S lie energetically lower. In the present case $S = 0$ or $S = 1$, and we see that the states $^1S_0$ and $^1D_2$ with $S = 0$ lie energetically above the states $^3P_{0,1,2}$ with $S = 1$. We can understand Hund's rule as follows: A higher value of S implies that any two electrons of the given configuration are more likely to have the same spin orientation (both up or both down) than otherwise and, therefore, their separation on the average will be larger than it would otherwise be (this owing in the final analysis to the Pauli principle[37] which tells us that two electrons cannot be at the same point in space with the same spin orientation); and this reduces the average Coulomb repulsion energy (a positive quantity) between the electrons (see also the last paragraph of Section B.8). We can say, looking at the diagram of Fig. 2.23, that the so-called exchange correlation interaction (the reduction of the Coulomb potential energy by the tendency of electrons of the same spin orientation to stay away from each other) is responsible for the energy difference between the average energy of the states with spin $S = 1$ and those with spin $S = 0$[49]. The energy difference between states of the same S but different L is due to residual electrostatic interaction between the electrons; as a rule the energy of the state is reduced as L increases; in our example the $^1D_2$ state ($L = 2$) is lower than that of the $^1S_0$ state ($L = 0$); one may say that in the state of higher L, two electrons of the given configuration are more likely to orbit in the same direction and therefore are less likely to come closer together, and so, by keeping apart, lower the repulsion energy between them. Finally the smaller energy differences between states of the same L and S but different J are due to spin–orbit interaction. The splitting of a level with given values of L and S into a number of levels with different values of J is called the fine structure or multiplet splitting of the level.

We can excite the carbon atom from its ground state, the $^3P_0$ state of Fig. 2.23, to one of the higher energy states shown in the same figure, by radiating it with photons of the appropriate frequency and in this way we can check the results of the theoretical calculation we have described. Needless to say, this has been done not only for carbon but for many other atoms as well with satisfactory agreement between theory and experiment. Of course the states shown in Fig. 2.23 are not the only excited states of the carbon atom. There are many others which we can introduce as follows. We note that the calculation which gives us the one-electron levels noted in Table 2.5 produces at the same time a number of higher one-electron energy levels corresponding to one electron states which are not occupied when the atom is in its ground state. Removing one electron from the noted in Table 2.5 shells to one of the higher energy shells produces a new configuration of the electrons which in turn leads (when the residual interaction between the electrons is taken into account) to a new collection of eigenstates of the atom. The corresponding energy level diagram is very rich indeed as can be seen from Fig. 2.24 which shows the lower energy levels of the carbon atom produced in this way. (Note that the multiplet splitting of an energy level can not be discerned on this diagram.) On the same diagram we show also the allowed transitions (with absorption or emission of photons) between these levels. It is worth noting that the allowed transitions obey very similar selection rules to those of the one-electron atom (see Section 2.13). We find that transitions usually occur only between configurations where only one electron changes its state; the l-value of the 'jumping' electron must change by $\Delta l = \pm 1$. For the atom as a whole the quantum numbers L, S, J and M_j must change as follows

$$\begin{gathered} \Delta S = 0 \\ \Delta L = 0, \pm 1 \\ \Delta J = 0, \pm 1, \text{ but } (J = 0) \text{ to } (J = 0) \text{ is not allowed.} \\ \Delta M_j = 0, \pm 1, \text{ but } (M_j = 0) \text{ to } (M_j = 0) \text{ is not allowed if } \Delta J = 0. \end{gathered} \tag{2.96}$$

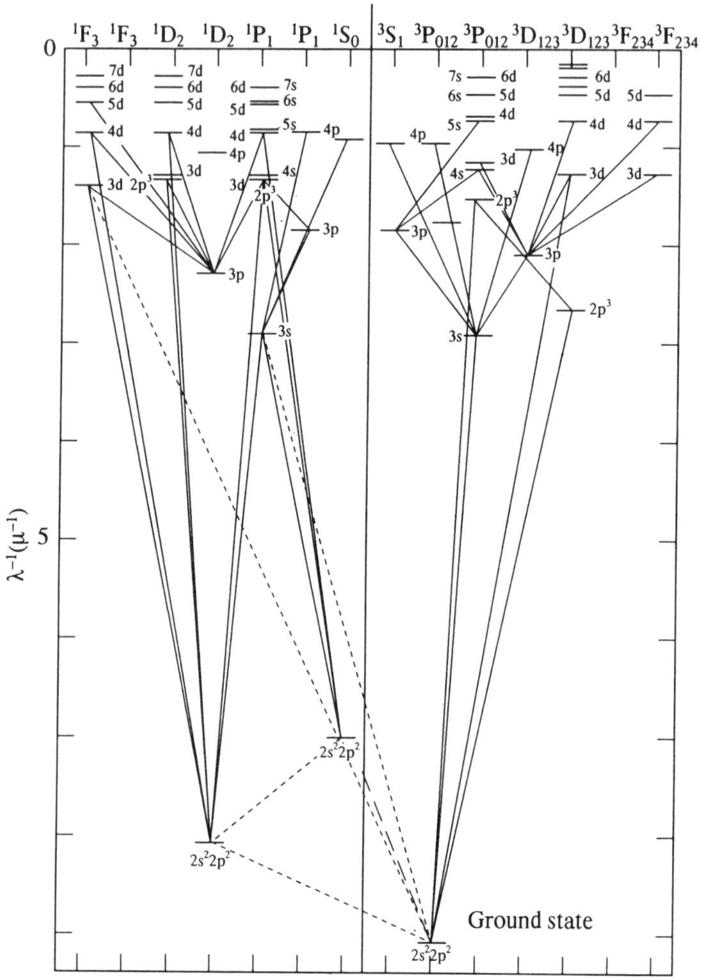

Figure 2.24. Energy-level diagram of the carbon atom. Configuration of electrons: Ground state: $1s^22s^22p^2$ ($3P_0$). Excited states: $1s^22s^22p^2$, $1s^22s^22p^3$, $1s^22s^22p$ (ms, mp, md). (Redrawn from R. B. Leighton, *Principles of Modern Physics*, (New York: McGraw-Hill) 1959. Reproduced with permission)

The above selection rules apply in the vast majority of transitions, but there are less common transitions (weak ones) which violate these rules. Transitions are possible, for example, in which more than one electron changes its state, or ΔS changes by one unit. The transitions which obey the above rules are shown by solid lines in Fig. 2.24; transitions which violate them are shown by broken lines.

We would like to conclude this section with a final observation concerning the atoms of the 3d-series ($Z = 21$ up to $Z = 29$). The reader may ask: why is it that the 4s shell is filled up with its complement of two electrons while the 3d shell, which corresponds to a lower energy remains partly empty? The answer to this question can be guessed from what we have said above about the ground and low-lying states of carbon. If we treat the one-electron states of the 3d shell and those of the 4s shell on the same footing, the ground-state energy of an atom of the 3d-series comes out (after taking into account the above-mentioned residual interactions) lower for the configuration noted in table 2.5. The ways of nature, it seems, are never too simple!

The above discussion of the atomic energy levels is based on the assumption that the orbital angular momenta of the electrons combine to give the total orbital angular momentum $\boldsymbol{L}$ of the atom, and their spins to give the total spin $\boldsymbol{S}$ of the atom, and that the magnitudes of $\boldsymbol{L}$ and $\boldsymbol{S}$ remain good quantum numbers. The above assumption is valid only when spin–orbit interaction and relativistic effects in general are small; more exactly the intervals in the fine structure of the energy spectrum must be small compared with the differences between levels of different L or S. This way of describing the atomic states is known as the Russel–Saunders approximation. The energy levels of the lighter atoms are described well by the Russel–Saunders scheme, but as the atomic number increases the relativistic interactions in the atom get stronger and this approximation becomes inapplicable. Alternative schemes for the description of the energy levels of heavy atoms do exist but we shall not discuss them in this volume.

2.15.3 THE RADII OF ATOMIC SHELLS

Let us now see what other useful information is to be had from the calculation of the eigenstates of an electron in the self-consistent field $U(r)$ of Section 2.15.1. We remember that, according to (2.92), the contribution to the electronic cloud in the region between r and $r + \mathrm{d}r$ from the shell nl is

$$q_{nl}[rR_{nl}(r)]^2\,\mathrm{d}r$$

where $R_{nl}(r)$ is the radial part of the eigenfunctions (2.91) of the shell. In Fig. 2.25 we show, schematically, the function of $[rR_{nl}(r)]^2$ for various shells. The radius $r_{\max}(nl)$—denoting the distance of the last maximum of the above function from the atomic nucleus—defines the spatial extension of the corresponding shell. In Table 2.6 we list the values of $r_{\max}(nl)$ for the occupied (fully or partly) shells of the lighter atoms calculated by Herman and Skillman and by Waber and Cromer. We note that, for given l, the radius $r_{\max}(nl)$ of a shell increases as the corresponding energy E_{nl} increases (we remember that Table 2.5 lists the absolute values of the negative E_{nl}).

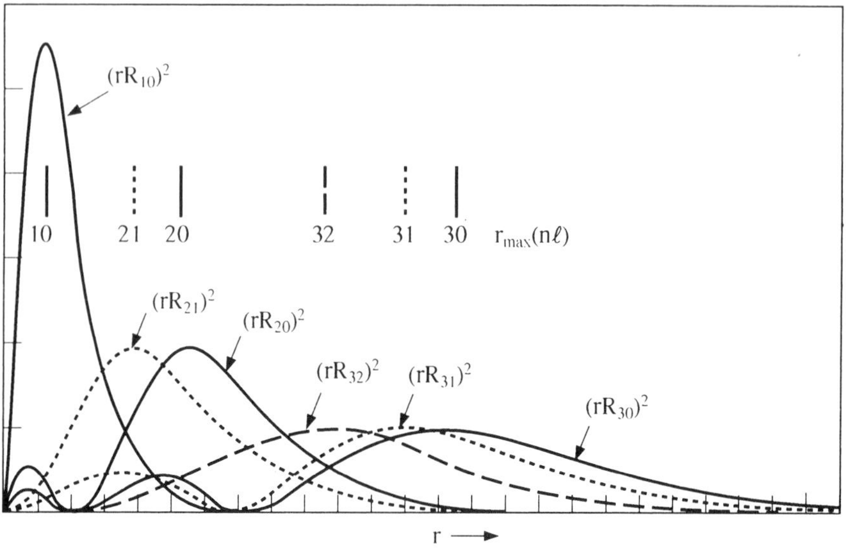

Figure 2.25. Schematic description of $[rR_{nl}(r)]^2$ for various shells nl. A vertical line marks the position of the last maximum, $r_{max}(nl)$, of $[rR_{nl}(r)]^2$

This is what we expect; it accords with our observations relating to the extension of the electronic wavefunction in the Coulomb field (see Section 2.5). However, it is worth noting that the extension (the radius r_{max}) of a higher-energy shell can be of the same magnitude or even a bit smaller than that of a lower-energy shell, when the orbital angular momentum of the higher-energy shell is larger than that of the lower-energy shell (see problem 2.17).

It is evident from Table 2.6 that we can always separate the shells of an atom into inner closed shells with small radii (the corresponding electronic clouds lie nearer to the atomic nucleus), and outer shells, with larger radii, the electronic clouds of which extend to the periphery of the atom. For example, in the case of sodium (Na) we have three inner shells (1s, 2s, 2p) and one outer shell (3s). In the case of carbon (C) we have one inner shell (1s) and two outer shells (2s, 2p). In the same manner we distinguish the inner shells (these are noted in parenthesis in the third column of Table 2.5) from the outer shells, of every atom listed in Table 2.5. We must be careful, however, not to include among the inner shells of an atom, shells which are only partly occupied (open shells), even when the radial extension of such shells appears to be rather small, as it is the case, for example, with the 3d-series of atoms ($Z = 21$ to $Z = 29$). These are correctly denoted as outer shells and put outside the parenthesis in column three of Table 2.5.

Table 2.6. $r_{max}(nl)$ of the one-electron orbitals of the atoms listed in Table 2.5 in angstroms (Herman and Skillman reproduced with permission from J. C. Slater, *Quantum Theory of Matter* (New York: McGraw Hill) 1968) and J. T. Waber and D. T. Cromer, *Journal of Chemical Physics*, **42** (1965) 4116).

	1s	2s	2p	3s	3p	3d	4s	4p
H	0.53							
He	0.291							
Li	0.186	1.586						
Be	0.138	1.040						
B	0.110	0.769	0.776					
C	0.091	0.620	0.596					
N	0.078	0.521	0.488					
O	0.068	0.450	0.413					
F	0.060	0.396	0.360					
Ne	0.054	0.354	0.318					
Na	0.049	0.32	0.278	1.173				
Mg	0.045	0.29	0.247	1.279				
Al	0.042	0.26	0.221	1.044	1.312			
Si	0.039	0.24	0.201	0.904	1.068			
P	0.036	0.22	0.184	0.803	0.918			
S	0.034	0.20	0.169	0.723	0.808			
Cl	0.032	0.19	0.157	0.660	0.723			
Ar	0.030	0.18	0.146	0.607	0.657			
K	0.029	0.17	0.137	0.55	0.593		2.162	
Ca	0.027	0.15	0.130	0.52	0.539		1.690	
Sc	0.026	0.146	0.124	0.48	0.500	0.539	1.570	
Ti	0.025	0.140	0.117	0.459	0.468	0.489	1.477	
V	0.024	0.134	0.110	0.44	0.439	0.449	1.401	
Cr	0.023	0.128	0.105	0.42	0.416	0.426	1.453	
Mn	0.022	0.122	0.100	0.40	0.392	0.389	1.278	
Fe	0.021	0.118	0.095	0.38	0.373	0.364	1.227	
Co	0.020	0.113	0.091	0.356	0.355	0.343	1.181	
Ni	0.019	0.109	0.088	0.34	0.339	0.324	1.139	
Cu	0.018	0.105	0.084	0.33	0.325	0.311	1.191	
Zn	0.018	0.100	0.081	0.32	0.311	0.292	1.065	
Ga	0.017	0.097	0.078	0.303	0.298	0.275	0.960	1.254
Ge	0.016	0.094	0.074	0.29	0.285	0.260	0.886	1.090
As	0.015	0.090	0.070	0.28	0.274	0.246	0.826	0.99
Se	0.015	0.087	0.068	0.27	0.263	0.234	0.775	0.91
Br	0.015	0.085	0.066	0.26	0.253	0.223	0.730	0.840
Kr	0.015	0.083	0.065	0.251	0.244	0.213	0.691	0.79

We expect, and this is confirmed by observation, that the chemical behaviour of an atom is determined by its outer shells: when two or more atoms bind together to form a molecule, the inner shells of the atoms are affected only slightly (to a first approximation they remain the same). We cannot add electrons to these shells because they are closed, and they remain closed because a lot of energy is required to remove an electron from them to a higher-energy state. Finally, interaction between neighbour atoms cannot affect significantly the electron states of the inner shells because of the small radial extension of the inner orbitals.

It is worth noting, in relation to Table 2.6, the following empirical rule. The value of r_{max} of the outermost shell of an atom determines the size of the atom and, indirectly, the size of the molecules which it forms with other atoms. Let AB be a diatomic molecule consisting of atoms A and B. Imagine two rigid spheres (like billiard balls) with radii equal, respectively, to the r_{max} of the outermost shell of A and to the r_{max} of the outermost shell of B. We find that the spatial extension of the molecule is obtained by having the above two spheres touching. A similar rule applies to polyatomic molecules and solids.

2.15.4 PERIODIC SYSTEM OF ELEMENTS

We find that atoms with similar outer shells (same number of electrons in shells of the same quantum number l) have similar chemical behaviour. For example, the atoms Ne, Ar, Kr, whose outer shells are closed p shells, are inert chemically. They do not form molecules; and the gases of these atoms liquefy at very low temperatures. Similarly, we find that the atoms Li, Na, K, all three of which have a single electron in their outer s shell, are chemically active in very much the same way, and so on. Because of this, the known atoms are traditionally listed in a table, known as the periodic table of the elements (Table 2.7), in such a way that atoms with similar outer shells, and therefore similar chemical behaviour, appear in the same column of the table[50]. On the other hand the atoms of a given row of the table make up a series of atoms with an increasing number of electrons in the same outer nl shell, which is noted at the beginning of the series. If there are more than one outer shells in a given atom, only the open (partly occupied) shell is noted at the beginning of the series. For example, we read that the first atom of the 3d series (the atom Sc) has one electron in its outer 3d shell. The closed shells of this atom $1s^2$, $2s^2$, $2p^6$, $3s^2$, $3p^6$ and $4s^2$, all of which, except the last one, are inner shells (see Table 2.5), are not noted in Table 2.7. The second atom of the 3d series (the atom Ti) has two electrons in the 3d shell, and so on. When the configuration of the electrons in the outer shells deviates from the above, the actual configuration is given in the box of the atom. In the case of Cr, for example, the actual configuration is $4s^1 3d^5$, as shown in the box of the atom, and not $4s^2 3d^4$. We should reiterate at this point that when the one-electron energies of the two outer shells of an atom are not very different, the ground state of the atom, which is described by a linear sum of mean-field states (Slater determinants), is obtainable from a configuration involving one-electron states from both

Table 2.7. Periodic table of the elements

1s	1 H																			2 He $1s^2$
2s	3 Li	4 Be												2p	5 B	6 C	7 N	8 O	9 F	10 Ne
3s	11 Na	12 Mg												3p	13 Al	14 Si	15 P	16 S	17 Cl	18 Ar
4s	19 K	20 Ca	3d	21 Sc	22 Ti	23 V	24 Cr $4s^1 3d^5$	25 Mn	26 Fe	27 Co	28 Ni	29 Cu $4s^1 3d^{10}$	30 Zn	4p	31 Ga	32 Ge	33 As	34 Se	35 Br	36 Kr
5s	37 Rb	38 Sr	4d	39 Y	40 Zr	41 Nb $5s^1 4d^4$	42 Mo $5s^1 4d^5$	43 Tc	44 Ru $5s^1 4d^7$	45 Rh $5s^1 4d^8$	46 Pd $5s^0 4d^{10}$	47 Ag $5s^1 4d^{10}$	48 Cd	5p	49 In	50 Sn	51 Sb	52 Te	53 I	54 Xe
6s	55 Cs	56 Ba	5d	57 La R.E.*	72 Hf	73 Ta	74 W	75 Re	76 Os	77 Ir	78 Pt $6s^1 5d^9$	79 Au $6s^1 5d^{10}$	80 Hg	6p	81 Tl	82 Pb	83 Bi	84 Po	85 At	86 Rn
7s	87 Fr	88 Ra	6d	89 Ac	90 Th H.E.+															
	s^1	s^2		d^1	d^2	d^3	d^4	d^5	d^6	d^7	d^8	d^9	d^{10}		p^1	p^2	p^3	p^4	p^5	p^6

* Rare earths	4f	58 Ce $5d^1$		59 Pr $5d^0$	60 Nd $5d^0$	61 Pm $5d^0$	62 Sm $5d^0$	63 Eu $5d^0$	64 Gd $5d^0 4f^7$	65 Tb $5d^0 4f^8$	66 Dy $5d^0$	67 Ho $5d^0$	68 Er $5d^0$	69 Tm $5d^0$	70 Yb $5d^0$	71 Lu $5d^1 4f^{14}$
		f^1	f^2	f^3	f^4	f^5	f^6	f^7	f^8	f^9	f^{10}	f^{11}	f^{12}	f^{13}	f^{14}	f^{14}
+ Heaviest elements	5f		91 Pa $6d^1$	92 U $6d^1$	93 Np $6d^1$	94 Pu $6d^1$	95 Am $6d^1$	96 Cm $6d^1$	97 Bk $6d^1$	98 Cf $6d^1$	99 Es	100 Fm	101 Md	102 No		

outer shells, as long as this leads to a lowering of the ground-state energy of the atom.

In the box of each atom, in Table 2.7, one also reads the atomic number (denoted by Z in Table 2.5) of the atom. We note that there may be a number of atoms with the same atomic number; we refer to these atoms as isotopes (of given Z). The nuclei of the isotopes have the same charge ($+Ze$) but differ from each other with respect to the mass, the size and possibly the spin of the nucleus. In the case of hydrogen, for example, we have the common hydrogen (its nucleus is a proton), and two more isotopes, the deuterium and the tritium, which are approximately two times and three times, respectively, heavier than the common hydrogen. The additional mass of their nuclei is due to neutrons; of which deuterium has one and tritium two. Here we need only say that the neutron is a neutral (it carries no electric charge) particle with a mass approximately equal to that of the proton. In general we may say that the nuclei of isotopes have the same number (Z) of protons (and therefore have the same charge $+Ze$), but each of them has a different number of neutrons (and therefore a different mass, a somewhat different size and possibly different spin and magnetic moment). The electronic and therefore the chemical properties of an atom depend, essentially, only on the charge of the nucleus and we can, for the purpose of studying these properties, replace the nucleus by a point charge ($+Ze$) as, indeed, we have already done. A very small dependence of the electronic structure of an atom on the size and spin of the atomic nucleus exists and leads in some cases to a hyperfine structure of the energy levels of an atom, but, as a rule, this has no significant consequences and can be disregarded.

2.15.5 TOTAL ENERGY OF AN ATOM

The total energy of an atom[51] is a negative quantity; it is the negative of the energy required to remove all electrons of the atom to infinite distance from the nucleus at rest. The mean-field value ε^0 of this quantity is given by

$$\varepsilon^0 = \sum_{nl} q_{nl} E_{nl} - \int U(r)\bar{n}(r)\,\mathrm{d}V - \frac{Ze^2}{4\pi\epsilon_O}\int \frac{\bar{n}(r)}{r}\,\mathrm{d}V$$
$$+ \frac{e^2}{8\pi\epsilon_O}\iint \frac{\bar{n}(r)\bar{n}(r')}{|\boldsymbol{r}-\boldsymbol{r}'|}\,\mathrm{d}V\,\mathrm{d}V' - \tfrac{9}{4}\alpha\left(\frac{e^2}{4\pi\epsilon_O}\right)\int\left(\frac{3}{8\pi}\bar{n}(r)\right)^{1/3}\bar{n}(r)\,\mathrm{d}V \tag{2.97}$$

where $\bar{n}(r)$, $U(r)$ and E_{nl} are the self-consistent values of these quantities calculated in the manner of Section 2.15.1. The first two terms of (2.97), taken together, give the kinetic energy of the electrons; the third term gives the electrostatic energy of attraction between the electrons and the nucleus;

and the fourth term gives the electrostatic energy of repulsion between the electrons. The last term relates (the relation is not straightforward and we shall not discuss it here) to the exchange interaction between the electrons and their Fermi holes.

In the second column of Table 2.8 we give the total energies of a few of the lighter atoms as calculated by the most accurate of the mean-field methods (the Hartree–Fock method); we note, however, that (2.97) would not give very different values. In the third column of the table we give the so-called correlation energy that has to be added to the mean field value to obtain the actual ground state energy of the atom. The correlation energy accounts for the difference between the real field (2.81) and the mean field (2.89); we have already introduced this in relation to the carbon atom (see Section 2.15.2). We see that correlation lowers the mean-field value of the energy by a small percentage (less than one percent for Li and smaller for the heavier atoms), which shows that the mean field approximation is a good one.

However, we must note that although the correlation energy is proportionally small, its magnitude can be quite large in comparison to such quantities as the dissociation energies of molecules and crystals. It is usually the case, for example, that the dissociation energy of a diatomic molecule (the energy required to remove the two atoms to infinite distance from each other at rest) is smaller than the correlation energies of the molecule and of the atoms; and similarly the energy required (per atom) to dissociate a crystal into its constituent atoms is generally smaller than the correlation energies of the crystal (per atom) and of the free atom. It is then obvious

Table 2.8. Hartree–Fock energy and correlation energy of atoms in their ground states, in Hartree units (From papers of E. Clementi and D. L. Raimondi, *Journal of Chemical Physics*, **38** (1963) 2686; and E. Clementi, *Journal of Chemical Physics*, **38** (1963) 2248, **39** (1963) 175. Reproduced with permission from J. C. Slater, *Quantum Theory of Matter* (New York: McGraw Hill) 1968.)

	ε^0(H-F)	Correlation energy	Correlation energy per electron
H	−0.50	0.00	0.00
Li	−7.432 726	−0.0453	0.0151
B	−24.529 05	−0.109	0.0218
N	−54.400 91	−0.188	0.0269
F	−99.409 29	−0.324	0.0360
Na	−161.858 89	−0.403	0.366
Al	−241.876 65	−0.482	0.0371
P	−340.718 71	−0.561	0.0374
Cl	−459.481 97	−0.712	0.0419
Ar	−526.817 43	−0.791	0.0439

that when we are considering the stability of a system in one form or the other, correlation should be taken into account. One would in particular like to know how the correlation energy in a molecule or a crystal differs from that in its constituent atoms. We shall not discuss this matter any further, except to note the rather interesting fact that the correlation energy per electron for the atoms listed in Table 2.8 appears to be changing very slowly, increasing by a factor of two from B to Ar while the correlation energy as a whole changes by a factor of about twenty or more.

Returning to (2.97) we should note that the total energy of the atom does not equal the sum of the one-electron energies E_{nl}. However, as we have stated in Section 2.15.1, these energies give a good first approximation of the ionisation energies of their respective shells. The above implies that the sum of the last four terms of (2.97) changes relatively little when one of the N electrons leaves the atom; it implies also that the one-electron energies themselves change relatively little when one electron is removed. The approximation is obviously better, the larger the atom ($N \gg 1$).

Problems

2.1 The interaction energy between a particle of mass m_1 and a particle of mass m_2 is given by $(U(|\boldsymbol{r}_1 - \boldsymbol{r}_2|)$, i.e. it depends only on the distance between the particles whose position vectors are denoted by $\boldsymbol{r}_1$ and $\boldsymbol{r}_2$ respectively. Show that the total energy of the system of the two particles can be written as:

$$E_{\text{tot}} = \tfrac{1}{2}MV_{\text{CM}}^2 + \tfrac{1}{2}m_r v^2 + U(r) \qquad (1)$$

where M and m_r are, respectively, the total mass and the reduced mass of the system, defined by

$$M = m_1 + m_2, \; m_r = \frac{m_1 m_2}{m_1 + m_2}$$

and $\boldsymbol{V}_{\text{CM}}$ and $\boldsymbol{v}$, the velocity of the centre of mass and the relative velocity respectively, are defined by

$$\boldsymbol{V}_{\text{CM}} = \frac{\mathrm{d}\boldsymbol{R}_{\text{CM}}}{\mathrm{d}t}, \; \boldsymbol{R}_{\text{CM}} \equiv (m_1\boldsymbol{r}_1 + m_2\boldsymbol{r}_2)/M$$

$$\boldsymbol{v} = \frac{\mathrm{d}\boldsymbol{r}}{\mathrm{d}t}, \; \boldsymbol{r} \equiv \boldsymbol{r}_1 - \boldsymbol{r}_2$$

Note that in the centre of mass system of coordinates ($V_{\text{CM}} = 0$), the system described by (1) is that of a particle of mass m_r moving in the centrally symmetric field $U(r)$. The picture simplifies further if $m_2 \gg m_1$; show that in this case, the massive particle (m_2) is practically stationary (at the origin of the CM system of coordinates) and about it, in the field $U(r)$, moves the lighter particle (m_1).

***2.2** A particle of mass m and energy $E = \hbar^2 k^2/2m$ is scattered by the spherical well potential

$$U(r) = -U_O,\ r < a$$
$$= 0,\ r > a$$

Find the effective cross section $d\sigma$ for scattering through angles in the range from θ to $\theta + d\theta$, and the effective total cross-section σ for scattering into any angle. Assume that $ka \ll 1$.
Hint. In the present case the wavefunction (2.9) for $r \geqslant a$ must be joined smoothly to the wavefunction inside the well ($r < a$) which should be finite at the origin ($r = 0$).

***2.3** (a) Show that there are infinite-many bound states of a particle of mass m in the spherical box defined by

$$U(r) = 0,\ r < a$$
$$= \infty,\ r > a$$

It will be sufficient to prove the above in relation to s-states ($l = 0$) and p-states ($l = 1$) using the data of Table 2.2.

(b) Obtain the first few energy levels of s-states and p-states of an electron in the above well ($a = 1$ Å), and write down the corresponding eigenfunctions.

***2.4** (a) Verify that the eigenfunctions listed in Table 2.3 satisfy Schrödinger's equation (B1):

$$-\frac{\hbar}{i}\frac{\partial\psi}{\partial t} = -\frac{\hbar^2}{2m}\nabla^2\psi + U(r)\psi$$

where $U(r)$ is given by (2.1).

(b) Using property (2.4a) of the spherical harmonics, show that the normalisation of the above eigenfunctions reduces to

$$\int_0^\infty [rR_{nl}(r)]^2\,dr = 1$$

and verify that the eigenfunctions listed in Table 2.3 do satisfy the above condition.

(c) Show that the eigenstates of the hydrogen atom (Table 2.3) have definite parity. The parity is even (odd) when the quantum-number l is even (odd).

2.5 Determine the classical turning point r_c of the electron in the ground state of the hydrogen atom (when $l = 0$, $U(r_c) = E$), and calculate the probability that the electron will be found in the classically forbidden region ($r > r_c$).

***2.6** (a) Determine, using first order perturbation theory, the correction to the ground-state energy of a hydrogen atom, when the latter is in a weak uniform static electric field (such as the one between the plates of a condenser).

(b) Consider, qualitatively, what might happen in an external field of large magnitude.

2.7 We have established that there are three linearly independent p-orbitals which have the form (see (2.3))

$$\psi_{1m} \equiv R(r)Y_{1m}(\theta, \varphi), \; m = -1, 0, 1 \tag{I}$$

(a) Show that the combinations of the above defined by

$$\begin{aligned} p_x &= -\frac{\psi_{1+1} - \psi_{1-1}}{\sqrt{2}} = \left(\frac{3}{4\pi}\right)^{1/2} \frac{x}{r} R(r) \\ p_y &= -\frac{\psi_{1+1} + \psi_{1-1}}{\sqrt{2}i} = \left(\frac{3}{4\pi}\right)^{1/2} \frac{y}{r} R(r) \\ p_z &= \psi_{10} = \left(\frac{3}{4\pi}\right)^{1/2} \frac{z}{r} R(r) \end{aligned} \tag{II}$$

are normalised and orthogonal orbitals; which means that we can choose (II) instead of (I) to represent the p-orbitals.

(b) Verify that the probability-densities associated with p_x, p_y, p_z are concentrated along the x, y and z directions respectively, as shown in Fig. 2.26.

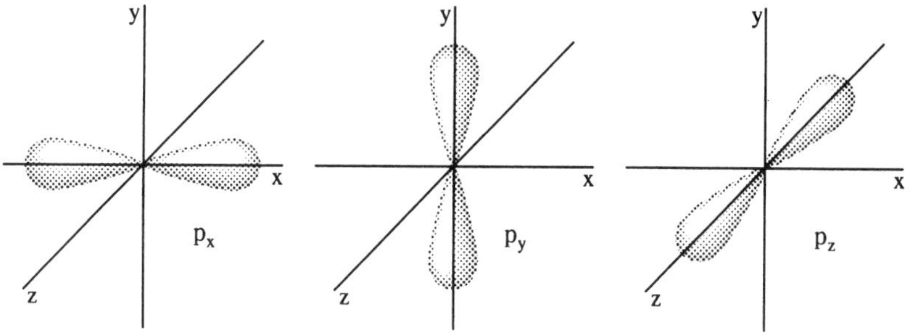

Figure 2.26. Probability densities for the p_x, p_y, p_z orbitals

2.8 (a) By integrating (2.31) over all wavelengths show that the energy radiated per unit time per unit surface of a black body is given by the so-called Stefan–Boltzmann law:

$$I(T) = \sigma T^4$$

$$\sigma = \frac{2\pi^5 k_B^4}{15c^2h^3} = 5.7 \times 10^{-8}\,\mathrm{W\,m^2\,K^{-1}}$$

Note:

$$\int_0^\infty \frac{x^3\,\mathrm{d}x}{e^x - 1} = \frac{\pi^4}{15}$$

(b) Estimate the temperature at the surface of the sun, if the radiated per unit time energy from the sum is 3.74×10^{26} W. The radius of the sun equals 6.96×10^8 m.

2.9 Assuming that the charge and mass of the electron are distributed uniformly over the circular orbit of Fig. 2.16, and letting the (classical) orbital angular momentum of the electron be l, show that the corresponding magnetic moment is given by (2.35).

2.10 Using the rules (2.58) determine the allowed transitions (accompanied by the absorption/emission of photons) between the states of the hydrogen atom corresponding to the energy levels of Fig. 2.15. Evaluate, at least for some of these transitions, the frequency of the absorbed/emitted photon.

2.11 (a) Using the orthogonality property of the one-particles states

$$\begin{aligned}\int u^*_{\alpha'}(\boldsymbol{r})u_\alpha(\boldsymbol{r})\,\mathrm{d}V &= 1, \text{ if } \alpha = \alpha' \\ &= 0, \text{ otherwise}\end{aligned} \tag{1}$$

verify that (2.69) is normalised, i.e.

$$\int |\psi(\boldsymbol{r}_1, \boldsymbol{r}_2)|^2\,\mathrm{d}^3r_1\,\mathrm{d}^3r_2 = 1$$

where $\mathrm{d}^3r = \mathrm{d}x\,\mathrm{d}y\,\mathrm{d}z$ and the integrations extend over all space.

(b) Verify the normalisation constant in (2.71) and that, $[(N_1!\ N_2!\ \ldots)/N!]^{-1/2}$, given in the text for the case when there are N_i particles in the one-particle state u_{α_i}. In each case show that

$$\int |\Psi(1, 2, \ldots N)|^2\,\mathrm{d}^3r_1 \ldots \mathrm{d}^3r_N = 1$$

We assume that the particles have zero spin.

***2.12** Using the orthogonality property of the one-particle states

$$\begin{aligned}\int u^*_{\alpha'}(q)u_\alpha(q)\,\mathrm{d}q &= 1, \text{ if } \alpha = \alpha' \\ &= 0, \text{ otherwise}\end{aligned}$$

where the $\mathrm{d}q$-integration involves spatial integration and summation over spin-indices as in (B196), show that (2.75) and (2.77) are normalised, i.e.

$$\int |\Psi(1, 2, \ldots, N)|^2\,\mathrm{d}q_1 \ldots \mathrm{d}q_N = 1$$

where 1, 2, ... stand for the spatial and spin coordinates of particles 1, 2, ..., and dq_i involves integration over the spatial coordinates and summation over the spin indices of particle i.

2.13 Evaluate the ground-state energy of a system of three non-interacting identical particles (the mass of a particle equals that of an electron) in a box of dimensions $a = 3$ Å, $b = 2$ Å, $c = 1$ Å and write down the corresponding eigenfunction for the system, when (*a*) the particles are fermions and (*b*) they are bosons.

2.14 (a) Show that the normalisation constant of the Boltzmann distribution (2.80) equals $(m/2\pi k_B T)^{3/2}$.

(b) Show that the average energy of a particle in a Boltzmann gas equals $3k_B T/2$.

2.15 Show that the potential energy of an electron at the centre of a uniformly charged sphere, when its total charge equals that of one electron, is $3e^2/(8\pi\epsilon_O R)$ where R is the radius of the sphere.

2.16 (a) Show that for large values of r the function $rR_{nl}(r)$, where R_{nl} denotes the radial part of the eigenfunction of an electron in a Coulomb field (see Table 2.3), is proportional to $r^n \exp(-Zr/na_0)$ and that, therefore, $r_{max}(nl) \simeq n^2 a_0/Z$.

(b) It is suggested by analogy to (a) above, that the radius of a shell of a many-electron atom can be approximately written as $r_{max}(nl) \simeq n^2 a_0/(Z - s)$ where s is a shielding constant (to be determined) which takes into account the screening of the nucleus by the electronic cloud. Use the results of Table 2.6 for the 1s and 2p shells to test the above proposition.

***2.17** The results listed in Tables 2.5 and 2.6 show that the radius $r_{max}(nl)$ of a shell depends on the energy and the angular momentum for the shell, and that it is possible for the radius of a shell of higher angular momentum (greater value of the quantum-number l) to be smaller than that of a lower angular momentum even if the energy of the former is higher than that of the latter (see, e.g., the entries for the 2s and 2p shells of the tables). Provide a qualitative interpretation of the above results on the basis of equation (B132) of Appendix B which determines the radial part of the wavefunction of the electron. To see the variation of the effective potential (B133) with r, approximate $U(r)$ interpolating roughly between its asymptotic values as $r \to 0$ and as $r \to \infty$. You may also consider the effective potential field when $U(r)$ is a bare Coulomb field, or a spherical potential well.

***2.18** Consider an atom with three electrons in the following configuration: 2p3p4d.

(a) Show, in relation to the total spin, that we have two sets of so-called doublet states ($S = \frac{1}{2}$) and one set of so-called quartet states ($S = \frac{3}{2}$). To prove the above combine the spins of two electrons to obtain S (for two electrons) $= 1, 0$, and then combine each of the above values with the spin of the third electron.

(b) To obtain the total orbital angular momentum, first combine the two p-states to obtain (for the two electrons) an S-state, a P-state and a D-state, corresponding respectively to $L = 0$, 1 and $L = 2$. Then combine each of the above values with the third (d-electron) to obtain the states for the three electrons:

$$\mathrm{S} + \mathrm{d} \rightarrow \mathrm{D}$$

$$\mathrm{P} + \mathrm{d} \rightarrow \mathrm{P}, \mathrm{D}, \mathrm{F}$$

$$\mathrm{D} + \mathrm{d} \rightarrow \mathrm{S}, \mathrm{P}, \mathrm{D}, \mathrm{F}, \mathrm{G}$$

(c) Finally, by combining (a) and (b) above find the total angular momentum states. Show that there are two sets of total angular momentum states, each containing a set of states described by

$$\begin{array}{c} {}^2\mathrm{D}_{3/2,5/2} \\ {}^2\mathrm{P}_{1/2,3/2}\ {}^2\mathrm{D}_{3/2,5/2}\ {}^2\mathrm{F}_{5/2,7/2} \\ {}^2\mathrm{S}_{1/2}\ {}^2\mathrm{P}_{1/2,3/2}\ {}^2\mathrm{D}_{3/2,5/2}\ {}^2\mathrm{F}_{5/2,7/2}\ {}^2\mathrm{G}_{7/2,9/2} \end{array} \qquad \text{(I)}$$

The above are obtained by combining the two $S = \frac{1}{2}$ states of the total spin with the total orbital angular momentum states. When spin–orbit interaction is taken into account the energy of every one of the above states will depend also on the angular momentum quantum-number J (lower index in the above notation), but since the two energy levels will not differ by much we refer to them as doublets. The states (I) of each of the above mentioned two sets correspond to eight doublets and one single level (${}^2\mathrm{S}_{1/2}$). We remember that for given J there are $(2J + 1)$ states of the same energy corresponding to the different values of the z-component of the total angular momentum.

Besides (I) we have a set of states of quartet levels described by

$$\begin{array}{c} {}^4\mathrm{D}_{1/2,3/2,5/2,7/2} \\ {}^4\mathrm{P}_{1/2,3/2,5/2}\ {}^4\mathrm{D}_{1/2,3/2,5/2,7/2}\ {}^4\mathrm{F}_{3/2,5/2,7/2,9/2} \\ {}^4\mathrm{S}_{3/2}\ {}^4\mathrm{P}_{1/2,3/2,5/2}\ {}^4\mathrm{D}_{1/2,3/2,5/2,7/2}\ {}^4\mathrm{F}_{3/2,5/2,7/2,9/2}\ {}^4\mathrm{G}_{5/2,7/2,9/2,11/2} \end{array} \qquad \text{(II)}$$

Obtain the above by combining the $S = \frac{3}{2}$ state of the total spin with the total orbital angular momentum states.

The reader should verify that the states in (I) and (II) together correspond to 65 distinct energy levels.

Footnotes of Chapter 2

1. We use the MKS system of units.
2. The orbiting of a planet about the sun is an example of such motion.
3. The angular momentum of a particle with respect to a centre is defined quite generally whether the potential field is centrally symmetric or not. In the general case when the angular momentum changes with time, Δt in the above definition must be very small ($\Delta t \rightarrow 0$) but in the case of a centrally symmetric field this is unimportant, because of the constancy of the angular momentum, and Δt may be taken as the unit of time.
4. It is always possible to write them in this way. There are, of course, linear combinations of these eigenfunctions which are eigenfunctions of the energy but not of the angular momentum (see, e.g., (2.6) of Section 2.3.1).

5. A complete list of the spherical Bessel and Hankel functions can be found in many books of mathematics and theoretical physics.
6. In practice the incident wavepacket has a wide cross-section (so that the scatterer sees effectively a plane wave) but, nevertheless, a finite one, so that there is no interference between the two terms of (2.10a) when $\theta > 0$ and $r \gg a$. The square modulus of ψ_E is taken at points where there is no incident wave (see Fig. 2.7).
7. In the present case one can actually evaluate, using (2.9a), some of the higher angular momentum terms and show that they are indeed very small compared to the $l = 0$ term.
8. Equation (2.12) which is exact, shows that we need not have done this approximation. By doing it we avoided some calculation (see problem 2.1).
9. This quantity is often referred to, in relation to electron states, as the density of the electronic cloud.
10. In the above discussion we disregarded the existence of hydrogen molecules for the sake of simplicity. In reality a hydrogen molecule has a lower energy than two independent hydrogen atoms and, therefore, at room temperature the gas consists of molecules, not hydrogen atoms. Only at higher temperatures ($k_B T > 4.476$ eV = dissociation energy of hydrogen molecule, see Chapter 3) there will be a significant number of atoms in the gas.
11. In terms of the frequency ν, this condition is written as follows: $2\pi\hbar\nu = (E_n - E_1)$.
12. Usually, the initial state is the ground state of the atom, as in (2.16). However, it may happen that the initial state is some other state of the atom, say $\psi_i = \psi_{n'l'm'}$.
13. It turns out (see Appendix B.11.2) that the transition rate is proportional to the quantity (2.17a) when the electric field is given by (2.15), except that in the general case ψ_{100} must be replaced by $\psi_i = \psi_{n'l'm'}$. The corresponding quantities for when the incident electric field is parallel to the x or the y direction are obtained from (2.17a) with z replaced by x and y respectively. One can show that the above quantities (all three of them) vanish when (2.19) are not satisfied.
14. The series are named after the spectroscopists who measured the above spectra at about the end of the 19th century.
15. When $V = L^3$ is large (of macroscopic dimensions), as we assume it to be, the results relating to observable quantities do not depend on the value of L.
16. $\hat{e}_1$ and $\hat{e}_2$ are the only linearly independent unit vectors normal to $\boldsymbol{k}$ in the sense that any other unit vector $\hat{e}$ normal to $\boldsymbol{k}$ can be written as: $\hat{e} = c_1\hat{e}_1 + c_2\hat{e}_2$. The two directions $\hat{e}_1$ and $\hat{e}_2$ correspond to the two linearly independent directions of the electric field associated with a plane electromagnetic wave of frequency $\nu_k = \omega_k/2\pi$ propagating in the direction of $\boldsymbol{k}$.
17. We can rewrite (2.26) as follows

$$\bar{E}_{k\hat{e}}(T) = -\frac{\partial \ln Z}{\partial \beta} \tag{i}$$

where $\beta = 1/k_B T$. So that the evaluation of (2.26) reduces to the evaluation of Z. Substituting (2.24) in (2.25a) we obtain

$$Z = \exp(-\hbar\omega_k\beta/2)\sum_{n=0}^{\infty}(\exp(-\beta\hbar\omega_k))^n$$

$$= \exp(-\hbar\omega_k\beta/2)/(1 - \exp(-\beta\hbar\omega_k)) \tag{ii}$$

which substituted in (i) gives (2.27).

18. It is customary to write (2.33a) as $\sigma^2 = s(s+1)\hbar^2$, $s = \frac{1}{2}$ and to say that the

electron has an intrinsic angular momentum, a spin $s = \frac{1}{2}$. By intrinsic we mean a property, like the mass or the charge of the electron, which is always there whether the electron is free or in some potential field.

19. Under the influence of an external magnetic field it behaves like a magnetic needle (a needle with magnetic moment).
20. The same is true in classical mechanics when the speed of a macroscopic particle becomes very large.
21. The theory was formulated by Albert Einstein in 1905. Relativistic corrections are large only when the speed of the particles involved becomes comparable to that of light.
22. It is also evident that the mean kinetic energy will be larger the more negative $U(r)$ is; which is the case when Z is larger. We infer from this that relativistic effects will be more important in heavy atoms.
23. Often we write $\psi_{n\uparrow}$ instead of ψ_{n+}, and $\psi_{n\downarrow}$ instead of ψ_{n-}.
24. We shall see in Section 2.14 that there are other particles besides the electron which have spin $s = \frac{1}{2}$.
25. Here and in what follows, when there is no danger of confusion, we refer to the spinor eigenstates of a particle, simply as eigenstates of the particle, for the sake of brevity.
26. In general we find that a weak perturbation of the potential field cannot change the number of eigenstates in any given energy region. It may split some energy levels by reducing the symmetry of the potential field, it may shift a bit some energy levels, but the number of eigenstates does not change (see also Section 1.3.3 and Appendix B.10).
27. To obtain the energy levels of the hydrogen atom one puts $Z = 1$ in (2.55).
28. They are called Clebsch–Gordan coefficients and can be found in books on atomic spectroscopy. We note that E_n in the time-factor of (2.57) cancels with the corresponding term of Ψ^0 (see Table 2.4) and one is left with a time-factor $\exp(-\mathrm{i}E_{nj}t/\hbar)$.
29. In our presentation we neglected the interaction of the angular momentum (orbital + spin) of the electron with the spin of the nucleus. If it were not for this interaction, the spin of the proton (see Section 2.14) would double the degeneracy of all energy levels of hydrogen (solid lines in Fig. 2.15). As it is, each solid line stands for two closely spaced levels; the separation between these two levels (due to the dipole–dipole interaction between the magnetic moments of the electron and proton) is hundreds of times smaller than the fine-structure splitting (due to spin–orbit coupling) and is called hyperfine structure.
30. Here by particle we do not mean only an elementary particle, such as the electron, which can not split into smaller particles, but any particle whose internal motion (i.e. with respect to its centre of mass) does not change in the phenomena under consideration. In a gas of atoms, at low temperatures, their internal state is their ground state (see Section 2.6) which is also an eigenstate of the total angular momentum of the atom (made up from the orbital and spin angular momenta of all the electrons and the nucleus of the atom). The atoms of the gas have, therefore, a definite (internal) energy and a total angular momentum of definite magnitude. In this section we call this, the total angular momentum of the atom, the spin of the atom. The thermodynamic behaviour of the gas is determined by the translational motion of the atoms and the collisions between them. When two atoms collide, they exchange translational energy but internally they do not change, they stay in their ground states.
31. We come to this conclusion by repeating the arguments of Section 1.3.2 relating to independent motions.

32. We disregard the Coulomb repulsion between the electrons; though it may seem surprising, it is often a good approximation to describe the conduction electrons of a metal in this way (see Section 4.9).
33. As usual, α denotes the set of quantum-numbers which determine the orbital and spin parts (if $s \neq 0$) of the one-particle eigenstates.
34. The formula gives the average value of n over an ensemble of a large number of macroscopically same systems (N particles in volume V at temperature T).
35. The eigenstates of the translational motion of one particle are those of a free particle in a box (described in Sections 1.3.3 and 2.12). In a large box there would be eigenstates of energy $E_\alpha < \mu$, if $\mu > 0$; putting $E_\alpha < \mu$, makes $n(\alpha)$ of (2.72) negative, which cannot be; therefore μ must be negative, at any temperature.
36. The reader who is familiar with determinants can show that the N-particle wavefunction (2.77) can be written in a determinantal form as follows

$$\Psi_{\alpha_1\alpha_2\ldots\alpha_N}(1, 2, \ldots, N) = \frac{1}{\sqrt{N!}} \begin{vmatrix} u_{\alpha_1}(1) & u_{\alpha_1}(2) & \ldots & u_{\alpha_1}(N) \\ u_{\alpha_2}(1) & u_{\alpha_2}(2) & \ldots & u_{\alpha_2}(N) \\ \ldots & \ldots & \ldots & \ldots \\ u_{\alpha_N}(1) & u_{\alpha_N}(2) & \ldots & u_{\alpha_N}(N) \end{vmatrix}$$
$$\times \exp(-\mathrm{i}(E_{\alpha_1} + \cdots + E_{\alpha_N})t/\hbar)$$

The above determinant is known as a Slater determinant.
37. One can also remark (this follows directly from (2.76)) that two fermions cannot be at the same point in space with the same spin orientation.
38. The atoms or molecules of an ordinary gas (at room temperature and under atmospheric pressure) behave as such as far as their translational motion is concerned.
39. We remember that in a macroscopic box ($V \to \infty$) the kinetic energy of the particle can, practically, take any value between zero and infinity.
40. Here, by density of the electronic cloud we mean, as in Section 2.5 (see in particular Fig. 2.10), a probability density. As we shall see, $n(\boldsymbol{r})$ can be written as a sum of independent contributions from the N electrons.
41. This is known as the $X\alpha$-model. See, J. C. Slater, *Quantum Theory of Matter*, (New York: McGraw-Hill) 1968; J. C. Slater, *The Self-consistent Field for Molecules and Solids: Quantum Theory of Molecules and Solids*, Volume 4 (New York: McGraw-Hill) 1974.
42. For large values of r, the energy $U(r)$ evaluated from (2.89) goes faster to zero than (2.90). This is due to the approximation involved in the calculation of $U_{\mathrm{ex}}(r)$.
43. In general the functions $R_{nl}(r)$ in (2.91) are known only numerically. We can represent them graphically, but, usually, we cannot obtain reliable analytic expressions of these functions. Their main features are similar to those of the hydrogen eigenfunctions discussed in Section 2.5.
44. The notation corresponds to the one used in the description of the eigenstates of an electron in a Coulomb field (see Sections 2.5 and 2.13). By convention the lowest eigenvalue for given l has $n = l + 1$.
45. The reader can see this by using (2.4a) to obtain an averaged spherically symmetric quantity (in the manner of Fig. 2.21) from the probability-density function $[R_{nl}(r)]^2 |Y_{lm}(\theta, \varphi)|^2$. It is also evident, from equation (2.4b), that the contribution to $n(\boldsymbol{r})$ from a closed shell *is* spherically symmetric.
46. F. Herman and S. Skillman, *Atomic Structure Calculations*, (New Jersey:

Prentice-Hall) 1963. Tables 2.5, 2.6 and 2.8 have been reproduced with permission from J. C. Slater, *Quantum Theory of Matter*, loc. cit..

47. One can easily see that the Coulomb repulsion between the electrons is not taken fully into account in the mean field approximation by noting that, in this approximation, the probability that two electrons '1' and '2' of different spin will be in the volume dV about $\boldsymbol{r}$ is given by the probability that '1' will be there, times the probability that the 'independent' electron '2' will be there. In reality if '1' is already there, it is more difficult for '2' to be there too, because of the Coulomb repulsion between the two electrons. The residual interaction between the electrons is often called the electron–electron correlation interaction, or simply the correlation interaction.
48. We disregard the spin of the nucleus. Its interaction with the angular momenta of the electrons is usually negligible.
49. Hund's rule, telling us that a spin-polarisation of the atom (say more electrons with spin up than down) is a likely feature of the ground state of an atom, suggests that we would have done a better mean-field calculation if we had allowed $n_{\uparrow}(\boldsymbol{r})$ and $n_{\downarrow}(\boldsymbol{r})$ to be different (see (2.85)), in which case the exchange potential (2.88) would be different for the two spins ($n(\boldsymbol{r})/2$ in (2.88) would be given by $n_{\uparrow}(\boldsymbol{r})$ and $n_{\downarrow}(\boldsymbol{r})$ respectively), and therefore the potential (2.89) and the corresponding one-electron levels and states (the orbital parts of them) would be different for spin up and spin down. The above is true and such calculations have been done. We note, however, that the energy-level diagram of the atom (as in Fig. 2.23) that one obtains when the residual interaction (and some such will exist even for a spin-polarised calculation) is taken into account, will not be different (except perhaps in the exact positions of the energy levels) from that we have described.
50. It is worth noting that the atom He has been removed from the top of the s^2 column to the top of the p^6 column. The reason is that He behaves chemically in very much the same way as the inert atoms of the latter column.
51. We assume always that the centre of the mass of the atom (the nucleus) is at rest at the origin of coordinates.

All these things being consider'd it seems probable to me, that God in the beginning form'd Matter in solid, massy, hard, impenetrable, moveable Particles of such Sizes and Figures, and with such other Properties, as most conduced to the End for which he formed them; . . . And therefore, that Nature may be lasting, the Changes of corporeal Things are to be placed only in the various Separations and new Associations and Motions of these permanent Particles.

ISAAC NEWTON, Opticks

3 MOLECULES

3.1 The Born–Oppenheimer approximation

A molecule consists of two or more nuclei and a complement of electrons. When no external forces are acting on the molecule the kinetic energy of its translational motion (this equals $MV_{CM}^2/2$, where M is the total mass of the molecule and V_{CM} the velocity of its centre mass) is constant, and in the following we disregard it. We assume, for simplicity, that the centre of mass of the molecule stays motionless at the origin of coordinates.[1]

We begin our discussion of the internal (i.e. with respect to a centre of mass system of coordinates) motion of the electrons and nuclei of the molecule by noting that the nuclei (which consist of protons and neutrons) are much heavier than the electrons. This means that forces of the same magnitude (and the forces acting on the electrons and the nuclei of a molecule are of the same magnitude because they derive from the same electrostatic interaction) produce a much larger acceleration of the electrons than of the nuclei. This means that for given total energy of the molecule, the average speed of the electrons is much larger than the average speed of the nuclei. We can say that, in the time taken by an electron to complete a revolution (an orbit) in the space of the molecule, the nuclei move relatively little, so that to a first approximation we may assume that they have not moved at all.

With the nuclei motionless at the positions $\boldsymbol{R}$, we can calculate the orbits of the electrons and the energy $E(\boldsymbol{R})$ of the molecule as a function of $\boldsymbol{R}$. Here and throughout this section, we denote by $\boldsymbol{R}$ the total of coordinates which determine the positions of all the nuclei of the molecule. We note that in a CM-system of coordinates we need $3\Lambda - 3$, and not 3Λ, coordinates to specify the positions of the Λ nuclei of the molecule. The energy $E(\boldsymbol{R})$ includes the kinetic energy of the electrons, the potential energy of attraction between the negatively charged electrons and the positively charged nuclei, the potential energy of repulsion between the electrons and, also, the potential energy of repulsion between the motionless nuclei.

Finally, we 'allow' the nuclei to move, always slowly in comparison with the motion of the electrons, and we remark: the total energy ε of the molecule, which is a constant of the motion (it does not change with time), is the sum of the kinetic energy K of the nuclei and the energy $E(\boldsymbol{R})$ as defined above:

$$\varepsilon = K + E(\boldsymbol{R}) \tag{3.1}$$

Equation (3.1) represents a system of Λ nuclei moving in the potential field $E(\boldsymbol{R})$. The classical motion of the nuclei in the field $E(\boldsymbol{R})$ is determined, at least in principle, in the same way as the motion of a particle in a potential well (see Section 1.4.1).

We know, of course, that the motion of the electrons and the nuclei of a molecule cannot be described by classical mechanics. However, the essence of the arguments leading to (3.1) remains valid, and is expressed in quantum-mechanical term as follows.

For given (fixed) positions $\boldsymbol{R}$ of the nuclei we calculate the eigenvalues $E_n(\boldsymbol{R})$ of the $E(\boldsymbol{R})$ defined above, and the corresponding eigenfunctions[2] $\psi_n\,(\boldsymbol{r}_1, \boldsymbol{r}_2, \ldots, \boldsymbol{r}_N; \boldsymbol{R})$ which describe the eigenstates of the system of the N electrons of the molecule for the given $\boldsymbol{R}$. We denote by $\boldsymbol{r}_i$ the position coordinates of electron i. And we emphasise that the position coordinates of the nuclei, which are denoted collectively by $\boldsymbol{R}$ in ψ_n, are not variables of ψ_n like the $\boldsymbol{r}_i$, but a parameter of the system which determines the potential field seen by the electrons. The index n denotes the set of quantum numbers required for a complete description of $E_n(\boldsymbol{R})$ and $\psi_n(\boldsymbol{r}_1, \ldots, \boldsymbol{r}_N; \boldsymbol{R})$. In Fig. 3.1 we show, schematically, the electronic terms (that is what the eigenvalues $E_n(\boldsymbol{R})$ are called) of a hypothetical molecule. If the muclei were indeed stationed at $\boldsymbol{R}$, the state of the molecule would be described by $\psi_n(\boldsymbol{r}_1, \ldots, \boldsymbol{r}_N; \boldsymbol{R}) \exp(-\mathrm{i}E_n(\boldsymbol{R})t/\hbar)$. But the nuclei are moving, however slowly, and therefore the energy $E_n(\boldsymbol{R})$ is not conserved (it is not a constant of the motion). Only the total energy ε of the molecule is conserved and this includes (see (3.1)), besides $E_n(\boldsymbol{R})$, the kinetic energy of the nuclei. Therefore, we identify the eigenvalues of ε, which correspond to a particular electronic term $E_n(\boldsymbol{R})$, with the eigenvalues of the energy of a physical system which consists of the Λ nuclei of the molecule moving under the influence of a mutual interaction $E_n(\boldsymbol{R})$. To these eigenvalues, which we shall denote by $\varepsilon_{n\mu}$, correspond, as usual, eigenstates of the nuclei described by wavefunctions[2], $\chi_{n\mu}(\boldsymbol{R}) \exp(-\mathrm{i}\varepsilon_{n\mu}t/\hbar)$, specified by an appropriate set of quantum numbers denoted by μ. And since the state of the electrons (when the nuclei are at $\boldsymbol{R}$) is given by $\psi_n(\boldsymbol{r}_1, \ldots, \boldsymbol{r}_N; \boldsymbol{R})$ and because in the approximation we are considering the electrons adjust rapidly (instantly) to the variation in the position of the nuclei, we can write the eigenfunction

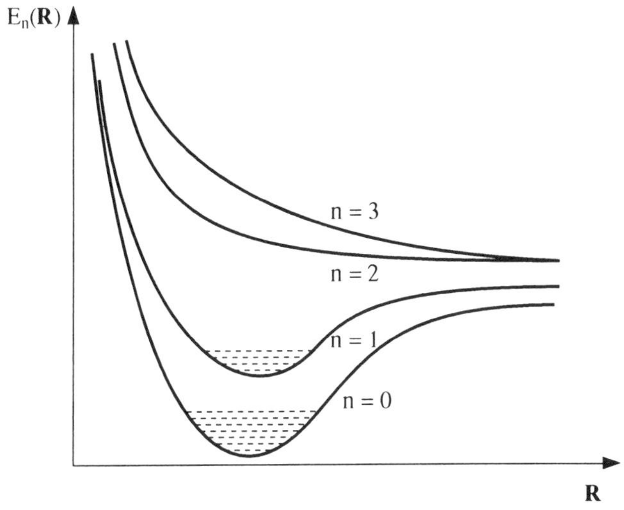

Figure 3.1. Electronic terms $E_n(\boldsymbol{R})$, $n = 0, 1, 2, 3$ of a hypothetical molecule. The broken lines represent total energies (eigenvalues of ε) corresponding to bound eigenstates of the molecule. Such states do not exist for the terms $n = 2, 3$

which describes the corresponding state of the molecule (electrons and nuclei) as follows

$$\Psi_{n\mu}(\boldsymbol{r}_1, \ldots, \boldsymbol{r}_N; \boldsymbol{R}; t) = \chi_{n\mu}(\boldsymbol{R})\psi_n(\boldsymbol{r}_1, \ldots, \boldsymbol{r}_N; \boldsymbol{R}) \exp(-i\varepsilon_{n\mu}t/\hbar) \quad (3.2)$$

The approximate description (3.2) of the eigenstates of a molecule is known as the Born–Oppenheimer approximation. Alternatively, we refer to it as the adiabatic approximation.

3.2 Hydrogen molecular ion—electronic terms

The positively charged molecular ion of hydrogen H_2^+ is the simplest molecule; it consists of one electron and two nuclei (protons). The potential energy of the system is

$$\begin{aligned} U(\boldsymbol{r}; R) &= \frac{1}{4\pi\epsilon_O}\left(\frac{e^2}{R} - \frac{e^2}{r_1} - \frac{e^2}{r_2}\right) \\ R &\equiv |\boldsymbol{R}_1 - \boldsymbol{R}_2| \\ r_1 &\equiv |\boldsymbol{r} - \boldsymbol{R}_1| \\ r_2 &\equiv |\boldsymbol{r} - \boldsymbol{R}_2| \end{aligned} \quad (3.3)$$

The position vectors $\boldsymbol{R}_1$ and $\boldsymbol{R}_2$ of the nuclei and $\boldsymbol{r}$ of the electron are shown in Fig. 3.2. The first term of (3.3) gives the potential energy of repulsion

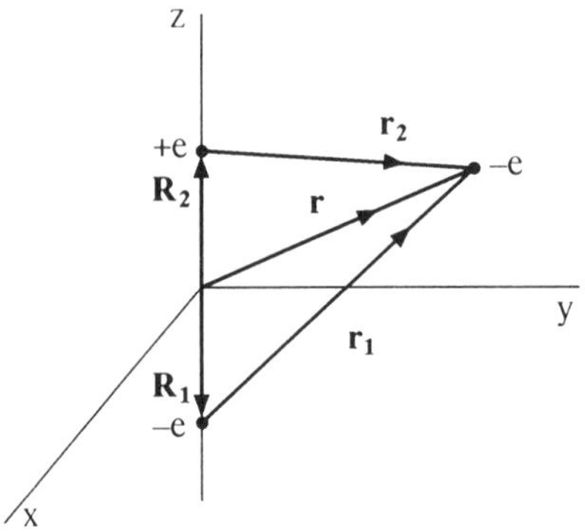

Figure 3.2. Molecular ion of hydrogen. The nuclei are stationed at $\boldsymbol{R}_1 = (0, 0, -R/2)$ and $\boldsymbol{R}_2 = (0, 0, R/2)$; we have taken the z-axis parallel to the axis of the molecule (the line joining the two nuclei) and we put the origin of coordinates at the centre of mass of the molecule

between the nuclei, and the second and third terms, the potential energy of attraction between the electron and the two nuclei.

With the nuclei motionless at $\boldsymbol{R}_1$ and $\boldsymbol{R}_2$, equation (3.3) defines the potential energy field in which the electron of the molecules moves. The ground state of the electron in this field is described approximately by the following wavefunction[3]

$$\psi_0(\boldsymbol{r}; R) = A_0\{e^{-r_1/a_0} + e^{-r_2/a_0}\} \tag{3.4}$$

where a_0 is the Bohr radius and A_0 is a normalisation constant. We note that the two terms of (3.4) are identical, apart from the normalisation constant, with the eigenfunction ψ_{100} which describes the ground state of an electron in the field of a proton when the latter is at $\boldsymbol{R}_1$ and $\boldsymbol{R}_2$ respectively[4]. We refer to these terms as atomic orbitals, and to a wavefunction which describes an electron state in a molecule as a molecular orbital. In the above approximation the molecular orbital ψ_0 is written as a linear sum of atomic orbitals. The idea is that the electron mostly sees the one or the other of the two nuclei and, therefore, its wavefunction can be approximated by a linear sum of corresponding atomic orbitals; and symmetry tells us that for the ground state this linear sum can only be (3.4) or (3.6). The normalisation constant A_0 is determined, as usual, by the requirement

$$\int |\psi_0(\boldsymbol{r}; R)|^2 \, dV = 1 \tag{3.5}$$

where $dV = dx\,dy\,dz$ and the integral extends over all space. The above approximation is very good asymptotically ($R \to \infty$), it is moderately good for $R \gtrsim 2$ a.u., is tolerable down to $R = 1$ a.u. and gets worse as $R \to 0$.

However, as we shall see, when the molecule is in its electronic ground state, the probability that the nuclei will be separated by $R < 2$ a.u. is very small and practically vanishes for $R \lesssim 1$ a.u., and in this respect we need not worry about the failure of (3.4) for small R.

In Fig. 3.3 we show the variation of ψ_0 along the axis of the molecule for three different values of the distance R between the nuclei of the molecule. In Fig. 3.4 we show the variation with R of the corresponding electronic term[5] $E_0(R)$. We see that when the distance between the nuclei is large enough ($R \to \infty$), $E_0(R) = -0.5$ Hartree, which is the ground energy of an electron in the field of a proton. We can understand this as follows: when the two nuclei stand away from each other ($R \to \infty$), the overlap between the two terms of (3.4) vanishes: the electron sees, at any given time, one or

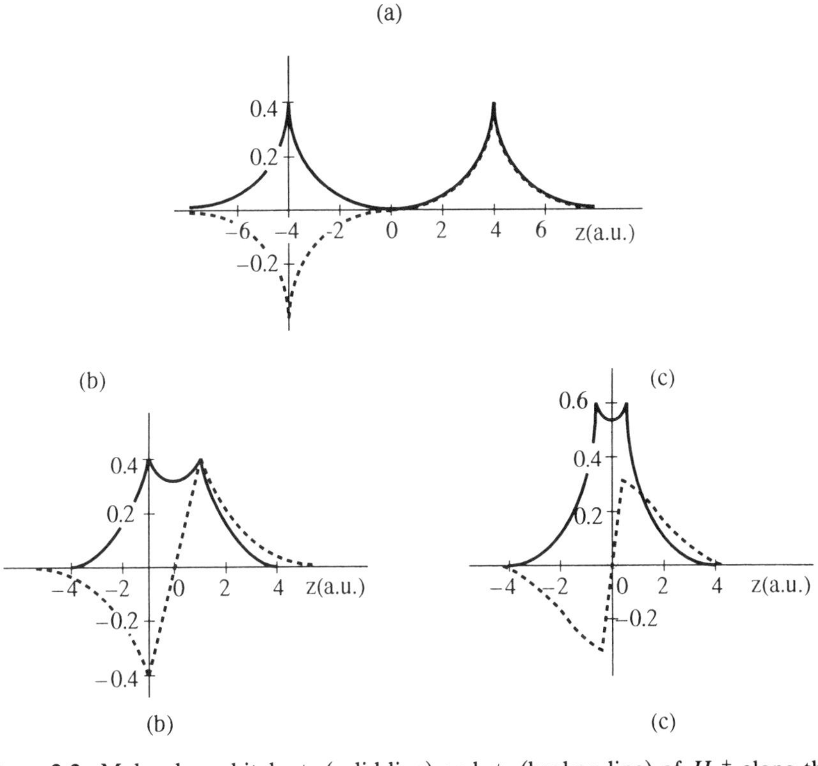

Figure 3.3. Molecular orbitals ψ_0 (solid line) and ψ_1 (broken line) of $H_2{}^+$ along the axis (z) of the molecule for three values of the internuclear separation (R): (a) $R = 8$ a.u.; (b) $R = 2$ a.u.; (c) $R = 1$ a.u. The graphs are based on exact wavefunctions, however these do not differ much from those one would obtain using the approximate expressions (3.4) and (3.6). 1 a.u. $= a_0 = 0.529$ Å). (Redrawn with permission from J. C. Slater, *Quantum Theory of Matter*, (New York: McGraw-Hill) 1968)

the other of the two nuclei, not both of them. We have then, in effect, a hydrogen atom in its ground state and a bare proton, the two of them not interacting with each other. As R decreases from infinity the potential field that the electron sees changes gradually, and the wavefunction ψ_0 also changes gradually due to the overlap of the two terms of (3.4). The resulting variation of $E_0(R)$ from infinite R to about 4 a.u. is relatively small and we shall not discuss it here. When the distance between the nuclei gets smaller than 4 a.u. or so, the overlap between the two terms of (3.4) leads to comparatively high values of ψ_0 in the region between and around the nuclei (see Fig. 3.3), which means, so to speak, that the electron is found there more often than elsewhere (the probability density being proportional to $|\psi_0|^2$). The concentration of the electronic cloud in the region between and around the nuclei leads, via the last two terms of (3.3), to a reduction of the mean potential energy of the system and, therefore, tends to reduce $E_0(R)$ as R decreases (see also comments following (3.7)). On the other hand, as the separation between the nuclei gets smaller than $R \simeq 4$ a.u., the mean kinetic energy of the electron begins to rise; this comes about from the confinement of the electron to a progressively smaller volume as R decreases. The situation is similar to that of an electron in a box; the smaller the box, the higher the (kinetic) energy of the electron (see (1.21)). Also, the energy of repulsion between the nuclei (first term of (3.3)) increases as the nuclear separation gets smaller. The increase of $E_0(R)$ due to the increase of the kinetic energy of the electron and of the nuclear repulsion as R gets smaller, competes with the reduction of $E_0(R)$ due to electronic cloud concentration between the nuclei (where the potential energy (3.7) is lower) as R gets smaller. The increase is smaller than the reduction as we move from $R \simeq 4$ a.u. to about 2 a.u. From then on ($R \lesssim 2$ a.u.) the opposite is true: the kinetic energy of the electron and the nuclear repulsion energy (especially the latter) increase very fast as R becomes smaller. The result is the $E_0(R)$ of Fig. 3.4 with a minimum at $R = 2a_0$. The existence of the minimum secures the existence of bound states of H_2^+, and the difference $E_0(R \to \infty) - E_0(R = 2a_0) \simeq 0.1$ Hartree determines the dissociation energy of the molecule (see Section 3.3.2). Obviously, if the total energy ε of H_2^+ exceeds $E_0(R \to \infty)$ the distance R between the nuclei can take any value up to infinity, which means that the molecule will dissociate: $H_2^+ \to H + H^+$. The energy spectrum of H_2^+ corresponding to the electronic term $E_0(R)$ is of course continuous in the energy region $\varepsilon > E_0(R \to \infty)$ as expected of non-bound states.

The electronic state of the system $H + H^+$, which corresponds to the electronic term $E_1(R)$ of Fig. 3.4, is described approximately by the wavefunction

$$\psi_1(\boldsymbol{r}; R) = A_1\{e^{-r_1/a_0} - e^{-r_2/a_0}\} \tag{3.6}$$

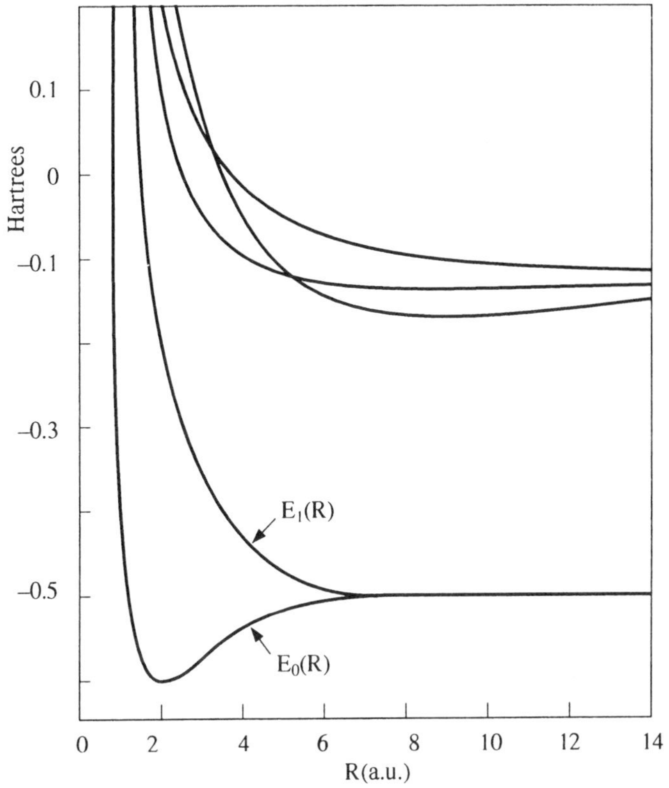

Figure 3.4. Electronic terms of H_2^+. The terms $E_0(R)$ and $E_1(R)$ correspond to the molecular orbitals ψ_0 and ψ_1 respectively (1 Hartree = 27.21 eV)

The above wavefunction, like (3.4), is very good for large values of R, moderately good for $R \gtrsim 2$ a.u. and not so good for $R < 2$ a.u.

For large values of R, ψ_1 does not differ significantly from ψ_0. But the two states are very different when the nuclear separation is smaller than $R \simeq 4$ a.u. and the overlap between the atomic orbitals is significant. In contrast to (3.4) the overlap between the two terms (atomic orbitals) of (3.6) results in a reduction of the probability-density in the region between the two nuclei (see Fig. 3.3) and, consequently, the reduction in the mean potential energy of the system from the last two terms of (3.3) is relatively small and cannot offset the increase from the kinetic energy of the electron and nuclear repulsion terms as R gets smaller and, therefore, the variation of the corresponding electronic term $E_1(R)$ with R, shown in Fig. 3.4, is very different from that of $E_0(R)$. We note in particular that $E_1(R)$ has no minimum and this means that there are no bound states of H_2^+ corresponding to this term. We often say, for this reason, that ψ_1 is an antibonding (molecular) orbital, in contrast to ψ_0 which is called a bonding orbital.

The other electronic terms shown in Fig. 3.4 correspond to molecular orbitals of H_2^+ which are linear sums of atomic orbitals representing excited states of the hydrogen atom. We shall not be concerned about these terms.

The solid lines in the two diagrams of Fig. 3.5 represent the potential energy

$$U_e(\boldsymbol{r}) = -\frac{1}{4\pi\epsilon_O}\left\{\frac{e^2}{|\boldsymbol{r}-\boldsymbol{R}_1|} + \frac{e^2}{|\boldsymbol{r}-\boldsymbol{R}_2|}\right\} \tag{3.7}$$

of the electron in the field of the two nuclei of H_2^+ along the axis of the molecule. $U_e(\boldsymbol{r})$ is the sum of the last two terms of (3.3). The reader will note the considerable reduction of $U_e(\boldsymbol{r})$ in the region between the nuclei

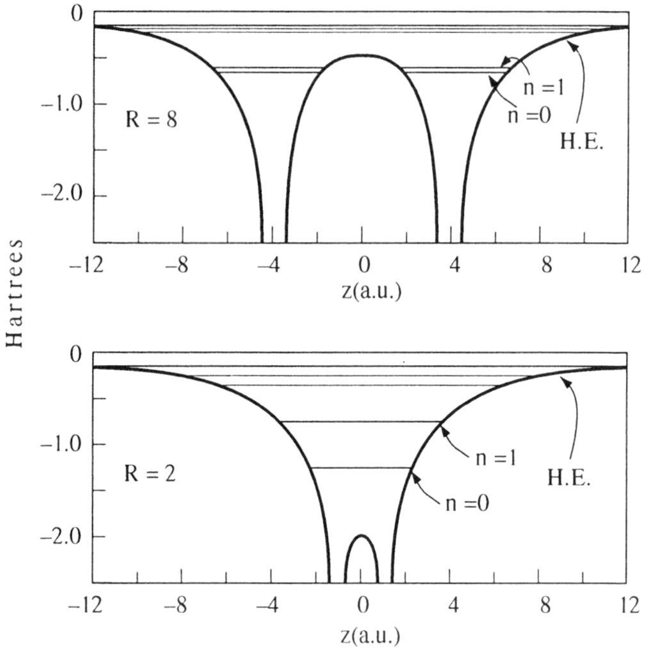

Figure 3.5. The potential energy (3.7) of the electron along the axis (z) of H_2^+ for two values, $R = 8$ a.u. and $R = 2$ a.u., of the internuclear separation. The zero of the z-axis coincides with the centre of mass of the molecule. The horizontal lines represent the eigenvalues of the energy, $\epsilon_n(R)$, of the electron in the potential field (3.7). The energy levels $n = 0$ and $n = 1$ correspond to the electronic terms $E_0(R)$ and $E_1(R)$ of Fig. 3.4, and those denoted by H. E. correspond to the higher electronic terms of Fig. 3.4. The separation between the energy levels has been exaggerated in the top diagram.

when the separation between the two changes from 8 a.u. to 2 a.u. The horizontal lines in the two diagrams represent the energy levels, $\epsilon_n(R)$, of the electron in the field (3.7). The energy levels ϵ_0 and ϵ_1 correspond to the molecular orbitals ψ_0 and ψ_1 described by (3.4) and (3.6) respectively; and evidently ϵ_0 is lower than ϵ_1 because $|\psi_0|^2$ is larger than $|\psi_1|^2$ in the region between the nuclei where $U_e(\boldsymbol{r})$ is smaller (more negative). We note that the difference $\epsilon_1(R) - \epsilon_0(R)$ between the two levels decreases as the separation R between the nuclei increases. It is much smaller at $R = 8$ a.u. than at $R = 2$ a.u. In the limit $R \rightarrow \infty$, $\epsilon_1 = \epsilon_0 = -0.5$ Hartree which is the ground-state energy of an electron in the field of *one* proton. To this energy-eigenvalue, the atomic level $\epsilon_1(R = \infty) = \epsilon_0(R = \infty)$, correspond two electron states localised about nucleus 1 and nucleus 2 respectively. We can say that in the first case the electron is found with a 100% probability about nucleus 1 and in the second case it is found with 100% probability about nucleus 2. As R gets smaller, the electron is able to move between the two nuclei and the above degeneracy is lifted. The atomic level is split into the two levels $\epsilon_0(R)$ and $\epsilon_1(R)$ corresponding to the molecular orbitals (3.4) and (3.6). We meet here for the first time the phenomenon which we shall later (in the treatment of one-electron states in crystals) call creation of a band of energy levels (and corresponding states).

It follows from the definition of $E_n(R)$ that this quantity is given, for the $H_2{}^+$ molecule, by the sum

$$E_n(R) = \epsilon_n(R) + \frac{e^2}{4\pi\epsilon_O R} \tag{3.8}$$

From the way it is calculated, the first term includes the kinetic energy of the electron and the electrostatic energy of attraction between it and the nuclei, while the second term of (3.8) represents the electrostatic energy of repulsion between the nuclei. It follows from (3.8) that

$$\epsilon_1(R) - \epsilon_0(R) = E_1(R) - E_0(R) \tag{3.9}$$

3.3 Nuclear motion

3.3.1 ROTATIONAL AND VIBRATIONAL MOTION OF A DIATOMIC MOLECULE

In the preceding section we have seen that the electronic term $E_0(R)$ is such that bound states of $H_2{}^+$ do exist for this term and we presented the electronic part (equation (3.4)) of the corresponding eigenstates of the molecule. We have now to determine the eigenvalues of the total energy of the molecule for the given electronic term and the corresponding nuclear wavefunctions.

The method of calculation of these quantities, denoted by $\varepsilon_{n\mu}$ and $\chi_{n\mu}(\boldsymbol{R})$

in the general formula (3.2), is the same essentially for all datomic molecules and therefore the exposition which follows is valid not only for $H_2{}^+$ but for all other diatomic molecules as well. We are interested in the bound eigenstates of a diatomic molecule of a given electronic term $E_n(R)$, which implies that this term has a minimum at a certain value of the internuclear separation R (as is the case with the $E_0(R)$ term of $H_2{}^+$) so that such states exist.

It is obvious that the positions of the nuclei of a diatomic molecule, when its centre of mass stays at the origin of coordinates, is fully determined by the direction of the axis of the molecule (angles Θ and Φ in Fig. 3.6) and the distance R between the two nuclei, and therefore the nuclear part of the wavefunction (3.2) takes the form: $\chi(R, \Theta, \Phi)$[6].

We remember (see Section 3.1) that the total energy ε of the molecule is the sum of the kinetic energy of the nuclei and the electronic term $E_n(R)$ which in the case of a diatomic molecule depends only on the internuclear separation. This fact allows us to write the nuclear part of the wavefunction as follows[7]

$$\chi(R, \Theta, \Phi) = (u(R)/R)Y_{lm}(\Theta, \Phi) \tag{3.10}$$

where $Y_{lm}(\Theta, \Phi)$ are the spherical harmonics introduced in Chapter 2. We recall that $l = 0, 1, 2, \ldots$, and that for given l the quantum-number m takes the values $m = -l, -l+1, \ldots, l-1, l$. Equation (3.10) tells us that the magnitude of the orbital angular momentum of the two atomic nuclei with respect to the centre of mass of the molecule may have one of the following values (and none other): $\hbar\sqrt{l(l+1)}$, $l = 0, 1, 2, \ldots$ We know that classically the kinetic energy of the nuclei due to rotational motion (variation of Θ and Φ with time) equals

$$K_{\text{rot}} = L^2/(2I) \tag{3.11}$$

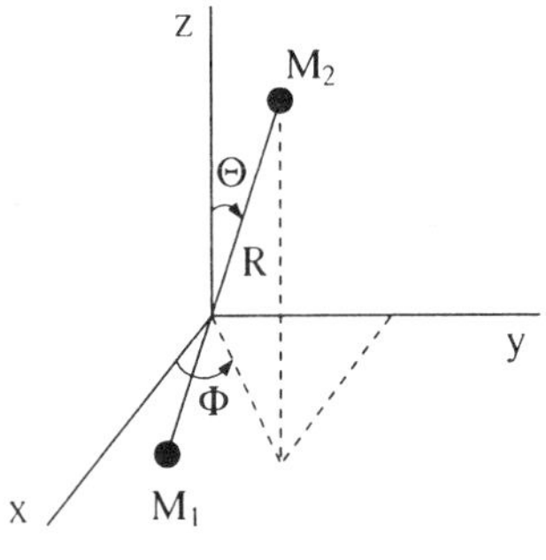

Figure 3.6. The position of the nuclei of a diatomic molecule, when its centre of mass stays at the origin of coordinates, is determined by the coordinates R, Θ, $\Phi[0 < R < +\infty, 0 \leq \Theta \leq \pi, 0 \leq \Phi < 2\pi]$

where L denotes the magnitude of the orbital angular momentum and I the moment of inertia of the two nuclei with respect to the centre of mass of the molecule. For a homonuclear molecule, such as H_2^+, we have $I = 2M(R/2)^2$ where M denotes the mass of one nucleus. We obtain the quantum-mechanical expression for the rotational energy of the molecule by substituting in (3.11) the allowed values of L:

$$K_{\text{rot}} = \frac{\hbar^2}{2I} l(l+1) = \frac{\hbar^2 l(l+1)}{2M_\text{r}R^2} \tag{3.12}$$

where $l = 0, 1, 2, \ldots$, and $M_\text{r} \equiv M/2$[8].

The total energy ε of the molecule, a conserved quantity, is the sum of $E_n(R)$ and the kinetic energy of the nuclei which consists of the rotational energy (3.12) and a vibrational term K_{vib} which derives from the variation of the internuclear separation R with the time. It can be shown (see problem 3.5) that the vibration of the nuclei, described by $u(R)$ in (3.10), can be identified with the motion of a particle of mass M_r moving in the one-dimensional potential field

$$U_{nl}(R) \equiv E_n(R) + \frac{\hbar^2 l(l+1)}{2M_\text{r}R^2} \tag{3.13}$$

The eigenvalues ε_{nlv} of the total energy of the molecule are given by the energy eigenvalues of a particle of mass M_r moving in the potential field (3.13); and the energy eigenfunctions of the particle in the above field give the vibrational parts $u_{nlv}(R)$ of the corresponding nuclear wavefunctions (3.10).

If we now put together the nuclear and electronic parts of the wavefunction (in the manner of Section 3.1) we obtain: to the eignevalue ε_{nlv} correspond the eigenstates

$$\Psi_{nlmv} = (u_{nlv}(R)/R)Y_{lm}(\Theta, \Phi)\psi_n(\boldsymbol{r}; R)\exp(-\text{i}\varepsilon_{nlv}t/\hbar) \tag{3.14}$$

where $m = -l, -l+1, \ldots, l-1, l$.[9] In the case of H_2^+, $\boldsymbol{r}$ denotes the position vector of the single electron of the molecule; in the case of a diatomic molecule of N electrons $\boldsymbol{r}$ stands for the coordinates $(\boldsymbol{r}_1, \boldsymbol{r}_2, \ldots, \boldsymbol{r}_N)$ of all the electrons of the molecule.

Let us, then, consider the energy levels and corresponding eigenstates of a particle of mass M_r in the one-dimensional potential field (3.13). This is shown schematically in Fig 3.7. We have assumed that the electronic term has a minimum at $R = R_{no}$. In the neighbourhood of the minimum one can write $E_n(R)$ as follows

$$E_n(R) = E_n(R_{no}) + \tfrac{1}{2}k_n(R - R_{no})^2 \tag{3.15}$$

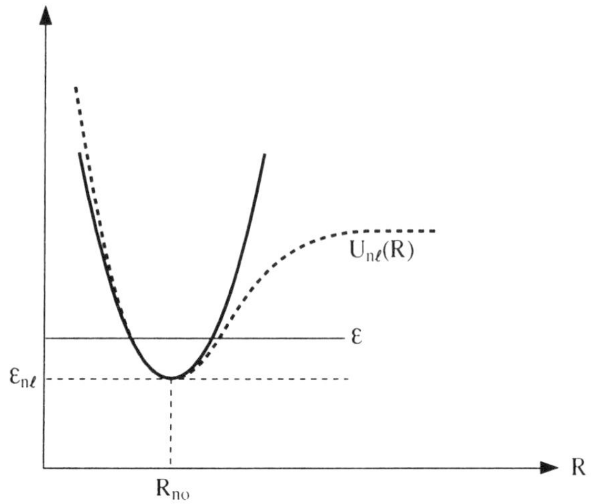

Figure 3.7. Schematic description (broken line) of the potential field (3.13). The solid line represents the harmonic approximation (3.17) of this field

where k_n is an appropriate constant. The potential energy (3.13) has a minimum at a position R_{nl} which lies close to R_{no}. We have:

$$R_{nl} = R_{no} + \frac{l(l+1)\hbar^2}{k_n M_r R_{no}^3} \simeq R_{no} \tag{3.16}$$

In the neighbourhood of R_{no} we can write the potential energy (3.13) as follows

$$U_{nl}(R) \simeq \varepsilon_{nl} + \tfrac{1}{2}k_n(R - R_{no})^2 \tag{3.17}$$

where

$$\varepsilon_{nl} = E_n(R_{no}) + \frac{\hbar^2 l(l+1)}{2M_r R_{no}^2} \tag{3.18}$$

The so-called harmonic approximation to the potential field, given by (3.17), is represented by the solid line in Fig. 3.7. We see that it is a very good approximation to the potential field (3.13), represented by the broken line in Fig. 3.7, in the neighbourhood of the potential minimum. This allows us to identify the low-energy eigenstates (ε near the bottom of the well $U_{nl}(R)$) of the particle in the field (3.13) with the low-energy eigenstates of the same particle moving in the field (3.17). A particle moving in the field (3.17) is often called a harmonic oscillator.

3.3.2 HARMONIC OSCILLATIONS

We consider a particle of mass m in the one-dimensional potential field

$$U(x) = U_O + \tfrac{1}{2}kx^2 \tag{3.19}$$

If classical mechanics were valid the particle could have any energy E above the potential minimum U_O and would, for the given energy, oscillate in perpetuity (see Section 1.4.1) between $-x_0$ and x_0, with period $2\pi/\omega$:

$$x(t) = x_0 \cos \omega t \tag{3.20}$$

where ω (the so-called natural frequency of the oscillator) is given below, and x_0 is determined from the equation

$$U_O + \tfrac{1}{2}kx_0^2 = E$$

In order to describe the harmonic oscillator quantum-mechanically we must define its energy eigenvalues and corresponding eigenfunctions. Its energy eigenvalues are:

$$E_v = U_O + \hbar\omega(v + \tfrac{1}{2}),\ v = 0, 1, 2, \ldots \tag{3.21}$$

where ω is the natural frequency of the oscillator given by

$$\omega = (k/m)^{1/2}$$

We note that the entire energy spectrum of a particle in the field (3.19) is a discrete one, as expected from the fact that whatever its energy, the particle remains bound in space, since $U(x) \to \infty$ as $x \to \pm\infty$. We note also that the energy difference between two successive energy levels is constant: $\hbar\omega$.

The energy levels (3.2.1) are non-degenerate (this is, as already mentioned in Section 1.4.2, a general property of bound states in one-dimensional fields). The corresponding eigenstates $u_v(x, t) = u_v(x) \exp(-iE_v t/\hbar)$, $v = 0, 1, 2, \ldots$ are described by analytic expressions. We present below $u_v(x)$ for $v = 0, 1, 2$ (see also Fig. 3.8). The interested reader will find a complete list and an exposition of the properties of the eigenstates of the harmonic oscillator in most books on quantum mechanics.

$$\begin{aligned} u_0(x) &= \frac{\alpha^{1/2}}{\pi^{1/4}} \exp(-\alpha^2x^2/2) \\ u_1(x) &= \frac{\alpha^{1/2}}{2^{1/2}\pi^{1/4}} 2\alpha x \exp(-\alpha^2x^2/2) \\ u_2(x) &= \frac{\alpha^{1/2}}{8^{1/2}\pi^{1/4}} (4\alpha^2x^2 - 2) \exp(-\alpha^2x^2/2) \end{aligned} \tag{3.22}$$

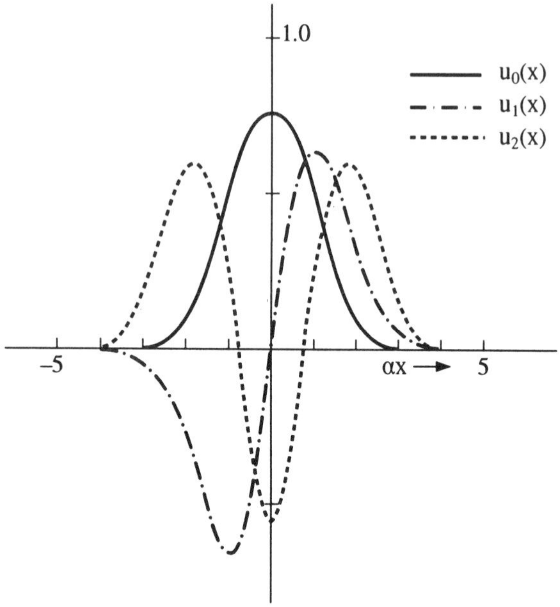

Figure 3.8. Eigenfunctions of a harmonic oscillator

where

$$\alpha \equiv (km/\hbar^2)^{1/4}$$

We are finally in a position to write down a formula for the eigenvalues of the total energy of a diatomic molecule. We have

$$\begin{aligned} \varepsilon_{nlv} &\simeq E_n(R_{no}) + \frac{\hbar^2 l(l+1)}{2M_r R_{no}^2} + \hbar\omega_n(v + \tfrac{1}{2}) \\ \omega_n &\equiv (k_n/M_r)^{1/2} \end{aligned} \tag{3.23}$$

where n, l, v are the quantum numbers defined above. The first term of (3.23) gives the electronic energy and the second and third terms give the rotational and vibrational energies of the molecule respectively. The above formula is obtained by comparison of (3.17) with (3.19), and using (3.21) with $k = k_n$, $m = M_r$ and $U_O = \varepsilon_{nl}$. The corresponding eigenfunctions are given by

$$\Psi_{nlmv} = [u_{nv}(R - R_{no})/R] Y_{lm}(\Theta, \Phi) \psi_n(\boldsymbol{r}; R) \exp(-i\varepsilon_{nlv} t/\hbar) \tag{3.24}$$

where $u_{nv}(R - R_{no})$ are obtained from the eigenfunctions (3.22) of the harmonic oscillator with the following substitutions: $x = R - R_{no}$, $\alpha = (k_n M_r/\hbar^2)^{1/4}$. We note that, in consequence of approximation (3.17), the functions $u_{nv}(R - R_{no})$ do not depend on the quantum number l. We note, also, that as $R \to 0$, $u_{nv}(R - R_{no})$ goes to zero faster than R, so that $u_{nv}(R - R_{no})/R$ vanishes as $R \to 0$.

Of particular importance is the ground state of the molecule ($n = l = v = 0$) with energy $E_0(R_{0o}) + \hbar\omega_0/2$. The difference between this energy and the limiting value ($R \to \infty$) of $U_{00}(R)$ of (3.13) is, by definition, the dissociation energy of the molecule; this is the minimum energy we must give to the molecule (in its ground state) to effect its dissociation into two atoms.

3.4 Diatomic molecules of *N* electrons

The calculation of the electronic terms $E_n(R)$ of a diatomic molecule XY, such as H_2, O_2, HCl, etc., which has more than one electron is in principle similar to the calculation of the eigenstates of a many-electron atom but technically more difficult because of the reduced symmetry of the molecular field. A first approximation to these terms and in particular of the ground-state term is obtained as follows. We assume that each of the N electrons of the molecule moves independently from the others in a mean potential field due to all electrons and the two nuclei of the molecule. The latter are, of course, assumed motionless at a distance R between them, as dictated by the Born–Oppenheimer approximation. The mean potential field the electron sees is calculated self-consistently, for given R, in essentially the same way as in the case of the N-electron atom, using a modified version (to take account of the second nucleus) of (2.83) to calculate the electrostatic field, and (2.88) to calculate the exchange-energy correction. Of course, the electronic cloud density $n(\boldsymbol{r})$ and the corresponding mean field $U(\boldsymbol{r})$ will be less symmetric (cylindrical instead of spherical) in the present case, and this means that the calculation of the one-electron states in $U(\boldsymbol{r})$ becomes more difficult; and of course the calculation of $n(\boldsymbol{r})$ from the occupied one-electron states and the whole self-consistency cycle becomes more elaborate than in the case of the atom. But it can be done and at the end of this calculation we have a description of the ground state of the N electrons in the field of the motionless nuclei: an N-electron wavefunction of the form (2.77) (a Slater determinant) where the $u_{\alpha_i}(i = 1, \ldots, N)$ are N different one-electron states corresponding to the lowest one-electron energy levels in the self-consistent mean field $U(\boldsymbol{r})$[10]. From the above we can calculate the energy of the system of the N electrons in the field of the motionless nuclei using a modified version (to take account of the second nucleus) of (2.97). To this energy we add the repulsion energy $e^2/(4\pi\epsilon_O R)$ of the nuclei to obtain the mean-field ground-state electronic term $E_0(R)$ of the molecule as a function of R. Given that the molecule *XY* exists, $E_0(R)$ must have a minimum and this, as a rule, occurs at

$R = (r_{max})_X + (r_{max})_Y$, where $(r_{max})_X$ and $(r_{max})_Y$ are respectively the radii of the outermost shells of the atoms X and Y as defined in Section 2.15.3. The above rule applies certainly when $X = Y$. In this case the binding of the atoms is effected in more or less the same way as in the case of the molecular ion H_2^+; electronic cloud accumulates to some degree between the nuclei where the mean-field potential energy $U(\boldsymbol{r})$ is more negative. A chemical bond of this kind is called covalent; the increased electronic cloud in the region between the nuclei 'belongs' to both atoms which in this way complete to a degree their outer shells. We should emphasise that for a reliable estimate of the dissociation energy of the molecule, one must go beyond the mean-field treatment sketched above; one must take into account electron–electron correlation (see also Section 2.15.5).

There is in fact another way of looking at covalent bonding which derives from the so-called Hellmann–Feynman theorem. We can say that the electronic charge accumulated between the two nuclei (or perhaps we should say between the two positive 'ions' consisting of the nuclei and the electrons in the inner shells of the respective atoms) attracts the nuclei towards it and therefore towards each other, while the repulsive force between them pushes them apart. At the equilibrium distance, corresponding to the minimum of the electronic term, the two forces balance each other. As R becomes smaller electronic charge is removed from the region between the nuclei to the left of the left-hand nucleus and the right of the right-hand nucleus and from there helps to pull the nuclei apart acting in the same direction as the repulsive force between the two nuclei.

In many heteronuclear molecules, such as HCl, NaCl, etc., the self-consistent electronic cloud corresponding to the ground state of the molecule is increased about one atomic nucleus and decreased (always relative to the situation of a free atom) about the other nucleus. For example in NaCl there is more electronic cloud around Cl and less around Na in comparison to the corresponding free atoms. We can approximately say that in the molecule NaCl the single electron in the outer (3s) shell of Na moves over to Cl occupying the one empty orbital in the outer (3p) shell of this atom. Consequently the positively charged Na^+ is attracted to the negatively charged Cl^- and in that way the total energy is reduced and the molecule NaCl is bound. Such a bond, where the main contribution to the binding energy (we may think of the binding energy as the negative of the dissociation energy) of the molecule comes from the electrostatic attraction between two oppositely charged ions, is called ionic. We note that the distance between the two nuclei (at the minimum of $E_0(R)$) is, in this case, given by $R = (r_{max})_{X^+} + (r_{max})_{Y^-}$, where $(r_{max})_{X^+}$ and $(r_{max})_{Y^-}$ are respectively the radii of the outermost shells of the ions X^+ and Y^-.

The above discussion had mostly to do with the ground-state electronic term of a diatomic molecule. In most diatomic molecules there are

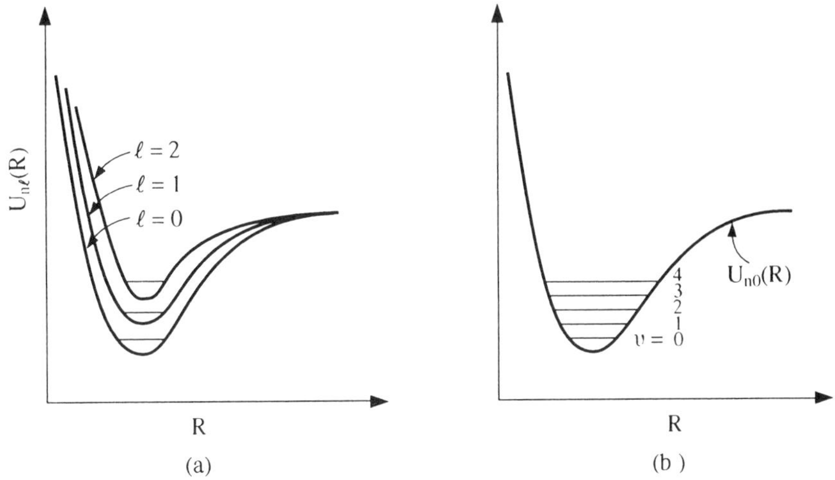

Figure 3.9. Energy eigenvalues ε_{nlv}. (a) The horizontal lines represent the energy levels ε_{nl0}, $l = 0, 1, 2$. (b) The horizontal lines represent the energy levels ε_{n0v}, $v = 0$, 1, 2, 3, 4. We note that the energy scale in diagram (*a*) is about ten times larger than in diagram (*b*)

additional electronic terms which can hold the molecule together at a higher energy as shown, schematically, in Fig. 3.1.

The (total) energy levels ε_{nlv} corresponding to a given electronic term (n) are described schematically in Fig. 3.9. The three curves of diagram (*a*) correspond to different values of the rotational (angular momentum) quantum number ($l = 0, 1, 2$). The horizontal lines represent the corresponding total energies of the molecule when the vibrational energy is the lowest possible ($v = 0$). Diagram (*b*) shows the (total) energy levels corresponding to the vibrational quantum numbers $v = 0, 1, 2, 3, 4$, for a given rotational quantum-number ($l = 0$). We note that the energy scale is about ten times larger in diagram (*a*) than in diagram (*b*). The difference $\hbar\omega_n$ between the two successive vibrational levels (of the same n and l) is about ten times larger than the average difference between two successive rotational levels (of the same n and v). We note also that $\hbar\omega_n$ is typically about ten times smaller than the difference between levels belonging to different electronic terms.

3.5 Absorption–emission of electromagnetic radiation

Absorption and emission of electromagnetic radiation by diatomic molecules occurs in three different regions of frequency.

A. *Region between the infrared and microwaves*. The molecular transitions responsible for absorption/emission of electromagnetic radiation (photons) in this region are of the following type

$$nlv \rightarrow nl'v \tag{3.25}$$

i.e. only the rotational state of the molecule changes in the transition. We find that such transitions are typical of molecules with an intrinsic electric dipole moment resulting from an asymmetric distribution of the electronic cloud in their ground states, as is the case, for example, with NaCl, MgO, etc.. They occur when the molecule is in its electronic ground state ($n = 0$) and, most often, in its vibrational ground state ($v = 0$) as well; and obey the following selection rule

$$\begin{aligned} \Delta l \equiv l' - l &= +1, \text{ absorption} \\ &= -1, \text{ emission} \end{aligned} \tag{3.26}$$

The frequently of the absorbed photon is given by

$$\nu = \frac{(l+1)\hbar}{2\pi M_r R_{0o}^2} \tag{3.27}$$

which is obtained from (3.23) with $n = 0$, using (3.25) and (3.26).

B. *Infrared region*. The spectral lines in this region are due to molecular transitions of the type

$$nlv \rightarrow nl'v' \tag{3.28}$$

in which the rotational and vibrational states of the molecule change while its electronic state stays the same. The above transitions occur, like those of (3.25), with the molecule in its electronic ground-state ($n = 0$), when the latter has an intrinsic electric dipode moment due to an asymmetric distribution of the electronic cloud, as it happens in NaCl, HF, MgO, etc.. They obey the following selection rules

$$\begin{aligned} \Delta l &\equiv l' - l = \pm 1 \\ \Delta v &\equiv v' - v = +1, \text{ absorption} \\ &\qquad\quad\;\; = -1, \text{ emission} \end{aligned} \tag{3.29}$$

The frequency of the absorbed photon is given by the following relations

$$\nu = \frac{\varepsilon_{0,l+1,v+1} - \varepsilon_{0,l,v}}{2\pi\hbar} = \frac{\omega_0}{2\pi} + \frac{(l+1)\hbar}{2\pi M_r R_{0o}^2} \qquad \Delta l = +1, \Delta v = +1 \tag{3.30a}$$

and

$$\nu = \frac{\varepsilon_{0,l-1,v+1} - \varepsilon_{0,l,v}}{2\pi\hbar} = \frac{\omega_0}{2\pi} - \frac{l\hbar}{2\pi M_r R_{0o}^2} \tag{3.30b}$$

$$\Delta l = -1, \Delta v = +1 \ (l \neq 0)$$

which are obtained from (3.23) with $n = 0$, and (3.29). Fig. 3.10 shows a typical infrared absorption spectrum.

We have noted that rotational and vibrational transitions are typical of molecules with an intrinsic electric dipole moment. Classically one can see that an electric dipole capable of rotation and/or vibration would couple easily with the oscillating electric field of the electromagnetic radiation. Quantum-mechanically we find that the dipole-moment matrix-elements between the initial and final states of these transitions (these are quantities, defined in essentially the same way as the coefficients f of (2.17a), which determine the strength of a transition) are relatively large when the electronic cloud in the ground state of the molecule is asymmetrically distributed (resulting in an intrinsic electric dipole moment of the molecule), and practically vanish for molecules, like H_2, O_2, Cl_2 etc., which have no electric dipole moment in their ground states.

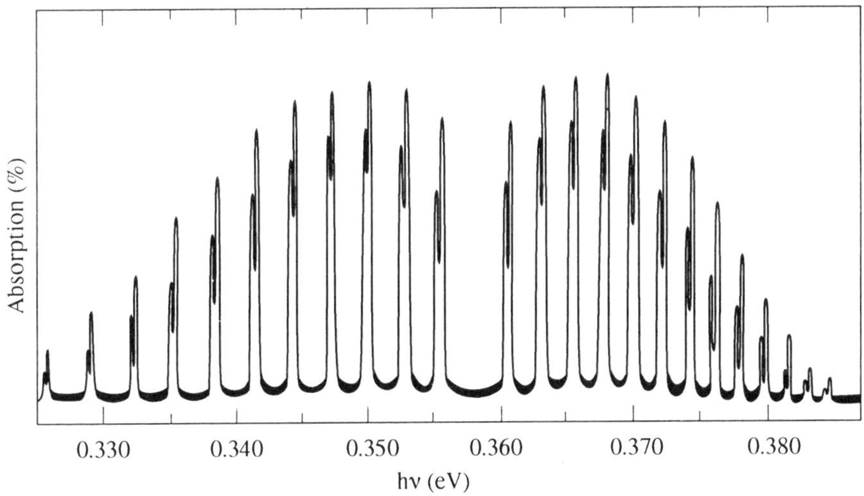

Figure 3.10. The absorption spectrum (infrared) of HCl. On the right of the centre of the spectrum the absorption lines correspond to transitions with $\Delta l = 1$ (equation (3.30a)), and to the left of the centre the absorption lines correspond to transitions with $\Delta l = -1$ (equation (3.30b)). The two peaks of each line are due to HCl molecules with different isotopes (^{35}Cl and ^{37}Cl) of Cl which have different nuclear masses (atomic weights: 35 and 37 respectively)

C. *Optical and ultraviolet region*. The spectral lines in this region are due to molecular transitions of the type

$$nlv \rightarrow n'l'v'$$

which means that the electronic state of the molecule changes and with it, as a rule, the vibrational and rotational states change too. The resulting absorption/emission spectra are usually very complicated, at least as much as the optical spectra of atoms discussed in Section 2.15.2, and for the same reasons, the situation is further complicated by changes in the rotational and vibrational motions of the nuclei.

3.6 Polyatomic molecules

These may have three or more atomic nuclei; examples are: H_2O (water), NH_3 (ammonia), CH_4 (methane), C_6H_6 (benzene), etc..

In Section 3.4 we sketched a mean-field calculation of the ground-state electronic term of a diatomic molecule which constitutes, in principle, a straightforward extension of the mean-field calculation of the ground state of an atom. In each case the mean-field ground-state is described by a Slater determinant (2.77). In the case of an atom, the one-particle states (the u_{α_i} in (2.77)) are atomic orbitals (2.91). In the case of a molecule the one-particle states are in general extended molecular orbitals: the electron does not 'move' about a specific atom but in the space of the entire molecule. We may, for our purposes, assume that an extended molecular orbital can be written as a linear sum of atomic orbitals, as in (3.4) of the molecular ion $H_2{}^+$, except that for a polyatomic molecule the sum will have not two but possibly as many terms as there are atoms in the molecule. Any symmetry in the spatial arrangement of the nuclei will be reflected in the mean-field the electron sees and this in general simplifies the evaluation of the coefficients which multiply the atomic orbitals in the linear sum which constitutes a particular molecular orbital. For example, in the molecular orbital (3.4) of $H_2{}^+$, the coefficients multiplying the two atomic orbitals are the same; the corresponding coefficients in the molecular orbital (3.6) have the same magnitude but opposite signs. Similar relations will apply in the case of molecular orbitals in polyatomic molecules depending on their symmetry. We shall see in the next chapter that in a crystalline solid (which is a polyatomic molecule) the periodicity of the mean potential field leads to molecular orbitals of the so-called Bloch type, which facilitates enormously the calculation of these orbitals.

There remains of course the question of how good a mean-field approximation is. It turns out that, for many purposes, the mean-field approximation works exceedingly well. On the other hand, we have seen that one must go beyond the mean-field treatment to understand the complexity of the optical spectra of atoms, and the same applies to the optical spectra of molecules; and we have noted that for a reliable estimate of the dissociation energy of a molecule, one must go beyond the mean-field approximation in evaluating the ground-state energies of the molecule and its constituent atoms in their free states (see Section 2.15.5) There follows a second question: is the mean-field treatment, and the associated with it extended molecular orbitals, always the best starting point? If electron–electron correlation is important, can there be a description of the state of the N electrons of the molecule which is nearer to the actual state, and yet is simple enough for us to be able to describe it in simple terms. This is a very difficult question to answer generally, and here we can only hint at what one might be able to do by looking at a specific case.

We shall describe qualitatively the ground state of CH_4 in terms of localised molecular orbitals. A CH_4 molecule consists of a carbon atom at the centre of a regular tetrahedron and four hydrogen atoms at the corners of the tetrahedron (these being the equilibrium positions of the nuclei corresponding to the minimum of the ground-state electronic term). The inner shell of carbon which accommodates two electrons is very little affected by, and does not contribute significantly to, the binding of the molecule, and we shall not be concerned with it. That would be the case also in the mean field treatment, but in that treatment all orbitals except those of the inner shell of carbon extend over all the atoms of the molecule. Now we proceed as follows. We begin by constructing four linearly independent combinations of the 2s and the three 2p orbitals of the outer shells of carbon (see Table 2.5) so that the corresponding wavefunctions are concentrated respectively along the lines from the carbon nucleus at the centre of the tetrahedron to its four corners. These hybridised orbitals, known as tetrahedral orbitals, are given by (see problem 3.7)

$$\begin{aligned} &\tfrac{1}{2}(s + p_x + p_y + p_z) \\ &\tfrac{1}{2}(s + p_x - p_y - p_z) \\ &\tfrac{1}{2}(s - p_x + p_y - p_z) \\ &\tfrac{1}{2}(s - p_x - p_y + p_z) \end{aligned} \tag{3.31}$$

where s stands for the 2s orbital which has the form $(4\pi)^{-1/2}R_{20}(r)$, and p_x, p_x, p_z, defined as in problem (2.7), have the form $(3/4\pi)^{1/2}R_{21}(r)x/r$, $(3/4\pi)^{1/2}R_{21}(r)y/r$ and $(3/4\pi)^{1/2}R_{21}(r)z/r$ respectively. Each of the above orbitals of carbon can be combined with the 1s orbital of the hydrogen on the respective corner into a bonding molecular orbital (one that builds up the wavefunction in between the two nuclei). We obtain two such localised molecular orbitals, one with spin up and one with spin down, between the carbon and each of the four hydrogen nuclei. Taken together they accommodate the eight electrons from the outer shells of the atoms of the molecule (four from carbon and one from each hydrogen). The occupation of these orbitals leads to accumulation of electronic charge and covalent bonding between the carbon and each of the hydrogen nuclei as shown schematically in Fig. 3.11.

It appears that, to a first approximation at least, each hydrogen is bonded to the carbon by a localised covalent bond. However, the accumulation of electronic charge between the carbon and the hydrogen nuclei shown in Fig. 3.11 need not be associated with localised bonds. It is also obtained in a mean-field approximation based on extended molecular orbitals. In this case the accumulation is the result of contributions from the totality of occupied extended molecular orbitals. We shall not attempt a comparison between the two approaches, except to say that one needs to improve on the ground-state wavefunction obtained by either of them in order to estimate accurately the ground-state energy of the molecule.

In what follows we need not be concerned with the way one calculates the electronic terms of a polyatomic molecule. We shall assume that by some means or other these have been obtained as functions of the positions of the nuclei, as shown schematically in Fig. 3.1. We remember that the minimum of a term (we assume that one exists) determines the equilibrium positions of the nuclei for this term. From the definition of the minimum, it follows that *small* displacements of any or all of the nuclei from their equilibrium positions will lead to an increase of $E_n(\boldsymbol{R})$ and, therefore, to restoring forces which tend to bring the nuclei back to their equilibrium positions. The restoring force in a diatomic molecule derives from the potential field (3.17) involving the separation R between the two nuclei of the molecule. We can visualise this force in terms of an elasting spring connecting the two atoms, the restoring force depending on the elastic constant k_n of the spring. The greater the value of k_n, the more difficult it is to extend the spring, for the restoring force is greater. Using the same language, we can say that in a polyatomic molecule each atom is connected via elastic springs with every other atom of the molecule, so that the displacement of an atomic nucleus

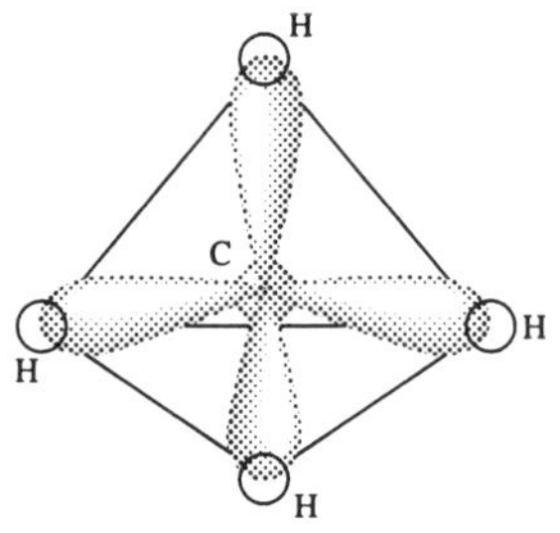

Figure 3.11. Covalent bonding in CH_4

from its equilibrium position along the x, y or z direction generates, in general, forces acting along all three directions not only on that atomic nucleus but on every other nucleus of the molcule as well. Classically one obtains[11], as a result of the above coupling, $3\Lambda - 6$ linearly independent coupled harmonic oscillations of the nuclei of the molecule, known as the normal modes of vibration of the molecule, where Λ is the number of nuclei in the molecule (we assume that the nuclei are not collinear, if the atomic nuclei lie on a straight line, the number of independent vibrations is $3N - 5$). When only one of the normal modes is excited, all the nuclei oscillate with the same frequency about their equilibrium positions tracing a certain pattern which is different for the different normal modes of vibration as is the frequency of the vibration. (Some normal frequencies may be degenerate in the sense that two different modes of vibration may have the same frequency because of symmetry.) The kth normal mode ($k = 1, 2, 3, \ldots, 3\Lambda - 6$) of frequency ω_k is described as follows

$$X_i^{(k)}(t) = X_{oi}^{(k)}\,\mathrm{e}^{\mathrm{i}\omega_k t},\ Y_i^{(k)}(t) = Y_{oi}^{(k)}\,\mathrm{e}^{\mathrm{i}\omega_k t},\ Z_i^{(k)}(t) = Z_{oi}^{(k)}\,\mathrm{e}^{\mathrm{i}\omega_k t} \quad (3.32)$$

where

$$X_{oi}^{(k)} = |X_{oi}^{(k)}|\exp(\mathrm{i}\varphi_{xi}^{(k)}),\ Y_{oi}^{(k)} = |Y_{oi}^{(k)}|\exp(\mathrm{i}\varphi_{yi}^{(k)}),\ \text{and}$$

$$Z_{oi}^{(k)} = |Z_{oi}^{(k)}|\exp(\mathrm{i}\varphi_{zi}^{(k)})$$

The x, y and z components of the actual displacement of the ith atomic nucleus are, of course, real quantities; they are given by

$$x\text{-component: } a_k|X_{oi}^{(k)}|\cos(\omega_k t + \varphi_{xi}^{(k)}) + b_k|X_{oi}^{(k)}|\sin(\omega_k t + \varphi_{xi}^{(k)})$$

which can be rewritten as

$$q_k|X_{oi}^{(k)}|\cos(\omega_k t + \varphi_{xi}^{(k)} + \delta_k) \quad (3.33a)$$

$$y\text{-component:} \quad q_k|Y_{oi}^{(k)}|\cos(\omega_k t + \varphi_{yi}^{(k)} + \delta_k) \quad (3.33b)$$

$$z\text{-component:} \quad q_k|Z_{oi}^{(k)}|\cos(\omega_k t + \varphi_{zi}^{(k)} + \delta_k) \quad (3.33c)$$

where a_k and b_k or, correspondingly, q_k and δ_k are arbitrary constants (see below). Equations (3.33) determine, for the kth mode, the direction a nucleus moves, the amplitude of its vibration and its phase. This is demonstrated schematically in Fig. 3.12 for a particular mode (k') of a square molecule. If each atom is displaced, at time $t = 0$, as shown by the vector attached to it and released with zero velocity, the subsequent motion will be given by (3.33) with $\omega_k = \omega_{k'}$ and $Z_{oi}^{(k')} = 0$ for all atoms ($i = 1, 2, 3, 4$); the remaining amplitudes and phases will be:

$$|X_{o1}^{(k')}| = |X_{o2}^{(k')}| = |Y_{o3}^{(k')}| = |Y_{o4}^{(k')}| = 0$$

$$|Y_{o1}^{(k')}| = |Y_{o2}^{(k')}| = |X_{o3}^{(k')}| = |X_{o4}^{(k')}| \neq 0$$

$$\varphi_{y1}^{(k')} = \varphi_{x3}^{(k')} = 0,\ \varphi_{y2}^{(k')} = \varphi_{x4}^{(k')} = \pi$$

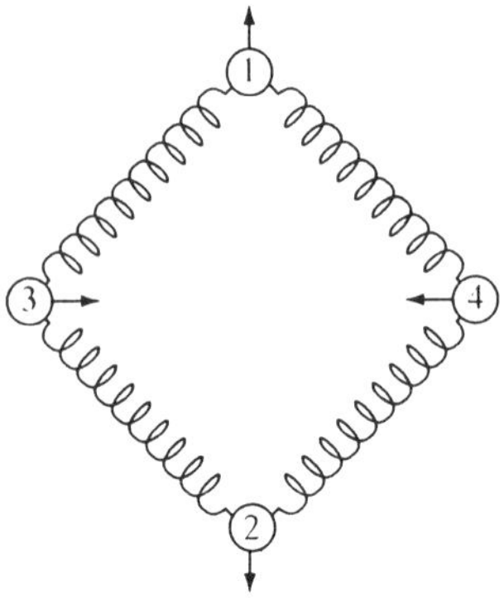

Figure 3.12. A vibrational mode of a square molecule

q_k is a positive number which depends on the magnitude of the initial displacement and $\delta_k = 0$.

The most general small-amplitude vibration of the molecule will be a superposition of (3.33); the displacement of the ith nucleus will be given by

$$\begin{aligned} \boldsymbol{R}_i(t) = \sum_{k=1}^{3\Lambda-6} q_k\{&|X_{oi}^{(k)}|\cos(\omega_k t + \varphi_{xi}^{(k)} + \delta_k)\hat{x} \\ &+ |Y_{oi}^{(k)}|\cos(\omega_k t + \varphi_{yi}^{(k)} + \delta_k)\hat{y} + |Z_{oi}^{(k)}|\cos(\omega_k t + \varphi_{zi}^{(k)} + \delta_k)\hat{z}\} \end{aligned} \tag{3.34}$$

where $\hat{x}$, $\hat{y}$, $\hat{z}$ denote unit vectors in the x, y, z directions. The q_k and δ_k coefficients are determined by the initial (at $t = 0$) displacements and initial velocities of the nuclei.

The quantisation of the normal modes (3.32), proceeds in exactly the same manner as the quantisation of the linear harmonic oscillator (3.20). To the kth normal mode corresponds a quantum-mechanical harmonic oscillator with energy eigenvalues

$$E_{kv} = \hbar\omega_k(v + \tfrac{1}{2}),\ v = 0, 1, 2 \ldots \tag{3.35}$$

The classical solution described above tells us that the displacement of the ith atom, in the kth mode, has the form:

$$\boldsymbol{R}_i^{(k)}(t) = q_k(t)\boldsymbol{R}_{oi}^{(k)}$$

where $\boldsymbol{R}_{oi}^{(k)}$ are certain vectors (and we have assumed that they are real quantities). We retain the above picture in quantum mechanics except in relation to q_k. In quantum mechanics we obtain the probability that it will have a certain value at time t. This is given, in the usual manner, by $|\psi(q_k, t)|^2$, where $\psi(q_k, t)$ is, in general, a linear sum of the eigenfunctions $u_{kv}(q_k)\exp(-\mathrm{i}E_{kv}t/\hbar)$ corresponding to the eigenvalues (3.35). (We need not write explicitly the eigenfunctions $u_{kv}(q_k)$, $v = 0, 1, 2, \ldots$, these have the same form as (3.22), except that α has a different definition in the present case.) We write

$$\psi(q_k, t) = \sum_v c_v u_{kv}(q_k) \exp(-iE_{kv}t/\hbar) \tag{3.36}$$

where c_v are to be determined by the initial (quantum-mechanical) state of the kth vibrator. When we know $\psi(q_k, t)$ we can calculate the mean values of q_k, q_k^2, etc., in the usual manner.

In a gas of molecules at thermal equilibrium, the different normal modes of vibration of a molecule behave as independent oscillators and, moreover, we can assume (see Section 2.6) that the kth oscillator is to be found in one or the other of its eigenstates, and not in a (3.36)-state. The probability that the kth oscillator will be in its vth state is given by

$$P(v) = \text{const} \cdot \exp(-E_{kv}/k_B T) \tag{3.37}$$

where E_{kv} is the energy (3.35) of the oscillator in that state. The normalisation is such that the sum of $P(v)$ over all v equals unity; therefore

$$\text{const} = \left(\sum_v \exp(-E_{kv}/k_B T)\right)^{-1}$$

We obtain the average energy of the kth oscillator at temperature T from the following formula

$$\bar{E}_k = \sum_v E_{kv} P(v) \tag{3.38}$$

Substituting (3.35) and (3.37) in (3.38) one obtains[12]

$$\bar{E}_k = \frac{\hbar\omega_k}{2} + \frac{\hbar\omega_k}{\exp(\hbar\omega_k/k_B T) - 1} \tag{3.39}$$

The first term of (3.39) represents the ground-state energy of the kth oscillator (normal mode) and does not depend on the temperature; the variation of $\bar{E}_k$ with temperature comes from the second term of (3.39). We can think of

$$n(k) \equiv \frac{1}{\exp(\hbar\omega_k/k_B T) - 1} \tag{3.40}$$

as the average number of vibration quanta of the kth type (mode of vibration). Equation (3.40) is seen to be a special case of the Bose–Einstein distribution (2.72), corresponding to $\mu = 0$. The vibration quanta are obviously bosons.

We conclude this section with a brief comment concerning the rotational motion of a polyatomic molecule about its centre of mass. If we disregard the coupling between the vibrational and rotational motions of the molecule, the rotational energy spectrum will be given by (3.12) with I representing the moment of inertia of the molecule with respect to its centre of mass. The coupling between the two motions introduces certain complications which we can not discuss in the present volume.

Problems

3.1 (a) Show that (3.22) are indeed eigenfunctions of the energy of the harmonic oscillator defined by (3.19).

(b) Show that the eigenfunctions (3.22) are normalised and orthogonal to each other.

(c) Discuss the similarities and differences between the energy spectrum of a particle in the potential field (3.19) and that of a particle in an one-dimensional box (given by (1.12)).

(d) Compare (3.22) with the eigenfunctions corresponding to the three-lower energy levels of a particle in an one-dimensional box (given by (1.13)).

3.2 An electronic term $E(R)$ of a diatomic molecule which has a minimum, such as the $E_0(R)$ term of Fig. 3.4, is described approximately by

$$E(R) = D\{\exp[-2\alpha(R - R_o)] - 2\exp[-\alpha(R - R_o)]\}$$

which is known as Morse's curve.

Show that (*i*) the minimum of $E(R)$ has the value $-D$ and occurs at $R = R_o$, and (ii) by an appropriate choice of α one can obtain the observed vibration frequency (given by (3.23)) for the term under consideration.

3.3 The dissociation energy of the H_2 molecule is 4.476 eV, the separation between the nuclei at equilibrium equals 0.75 Å and the vibration frequency for the ground-state electronic term is $(\omega/2\pi) = 1.30 \times 10^{14}$ cycles per second. Determine D and R_o of the corresponding Morse curve (see problem 3.2).

3.4 The space lattice of crystalline LiH is cubic with hydrogen atoms at the corners and at the centres of the faces of the cube (a unit cell) and lithium atoms at the centre of the cube and at the centres of its edges. Estimate the equilibrium separation between the atoms of the molecule LiH assuming it is the same with that between the atoms of crystalline LiH. We know that the density of crystalline LiH is 0.83×10^{13} kg/m^3. Using the above data estimate the wavelength of the transition (3.25): $l = 1 \rightarrow l = 0$ $(n = v = 0)$.

***3.5** Starting from the Hamiltonian for a system of two particles interacting through a potential energy term $E_n(R)$, where R is the separation between the particles,

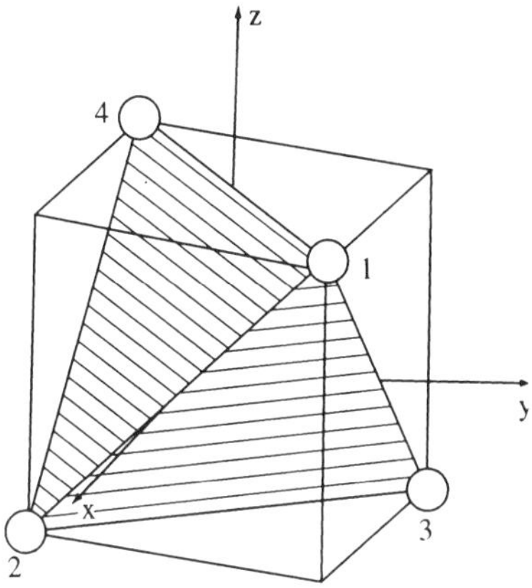

Figure 3.13. Positions of the hydrogen nuclei in CH_4. The carbon nucleus is at the centre of the tetrahedron

show that in the centre-of-mass system of coordinates the nuclear motion of a diatomic molecule is described by (3.10), and that $u(R)$ in (3.10) can be identified with the motion of a particle of mass M_r (the reduced mass of the two nuclei) in the field (3.13). [See problem 2.1 and Section B.7.]

3.6 By a careful analysis of the absorption spectrum of HCl, shown in Fig. 3.10, estimate (a) the constant k_0 of the corresponding harmonic oscillator (see (3.23)) and (b) the equilibrium separation R_{0_o} between the atoms of the molecule.

3.7 Show that the tetrahedral orbitals defined by (3.31) are normalised and orthogonal with respect to each other; and verify that the probability-densities corresponding to them are concentrated along the lines from the carbon nucleus at the centre of the tetrahedron to its four corners (1, 2, 3 and 4 in Fig. 3.13).

Footnotes of Chapter 3

1. Sometimes we say this in a different way: we say that the motion is described in a centre of mass (CM) system of coordinates.
2. For the sake of simplicity we suppress the spin part of ψ_n (i.e. that part which would tell us the spin state of the electrons); and similarly we suppress the spin part of the nuclear wavefunction $\chi_{n\mu}$, which would tell us the spin state of the nuclei (see (3.2)).
3. For simplicity we write down only the orbital part of the wavefunction. The complete spinor-eigenfunction is

$$\begin{pmatrix} \psi_0(\boldsymbol{r};\, R) \\ 0 \end{pmatrix} \text{ or } \begin{pmatrix} 0 \\ \psi_0(\boldsymbol{r};\, R) \end{pmatrix}$$

depending on the spin (up or down) of the electron. The time-factor $\exp(-iE_0(R)t/\hbar)$, which would describe the time development of the eigenfunction if the nuclei were actually motionless, is omitted (see Section 3.1).

4. ψ_{100} is obtained from table 2.3 with $Z = 1$. We disregard the slight difference between a_0' and a_0.
5. $E_n(R)$, $n = 0, 1$ is calculated from the following formula (see Appendix B.10)

$$E_n(R) = \frac{-\hbar^2}{2m}\int \psi_n(\boldsymbol{r};\, R)\nabla^2\psi_n(\boldsymbol{r};\, R)\,\mathrm{d}V + \int U_e(\boldsymbol{r})|\psi_n(\boldsymbol{r};\, R)|^2\,\mathrm{d}V + \frac{e^2}{4\pi\epsilon_O|\boldsymbol{R}_1 - \boldsymbol{R}_2|}$$

The first and second terms in the above equation give respectively the mean kinetic energy and the mean potential energy of the electron in the field (3.7) of the two nuclei, and the last term gives the repulsion energy between the nuclei. We note that $E_n(R)$ does not depend on the direction of the molecular axis.
6. The variables R, Θ, Φ define the vector $\boldsymbol{R} = \boldsymbol{R}_2 - \boldsymbol{R}_1$ where $\boldsymbol{R}_2$ and $\boldsymbol{R}_1$ are the position vectors of the nuclei; their masses: M_2 and M_1. With the centre of mass of the molecule at the origin of coordinates, we obtain

$$\boldsymbol{R}_2 = \left(\frac{M_1}{M_1 + M_2}\right)\boldsymbol{R} \text{ and } \boldsymbol{R}_1 = -\left(\frac{M_2}{M_1 + M_2}\right)\boldsymbol{R}$$

7. This form, which implies that the orbital angular momentum of the nuclei with respect to the centre of mass of the molecule (origin of coordinates) is a constant of the motion, is rigorously valid only when the total orbital angular momentum of the electrons of the molecule is zero. We note also that the spin part of the nuclear wavefunction has been dropped for the sake of simplicity.
8. The reduced mass M_r of a diatomic molecule whose nuclei have masses M_1 and M_2 respectively is defined by the relation $M_r = (M_1M_2)/(M_1 + M_2)$. Equation (3.12) written in that form is valid for any diatomic molecule. When $M_1 = M_2 = M$ (homonuclear molecule): $M_r = M/2$.
9. We note that for the determination of the degeneracy of a particular level one must also take into account the spin of both the electrons and the nuclei of the molecule. The spin of the nuclei is important also in another respect. In a homonuclear molecule the nuclear part of the wavefunction must be symmetric or antisymmetric with respect to the interchange of the position and spin coordinates of the nuclei depending on the spin (integer or half-integer) of the individual nucleus. We assume that the (suppressed) spin part of the nuclear wavefunction is such that this requirement is satisfied.
10. We remember that α_i denotes a set of quantum numbers which defines the orbital and spin parts of the one-electron state. In the absence of spin–orbit interactions we have two such states (spin up and spin down) with the same orbital part and the same energy.
11. The classical equation of motion (mass times the acceleration of a particle equals the force acting on it) takes the following form when applied to the ith nucleus of the molecule (we assume that all displacements are sufficiently small so that the forces resulting from them vary linearly with them; this is the essence of the harmonic approximation discussed in Sections 3.3.1 and 3.3.2; we note that the force corresponding to (3.19) equals, according to (1.33), $F = -kx$):

$$M_i \frac{d}{dt}\left(\frac{dX_i}{dt}\right) = -\sum_{j=1}^{\Lambda} \{k_{ij;xx} X_j + k_{ij;xy} Y_j + k_{ij;xz} Z_j\}$$
$$M_i \frac{d}{dt}\left(\frac{dY_i}{dt}\right) = -\sum_{j=1}^{\Lambda} \{k_{ij;yx} X_j + k_{ij;yy} Y_j + k_{ij;yz} Z_j\} \qquad \text{(I)}$$
$$M_i \frac{d}{dt}\left(\frac{dZ_i}{dt}\right) = -\sum_{j}^{\Lambda} \{k_{ij;zx} X_j + k_{ij;zy} Y_j + k_{ij;zz} Z_j\}$$

where M_i is the mass of the ith nucleus; X_i, Y_i, Z_i are the x, y, z components respectively of the displacement of the ith nucleus from its equilibrium position; and $k_{i,j;xx}$, $k_{ij;xy}$, etc. are appropriate elastic constants. The sum over j in the above equations includes all the nuclei of the molecule. We have 3Λ such equations ($i = 1, 2, \ldots, \Lambda$). Putting

$$X_i = X_{oi}\,e^{i\omega t},\ Y_i = Y_{oi}\,e^{i\omega t},\ Z_i = Z_{oi}\,e^{i\omega t} \qquad \text{(II)}$$

in the equations (I) we obtain the following system of 3Λ algebraic equations, linear and homogeneous, with 3Λ unknowns: X_{oi}, Y_{oi}, Z_{oi}, $i = 1, 2, \ldots, \Lambda$

$$\omega^2 M_i X_{oi} = \sum_{j=1}^{\Lambda} \{k_{ij;xx} X_{oj} + k_{ij;xy} Y_{oj} + k_{ij;xz} Z_{oj}\}$$
$$\omega^2 M_i Y_{oi} = \sum_{j=1}^{\Lambda} \{k_{ij;yx} X_{oj} + k_{ij;yy} Y_{oj} + k_{ij;yz} Z_{oj}\} \qquad \text{(III)}$$
$$\omega^2 M_i Z_{oi} = \sum_{j=1}^{\Lambda} \{k_{ij;zx} X_{oj} + k_{ij;zy} Y_{oj} + k_{ij;zz} Z_{oj}\}$$

We know that a system of linear homogeneous equations has non-zero solutions (i.e. solutions where at least one of the unknowns is not zero) when the so-called determinant of the coefficients vanishes. This condition of solvability is satisfied only for certain values of ω^2; we refer to the corresponding values of ω as the eigenfrequencies (also called natural or normal frequencies) of the system. To each eigenfrequency corresponds one or more than one (for a degenerate eigenfrequency) solution (II), and there are in all 3Λ linearly independent solutions of (III). In the present case six of these solutions correspond to $\omega = 0$; three of them describe uniform displacements of the molecule along the three coordinate axes and the other three uniform rotations about these axes. We disregard these solutions. The remaining $3\Lambda - 6$ solutions describe the normal modes of vibration (3.32). (In linear molecules we obtain only two rotations, for in this case a rotation about the axis of the molecule has no physical meaning. Therefore in linear molecules we obtain $3\Lambda - 5$ normal vibrations.) The solution of (III) for a given normal frequency ω_k determines $X_{oi}^{(k)}$, $Y_{oi}^{(k)}$ and $Z_{oi}^{(k)}$ apart from a common multiplicative constant which we can fix by an appropriate normalisation.

The fact that the coefficients in (I) are real quantities implies that the real and imaginary parts of a complex normal mode (3.32) are themselves solutions of (I), and so is a linear sum (3.33) of the two parts.

12. The calculation proceeds as in footnote 17 of Chapter 2.

Be bold and resolute in interpretation!
If you can't work it out, then work something into it.
JOHANN WOLFGANG GOETHE, *Gesellschaft*

4 SOLIDS

4.1 Space lattices

The main characteristic of a crystalline solid is its periodic structure: unit cells of the same volume, shape and content fill up all the space occupied by the solid. Fig. 4.1 is a photograph of a crystal of touching spheres. The spheres are centred on the sites (points) of a body-centred cubic lattice.[1] We can describe the bcc lattice as follows: we imagine the entire space divided into identical cubes of edge *a* as shown in Fig. 4.2. The corners of the cubes (open circles in the figure) and their centres (full circles in the figure) define

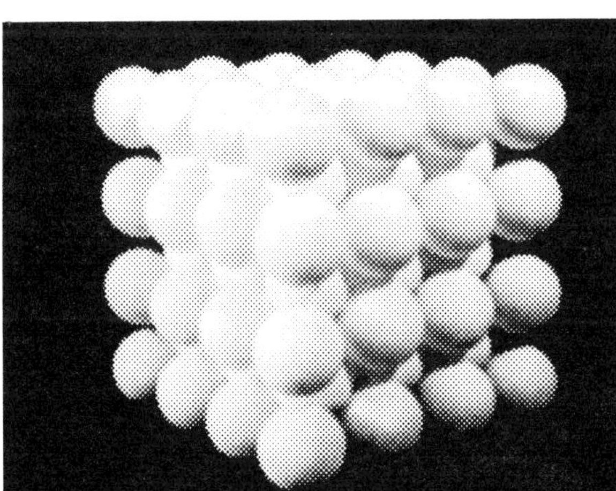

Figure 4.1. A crystal of touching spheres. The spheres are centred on the sites of a body-centred cubic lattice. (Reproduced from K. J. Pascoe, *Properties of Materials for Electrical Engineers* (New York: Wiley) 1973. Reprinted by permission of John Wiley & Sons Ltd)

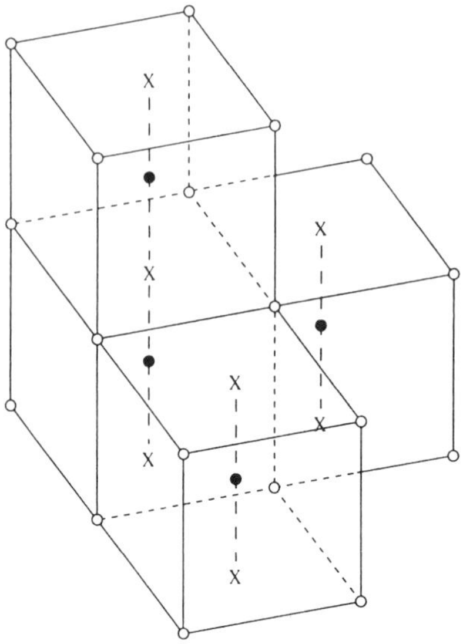

Figure 4.2. Body-centred cubic lattice. The lattice sites (points) coincide with the corners (open circles) and the centres (full circles) of the cubes. The points are completely equivalent and are denoted by open and full circles only for the sake of clarity of presentation

the sites of a bcc lattice. We refer to the edge a of the cube as the lattice constant. We observe that to the unit cell we have chosen (cube of volume a^3) correspond two lattice sites: one at the centre of the cube and a second one which is counted as follows: a corner site is shared by eight cubes; there are eight corners to a cube and, therefore, a lattice site per cube.

We can easily construct a unit cell (we shall call it a primitive cell), with volume $a^3/2$, corresponding to one lattice point. Let t_1, t_2, t_3 be the vectors from a lattice point (which we take as the origin of coordinates x, y, z) to three of its nearest points (see Fig. 4.3). Denoting by $\hat{x}$, $\hat{y}$, $\hat{z}$ the unit vectors in the directions x, y, z respectively, we can write t_1, t_2 and t_3 as follows

$$
\begin{aligned}
t_1 &= \frac{a}{2}(-\hat{x} + \hat{y} + \hat{z}) \\
t_2 &= \frac{a}{2}(\hat{x} - \hat{y} + \hat{z}) \qquad (4.1) \\
t_3 &= \frac{a}{2}(\hat{x} + \hat{y} - \hat{z})
\end{aligned}
$$

One can see that the parallelepiped defined by the above vectors (see Fig.

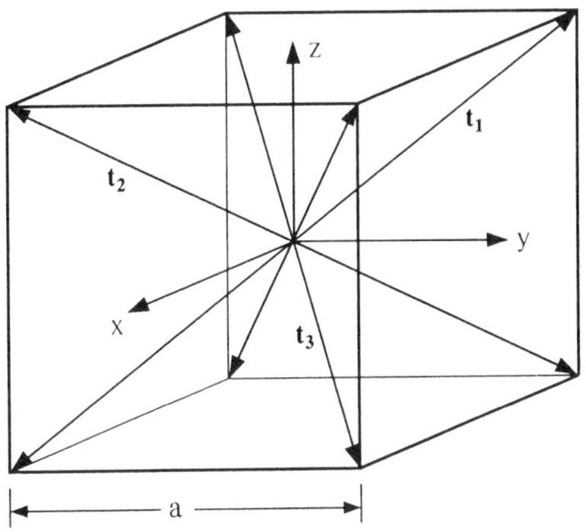

Figure 4.3. Unit cell and primitive vectors t_1, t_2, t_3 of the body-centred cubic lattice

4.4) is a primitive cell. It has the right volume: $a^3/2$. Moreover, one can easily see that successive translations t_1, t_2 and t_3 of the above parallelepiped reproduce the body-centred cubic lattice. The entire lattice is described by the following formula

$$\boldsymbol{R}_n = n_1\boldsymbol{t}_1 + n_2\boldsymbol{t}_2 + n_3\boldsymbol{t}_3 \quad n_1, n_2, n_3 = 0, \pm 1, \pm 2, \pm 3, \ldots \tag{4.2}$$

Three integers define through (4.2) a particular lattice site (we may think of the lattice vector $\boldsymbol{R}_n$ as the position vector of the corresponding lattice site); and to every site of the lattice corresponds one of the lattice vectors (4.2).

The parallelepiped of Fig. 4.4 is not the only possible primitive cell. We can construct a more symmetric primitive cell as follows. We draw straight lines from a lattice point to all its neighbour points, and through the midpoint of each line and normal to it we draw a plane. In this way we

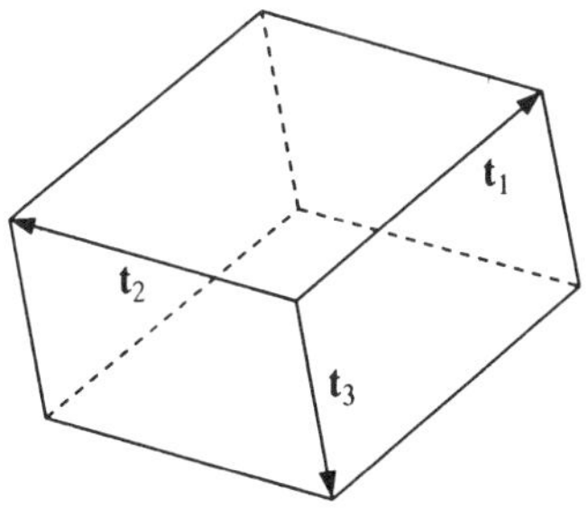

Figure 4.4. Primitive unit cell of the body-centred cubic lattice

obtain a closed volume, a cell of fourteen faces, about the central point as shown in Fig. 4.5. By the way it was constructed, the volume of this cell is that corresponding to one lattice point (and therefore equals $a^3/2$). It is also obvious (see Fig. 4.6) that successive translations of this cell, which is known as the Wigner–Seitz cell, will cover all space. The Wigner–Seitz cell of Fig. 4.5 is, therefore, a primitive cell of the bcc lattice. What makes it special is its symmetry: rotations about symmetry axes, mirror reflections with respect to symmetry planes and other operations which leave a point in space (the

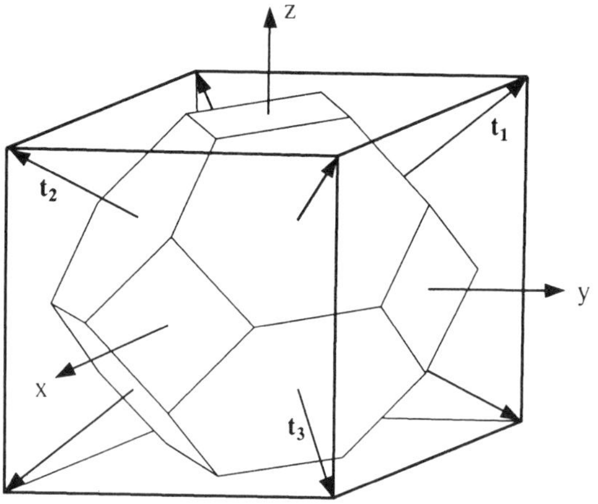

Figure 4.5. Wigner–Seitz cell of the body-centred cubic lattice

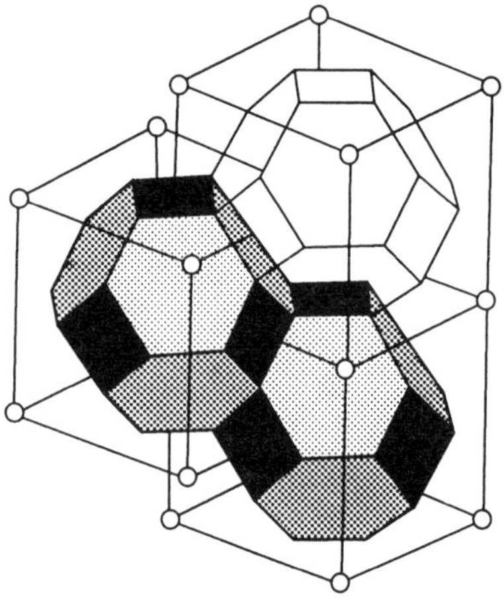

Figure 4.6. Wigner–Seitz cells cover all space

origin of coordinates) fixed and bring the lattice upon itself, bring the Wigner–Seitz cell upon itself as well. The Wigner–Seitz cell has the cubic symmetry of the bcc lattice. No other primitive cell has this property.

We have seen that the body-centred cubic lattice is completely determined by equations (4.1) and (4.2), and in particular by the primitive vectors $\boldsymbol{t}_1$, $\boldsymbol{t}_2$, $\boldsymbol{t}_3$. All space lattices can be described in the same way. In Fig. 4.7 we show unit cells (not primitive) of the fourteen known lattices (they are called Bravais lattices). Each unit cell is characterised by the lengths a, b, c along three axes, by the angles α, β, γ between these axes and by extra lattice points within the unit cell. In each case the lattice can be decribed by a set of three primitive vectors as in the case of the bcc lattice, and a symmetric primitive cell obtained by the method of Wigner and Seitz.

We shall describe, in detail, one more space lattice: the face-centred cubic[2]. This is defined by

$$\boldsymbol{R}_n = n_1\boldsymbol{t}_1 + n_2\boldsymbol{t}_2 + n_3\boldsymbol{t}_3$$

$$n_1, n_2, n_3 = 0, \pm 1, \pm 2, \ldots$$

$$\begin{aligned} \boldsymbol{t}_1 &= \frac{a}{2}(\hat{y} + \hat{z}) \\ \boldsymbol{t}_2 &= \frac{a}{2}(\hat{x} + \hat{z}) \\ \boldsymbol{t}_3 &= \frac{a}{2}(\hat{x} + \hat{y}) \end{aligned} \tag{4.3}$$

where a is the lattice constant. Fig. 4.8 shows a unit cell of the lattice, a cube of edge a, with primitive vectors $\boldsymbol{t}_1$, $\boldsymbol{t}_2$, $\boldsymbol{t}_3$ and others leading from the lattice site at the centre of the cube to its twelve neighbours at the midpoints of the edges of the cube[3]. The volume about the central point enclosed by planes normal to the above vectors, at the midpoints between neighbours, defines the Wigner–Seitz cell of the fcc lattice. This, a dodecahedron of volume $a^3/4$, is shown in Fig. 4.9, and obviously has the cubic symmetry expected of it. (We determine the volume of the Wigner–Seitz cell (volume per lattice site) by noting that there are four lattice sites corresponding to a unit cell (Fig. 4.8) of volume a^3; we have a lattice site at the centre of the cube and 1/4 from every one of the twelve edges of the cube, since an edge-site is shared by four cubes.)

We must now point out that for a complete description of a crystalline solid we must know, besides the space lattice, the content of a primitive cell. In many instances we have only one atomic nucleus in a primitive cell (one atom per lattice site). We can, then, identify the 'equilibrium positions' (we shall explain the term below) of the atomic nuclei of the crystal with the sites of the lattice. For example, copper has an fcc lattice with one atom per

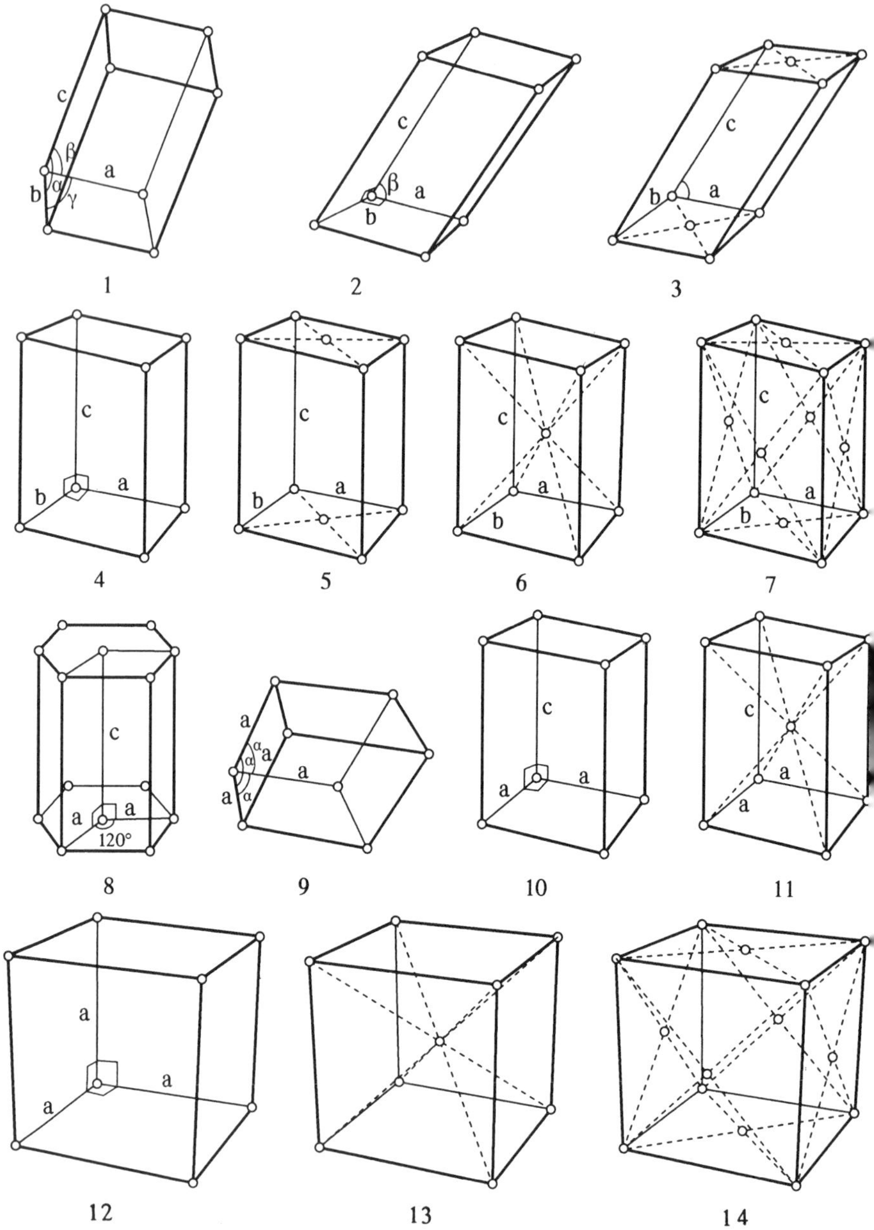

Figure 4.7. Unit cells of the 14 Bravais lattices. Triclinic ($a \neq b \neq c$, $\alpha \neq \beta \neq \gamma$): 1; Monoclinic ($a \neq b \neq c$, $\alpha = \gamma = 90° \neq \beta$), simple:2, base-centred:3; Orthorhombic ($a \neq b \neq c$, $\alpha = \beta = \gamma = 90°$), simple:4, base-centred:5, body-centred:6, face-centred:7; Hexagonal ($a = b \neq c$, $\alpha = \beta = 90°$, $\gamma = 120°$):8; Rhombohedral ($a = b = c$, $\alpha = \beta = \gamma \neq 90°$):9; Tetragonal ($a = b \neq c$, $\alpha = \beta = \gamma = 90°$), simple:10, body-centred:11; Cubic ($a = b = c$, $\alpha = \beta = \gamma = 90°$), simple:12, body-centred:13, face-centred:14

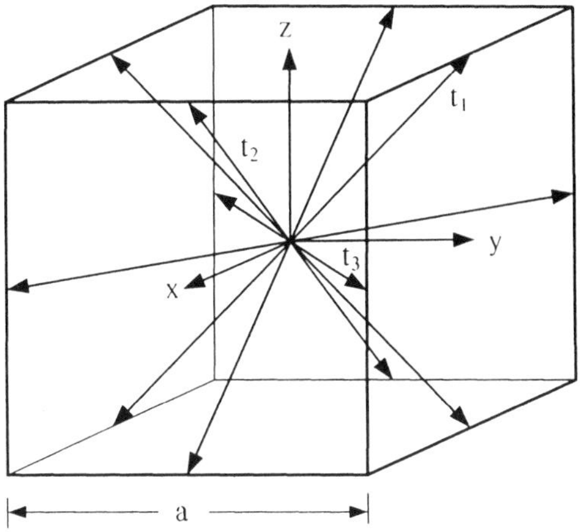

Figure 4.8. Unit cell and primitive vectors $\boldsymbol{t}_1$, $\boldsymbol{t}_2$, $\boldsymbol{t}_3$ of the face-centred cubic lattice

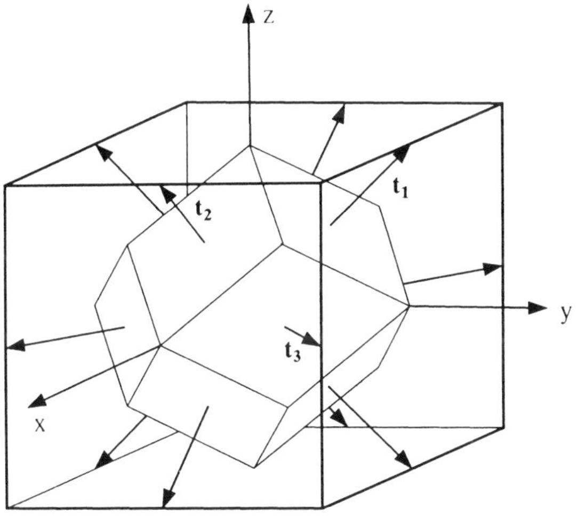

Figure 4.9. Wigner–Seitz cell of the face-centred cubic lattice

lattice site, and tungsten has a bcc lattice also with one atom per lattice site. Diamond and the semiconductors germanium and silicon have fcc lattices with two atoms per lattice site. Fig. 4.10 shows a unit cell of the diamond structure. This (a cube of edge a) is not of course a primitive cell; it has, according to what we said above in connection to the fcc lattice, four lattice points corresponding to it. And we see from the figure that there are two atoms per lattice site. The diamond structure can be described as two interpenetrating fcc structures which are displaced relative to each other

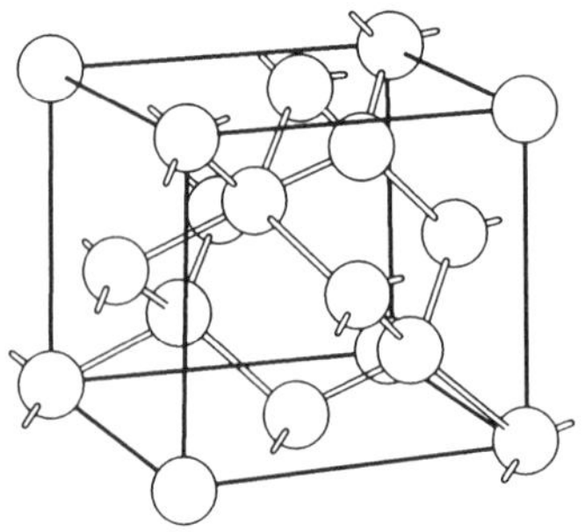

Figure 4.10. Unit cell of diamond (a cube of edge a): the space lattice is face-centred cubic with two atoms (spheres in the figure) per lattice site. Germanium and silicon have the same structure. (Redrawn from K. J. Pascoe, *Properties of Materials for Electrical Engineers*, (New York: Wiley) 1973. Reprinted by permission of John Wiley & Sons Ltd)

along the main diagonal of the cube. The position of the origin of the second fcc structure relative to that of the first is $(a/4, a/4, a/4)$. We note that each atom is surrounded by four nearest neighbours in a tetrahedral configuration.

The above description of a crystal assumes that the atomic nuclei are motionless. We must justify this picture. Thinking as in the case of molecules (Section 3.1), we assume that because the atomic nuclei are much heavier than the electrons we can, in the spirit of the Born–Oppenheimer approximation, consider the motion of the electrons in the field of motionless nuclei, stationed at the positions $\boldsymbol{R}_1, \boldsymbol{R}_2, \ldots, \boldsymbol{R}_N$ where N denotes the number of nuclei in the crystal. The electronic energy of the crystal, which includes the kinetic energy of the electrons, the potential energy of attraction between the electrons and the nuclei, the potential energy of repulsion between the electrons and, also, the potential energy of repulsion between the motionless nuclei of the crystal, is then a function $E(\boldsymbol{R}_1, \boldsymbol{R}_2, \ldots, \boldsymbol{R}_N)$ of the positions of the nuclei. It has a minimum when the nuclei are at certain positions which are, for a crystal, periodically arranged in space. Displacement of the nuclei of the crystal from the positions which minimise $E(\boldsymbol{R}_1, \boldsymbol{R}_2, \ldots, \boldsymbol{R}_N)$ generates forces which tend to bring back the nuclei to the positions of minimum energy. This is why we referred to them as the 'equilibrium positions'. In reality the nuclei vibrate about these positions in much the same way as the atomic nuclei of a polyatomic molecule vibrate about their equilibrium positions (see Section 3.6). In a first approximation, we may assume that each atomic nucleus vibrates independently of the others[4]. In any case the total energy of the crystal is the sum of its electronic energy and the vibrational energy of the nuclei. At low temperatures the vibrators are in low-energy states and, therefore, the nuclei stay very near their equilibrium positions; so much so that we can assume them stationary[5]. A reliable calculation of the atomic structure of a crystal (by which we mean the space lattice and the positions of the atomic nuclei, if there is more than one, in the primitive cell) is very difficult. One usually determines this structure experimentally, by scattering X-rays (electromagnetic radiation of a wavelength comparable to inter-atomic distances in the solid) off the crystal. Scattered rays (diffracted beams) propagate only along certain directions determined by the space lattice; this fact and a careful analysis of the variation of the intensity of the diffracted beams with the frequency of the radiation make it possible to determine accurately the atomic structure of the crystal.

4.2 Reciprocal lattice

Before we can describe the electronic states of a crystal we need to introduce the concept of the reciprocal lattice. We have seen that a space lattice $\{\boldsymbol{R}_n\}$ is defined by three primitive vectors $\boldsymbol{t}_1$, $\boldsymbol{t}_2$, $\boldsymbol{t}_3$. We have

$$\boldsymbol{R}_n = n_1\boldsymbol{t}_1 + n_2\boldsymbol{t}_2 + n_3\boldsymbol{t}_3$$
$$n_1, n_2, n_3 = 0, \pm 1, \pm 2, \pm 3, \ldots \tag{4.4}$$

We define the primitive vectors $\boldsymbol{b}_1$, $\boldsymbol{b}_2$, $\boldsymbol{b}_3$ of the reciprocal (to (4.4)) lattice $\{\boldsymbol{K}_h\}$ as follows:

$\boldsymbol{b}_1$: it is normal to the plane defined by $\boldsymbol{t}_2$ and $\boldsymbol{t}_3$ and has magnitude $b_1 = 2\pi/(t_1 \cos\theta_1)$, where θ_1 is the angle between $\boldsymbol{b}_1$ and $\boldsymbol{t}_1$ (see Fig. 4.11).

$\boldsymbol{b}_2$: it is normal to the plane defined by $\boldsymbol{t}_1$ and $\boldsymbol{t}_3$ and has magnitude $b_2 = 2\pi/(t_2 \cos\theta_2)$, where θ_2 is the angle between $\boldsymbol{b}_2$ and $\boldsymbol{t}_2$.

$\boldsymbol{b}_3$: it is normal to the plane defined by $\boldsymbol{t}_1$ and $\boldsymbol{t}_2$ and has magnitude $b_3 = 2\pi/(t_3 \cos\theta_3)$, where θ_3 is the angle between $\boldsymbol{b}_3$ and $\boldsymbol{t}_3$.

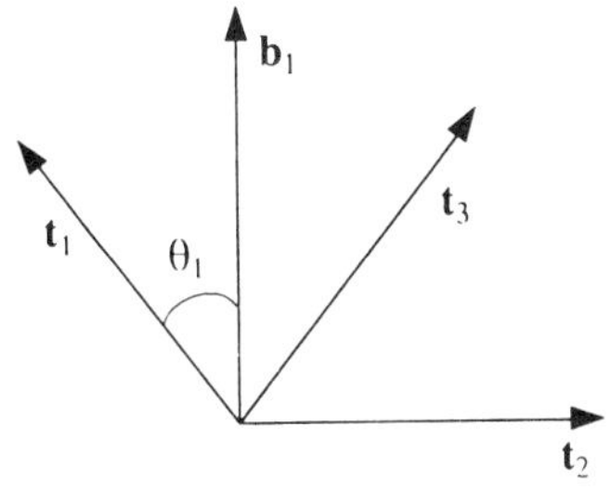

Figure 4.11. Primitive vector $\boldsymbol{b}_1$ of the reciprocal lattice. $\boldsymbol{b}_1$ is normal to the plane defined by $\boldsymbol{t}_2$ and $\boldsymbol{t}_3$, and has magnitude $b_1 = 2\pi/(t_1 \cos\theta_1)$

The reciprocal (to $\{\boldsymbol{R}_n\}$) lattice is given by

$$\boldsymbol{K}_h = h_1\boldsymbol{b}_1 + h_2\boldsymbol{b}_2 + h_3\boldsymbol{b}_3$$
$$h_1, h_2, h_3 = 0, \pm 1, \pm 2, \pm 3, \ldots \tag{4.5}$$

It is a lattice in wavevector space ($\boldsymbol{k}$-space), also called reciprocal space.

We note the following properties of the reciprocal lattice (which follow from the definition of $\boldsymbol{b}_1$, $\boldsymbol{b}_2$, $\boldsymbol{b}_3$):

$$\begin{aligned} \boldsymbol{b}_i \cdot \boldsymbol{t}_j &= 2\pi, \qquad \text{if } i = j \\ &= 0, \qquad \text{if } i \neq j \\ i, j &= 1, 2, 3 \end{aligned} \tag{4.6}$$

Using (4.6) and the definitions (4.4) and (4.5) of $\{\boldsymbol{R}_n\}$ and $\{\boldsymbol{K}_h\}$, we obtain

$$\boldsymbol{K}_h \cdot \boldsymbol{R}_n = (\text{integer number}) \times 2\pi \tag{4.7}$$

Using (4.7) and the mathematical identity (see Appendix A.2)

$$\begin{aligned} \exp(\mathrm{i}2\pi l) &= \cos 2\pi l + \mathrm{i}\sin 2\pi l \\ &= 1, \text{ if } l = \text{integer number} \end{aligned} \tag{4.8}$$

we obtain the following important property

$$\exp(\mathrm{i}\boldsymbol{K}_h \cdot \boldsymbol{R}_n) = 1 \tag{4.9}$$

which means that the functions

$$\exp(\mathrm{i}\boldsymbol{K}_h \cdot \boldsymbol{r}) \tag{4.10}$$

where $\boldsymbol{K}_h$ is a reciprocal vector (4.5), have the following property

$$\exp(\mathrm{i}\boldsymbol{K}_h \cdot (\boldsymbol{r} + \boldsymbol{R}_n)) = \exp(\mathrm{i}\boldsymbol{K}_h \cdot \boldsymbol{r}) \tag{4.11}$$

Equation (4.11) tells us that the functions (4.10) are periodic functions of $\boldsymbol{r}$ and that their periodicity is that of the lattice (4.4). The usefulness of these functions derives from the following theorem: if a function $\psi(\boldsymbol{r})$ is periodic, i.e. if

$$\psi(\boldsymbol{r} + \boldsymbol{R}_n) = \psi(\boldsymbol{r}) \tag{4.12}$$

for every vector $\boldsymbol{R}_n$ of a space lattice (4.4), then $\psi(\boldsymbol{r})$ can be written as a linear sum of the functions (4.10):

$$\psi(\boldsymbol{r}) = \sum_{\boldsymbol{K}_h} C(\boldsymbol{K}_h) \exp(\mathrm{i}\boldsymbol{K}_h \cdot \boldsymbol{r}) \tag{4.13}$$

where $C(\boldsymbol{K}_h)$ are coefficients to be determined. In principle, the sum (4.13) has infinitely many terms, one for every vector of the reciprocal lattice (4.5);

in practice we find that $C(\boldsymbol{K}_h) = 0$ when the magnitude of $\boldsymbol{K}_h$ exceeds a certain value.

We conclude this section with a description of the reciprocal lattices of the bcc and fcc space lattices.

We find that the reciprocal lattice of the bcc lattice (4.1, 4.2) is given by (4.5) with

$$\begin{aligned} \boldsymbol{b}_1 &= \frac{2\pi}{a}(\hat{y} + \hat{z}) \\ \boldsymbol{b}_2 &= \frac{2\pi}{a}(\hat{x} + \hat{z}) \\ \boldsymbol{b}_3 &= \frac{2\pi}{a}(\hat{x} + \hat{y}) \end{aligned} \tag{4.14}$$

A comparison of (4.14) with (4.3) shows that the reciprocal of a body-centred cubic space lattice is (in $\boldsymbol{k}$-space) a face-centred cubic lattice. The volume (in $\boldsymbol{k}$-space) of its primitive cell, which equals the volume of the parallelepiped defined by the primitive vectors (4.14), is $(2\pi)^3/(a^3/2)$; it is inversely proportional to the volume of a primitive cell of the space lattice.

Similarly we find that the reciprocal of the fcc lattice (4.3) is given by (4.5) with

$$\begin{aligned} \boldsymbol{b}_1 &= \frac{2\pi}{a}(-\hat{x} + \hat{y} + \hat{z}) \\ \boldsymbol{b}_2 &= \frac{2\pi}{a}(\hat{x} - \hat{y} + \hat{z}) \\ \boldsymbol{b}_3 &= \frac{2\pi}{a}(\hat{x} + \hat{y} - \hat{z}) \end{aligned} \tag{4.15}$$

A comparison of (4.15) with (4.1) shows that the reciprocal of a face-centred cubic space lattice is (in $\boldsymbol{k}$-space) a body-centred cubic lattice. The volume (in $\boldsymbol{k}$-space) of its primitive cell, which equals the volume of the parallelepiped defined by the vectors (4.15), is $(2\pi)^3/(a^3/4)$, inversely proportional to the volume of the primitive cell of the space lattice.

We can construct, using the method of Wigner and Seitz, a primitive cell of the reciprocal lattice which exhibits the symmetry of this lattice. The equivalent of the Wigner–Seitz cell in reciprocal space is known as the first Brillouin zone or, simply, as the Brillouin zone and is denoted by BZ. It follows from what we said above, that the BZ of (the reciprocal of) a bcc lattice will have the shape (Fig. 4.9) of the Wigner–Seitz cell of an fcc space lattice, and the BZ of (the reciprocal of) an fcc lattice will have the shape (Fig. 4.5) of the Wigner–Seitz cell of a bcc space lattice.

4.3 One-electron eigenstates in a periodic potential field

We obtain a good approximation to the electronic properties of a crystal if we assume that every electron of the crystal moves, independently of all the others, in a mean potential field $U(\boldsymbol{r})$ due to the totality of the electrons and stationary nuclei of the crystal. The mean field has, of course, the periodicity of the crystal lattice:

$$U(\boldsymbol{r} + \boldsymbol{R}_n) = U(\boldsymbol{r}) \tag{4.16}$$

where $\boldsymbol{R}_n$ is any vector of the given (experimentally determined) space lattice. We note that (4.16) is strictly valid for an infinite crystal, one which extends over all space: $-\infty < x, y, z < +\infty$. A real crystal does not extend over all space; we shall, however, maintain the assumption of an infinite crystal so as to able to use (4.16). It turns out that this is not a bad approximation when one deals with the bulk properties of the crystal, and that one can take into account the finite size of the crystal by means of appropriate periodic boundary conditions on the one-electron eigenfunctions (see below).

An important theorem (Bloch's theorem) tells us that the eigenstates of the energy of an electron in the potential field (4.16) can be described by eigenfunctions of the following form[6]

$$\psi_{\boldsymbol{k}\alpha}(\boldsymbol{r}, t) = A \exp(\mathrm{i}\boldsymbol{k} \cdot \boldsymbol{r}) u_{\boldsymbol{k}\alpha}(\boldsymbol{r}) \exp(-\mathrm{i}E_\alpha(\boldsymbol{k})t/\hbar) \tag{4.17}$$

where $u_{\boldsymbol{k}\alpha}$ is a periodic function:

$$u_{\boldsymbol{k}\alpha}(\boldsymbol{r} + \boldsymbol{R}_n) = u_{\boldsymbol{k}\alpha}(\boldsymbol{r}) \tag{4.18}$$

which we can write, using formula (4.13), as follows

$$u_{\boldsymbol{k}\alpha}(\boldsymbol{r}) = \sum_{\boldsymbol{K}_h} C(\boldsymbol{K}_h) \exp(\mathrm{i}\boldsymbol{K}_h \cdot \boldsymbol{r}) \tag{4.18a}$$

$\boldsymbol{K}_h$ are the vectors of the reciprocal lattice (4.5) and $C(\boldsymbol{K}_h)$ are coefficients which depend on the so-called wavevector $\boldsymbol{k}$ and additional quantum-numbers, denoted collectively by α, which are needed for a complete description of an eigenstate (4.17). $E_\alpha(\boldsymbol{k})$ denotes the energy of the eigenstate and A is a normalisation constant such that

$$\int_V |\psi_{\boldsymbol{k}\alpha}(\boldsymbol{r}, t)|^2 \, \mathrm{d}V = \int_V |A|^2 |u_{\boldsymbol{k}\alpha}(\boldsymbol{r})|^2 \, \mathrm{d}V = 1 \tag{4.19}$$

where V denotes the volume of the crystal. Wavefunctions of the form (4.17) are called Bloch waves. In Fig. 4.12 we show the variation of such a wave along a line passing through the centres of a row of atoms, together with the variation of the potential field along the same line.

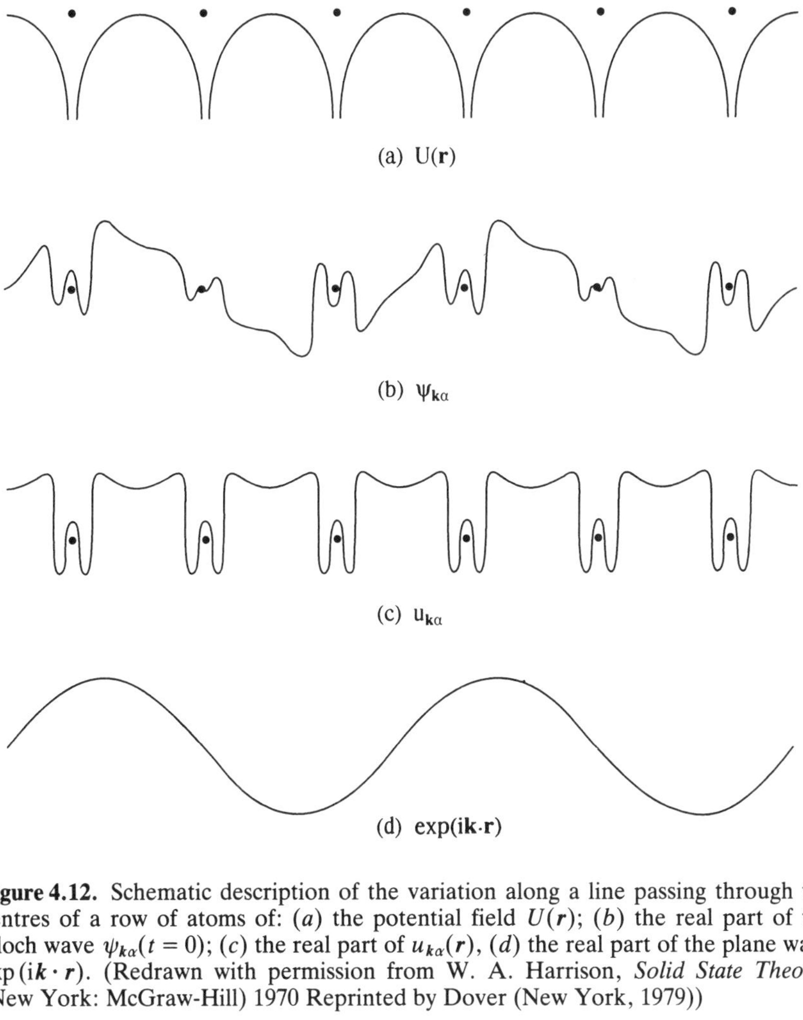

Figure 4.12. Schematic description of the variation along a line passing through the centres of a row of atoms of: (*a*) the potential field $U(\boldsymbol{r})$; (*b*) the real part of the Bloch wave $\psi_{\boldsymbol{k}\alpha}(t=0)$; (*c*) the real part of $u_{\boldsymbol{k}\alpha}(\boldsymbol{r})$, (*d*) the real part of the plane wave $\exp(\mathrm{i}\boldsymbol{k}\cdot\boldsymbol{r})$. (Redrawn with permission from W. A. Harrison, *Solid State Theory*, (New York: McGraw-Hill) 1970 Reprinted by Dover (New York, 1979))

The characteristic property of a Bloch wave (a direct consequence of (4.17) and (4.18)) is:

$$\psi_{\boldsymbol{k}\alpha}(\boldsymbol{r}+\boldsymbol{R}_n,\, t) = \exp(\mathrm{i}\boldsymbol{k}\cdot\boldsymbol{R}_n)\psi_{\boldsymbol{k}\alpha}(\boldsymbol{r},\, t) \tag{4.20}$$

The wavevector $\boldsymbol{k}$ of a Bloch wave can have any value ($-\infty < k_x, k_y, k_z < +\infty$). We observe, however, that values of $\boldsymbol{k}$ differing by a reciprocal vector $\boldsymbol{K}_{h'}$ lead to an identical set of Bloch waves. (The Bloch waves corresponding to $\boldsymbol{k}+\boldsymbol{K}_{h'}$ are not different from the ones corresponding to

$\boldsymbol{k}$.) We can see that this is so, by writing

$$\exp[\mathrm{i}(\boldsymbol{k} + \boldsymbol{K}_{h'}) \cdot \boldsymbol{r}] u_{\boldsymbol{k}+\boldsymbol{K}_{h'},\alpha}(\boldsymbol{r}) = \exp(\mathrm{i}\boldsymbol{k} \cdot \boldsymbol{r}) \sum_{\boldsymbol{K}_h} C(\boldsymbol{K}_h) \exp[\mathrm{i}(\boldsymbol{K}_h + \boldsymbol{K}_{h'}) \cdot \boldsymbol{r}] \tag{4.21}$$

where we made use of (4.18a). Given that $\boldsymbol{K}_h + \boldsymbol{K}_{h'}$ is also a vector of the reciprocal lattice and since the sum (4.21) includes all the vectors of this lattice, we can rewrite (4.21) as follows

$$(4.21) = \exp(\mathrm{i}\boldsymbol{k} \cdot \boldsymbol{r}) \sum_{\boldsymbol{K}_h} C'(\boldsymbol{K}_h) \exp(\mathrm{i}\boldsymbol{K}_h \cdot \boldsymbol{r}) \tag{4.22}$$

Now, since the coefficients $C'(\boldsymbol{K}_h)$ remain to be determined, as those of (4.18a), the two expressions, $\exp(\mathrm{i}\boldsymbol{k} \cdot \boldsymbol{r}) u_{\boldsymbol{k}\alpha}(\boldsymbol{r})$ and $\exp(\mathrm{i}[\boldsymbol{k} + \boldsymbol{K}_{h'}] \cdot \boldsymbol{r}) u_{\boldsymbol{k}+\boldsymbol{K}_{h'},\alpha}(\boldsymbol{r})$ are equivalent. This means that when we seek the eigenstates (4.17) of an electron in the field (4.16), we need only consider wavevectors $\boldsymbol{k}$ (one may say points $\boldsymbol{k} = (k_x, k_y, k_z)$) within and on the boundary of a primitive cell of the reciprocal lattice. We may refer to this cell as the reduced $\boldsymbol{k}$-zone of the crystal. The most appropriate choice of the reduced $\boldsymbol{k}$-zone, when we study the electronic properties in the interior of the crystal, is the BZ defined in the previous section. Therefore, we assume in what follows that $\boldsymbol{k}$ lies within or on the boundary of the BZ of the crystal under consideration.

At this stage we are ready to take into account the finite volume V of the crystal. We do so by imposing periodic boundary conditions on the eigenfunctions (4.17). We demand that

$$\begin{aligned} \psi_{\boldsymbol{k}\alpha}(x + L, y, z, t) = \psi_{\boldsymbol{k}\alpha}(x, y + L, z, t) &= \psi_{\boldsymbol{k}\alpha}(x, y, z + L, t) \\ &= \psi_{\boldsymbol{k}\alpha}(x, y, z, t) \end{aligned} \tag{4.23}$$

where $L = V^{1/3}$. (We have assumed for the sake of simplicity that the crystal has the shape of a cube (as in Fig. 4.1); similar boundary conditions can be applied to any reasonable shape of the crystal, and the results are the same; they are independent of the shape of the crystal as long as the volume of the crystal is sufficiently large. We may add that 'sufficiently large' turns out to mean that every dimension of the crystal (length, width and thickness) exceeds a few thousand atomic units.) The conditions (4.23) are satisfied if (and only if)

$$\begin{aligned} \boldsymbol{k} = (k_x, k_y, k_z) &= (2\pi/L)(n_x, n_y, n_z) \\ n_x, n_y, n_z &= 0, \pm 1, \pm 2, \pm 3, \ldots \end{aligned} \tag{4.24}$$

When $\boldsymbol{k}$ has one of the above values, $\exp(\mathrm{i}\boldsymbol{k} \cdot \boldsymbol{r})$ has the required periodicity $[\exp(\mathrm{i}k_x(x + L)) = \exp(\mathrm{i}k_x x),\ \exp(\mathrm{i}k_y(y + L)) = \exp(\mathrm{i}k_y y),\ \exp(\mathrm{i}k_z(z + L)) = \exp(\mathrm{i}k_z z)]$ because of (4.8), and since $u_{\boldsymbol{k}\alpha}(\boldsymbol{r})$ is by construction a periodic function, we may conclude that $\psi_{k\alpha} \propto \exp(\mathrm{i}\boldsymbol{k}\cdot\boldsymbol{r})u_{k\alpha}(\boldsymbol{r})$ is also a periodic function: it satisfies (4.23). By restricting, via the conditions (4.23), the values of the wavevector $\boldsymbol{k}$ of a Bloch wave to those in the BZ which satisfy (4.24), we make certain that the number of eigenstates (4.17) of an electron in the crystal under consideration (of volume V) in any energy interval is the correct one, and at the same time we maintain our assumption of an effective infinite crystal, which is necessary if we are to use Bloch's theorem. It turns out that the electronic properties of the bulk of the crystal (specific heat, electrical conduction, etc.) which are determined by the behaviour of the electrons in the interior of the crystal are described perfectly well within the above scheme. The final results are the same with those one would obtain using more realistic boundary conditions (e.g., the eigenfunctions of the electron must decay exponentially to zero outside the crystal). We note that the electronic cloud associated with the electron states of energy between E and $E + \mathrm{d}E$ obtained with the periodic boundary conditions (4.23) is correct in the interior of the crystal up to a distance from the surface of a few atomic units, if the crystal happens to be a metal; and up to a distance from the surface of, at most, a few tens of atomic units, if the crystal is a semiconductor or an insulator (see also problem 1.8 and Section 4.13.2).

We accept, therefore, that only wavevectors $\boldsymbol{k}$, in the BZ, which satisfy (4.24) are to be allowed in (4.17). To every one of the allowed $\boldsymbol{k}$-points corresponds, according to (4.24), a volume (in $\boldsymbol{k}$-space) equal to $(2\pi)^3/L^3 = (2\pi)^3/(NV_o)$, where V_o denotes the volume of a primitive cell of the crystal and N is the total number of such cells in the crystal under consideration. Given that the BZ has a volume of $(2\pi)^3/V_o$ (see Section 4.2), we conclude that there are N allowed values of $\boldsymbol{k}$ in the BZ, as many as the number of primitive cells in the crystal.

In Appendix B.9 we show that the eigenstates (4.17) of an electron, for given wavevector $\boldsymbol{k}$ (one of the allowed values (4.24)), correspond to a discrete spectrum of energy levels. In the simplest, and most common, case (no degeneracy except that due to the spin[6]) we obtain a series of eigenstates[7]

$$\psi_{k\alpha}(\boldsymbol{r}, t) = \frac{1}{\sqrt{V}} \exp(\mathrm{i}\boldsymbol{k}\cdot\boldsymbol{r})u_{k\alpha}(\boldsymbol{r}) \exp(-\mathrm{i}E_\alpha(\boldsymbol{k})t/\hbar) \tag{4.25}$$

which correspond to different eigenvalues of the energy

$$E_\alpha(\boldsymbol{k}), \quad \alpha = 1, 2, 3, \ldots \tag{4.26}$$

Fig. 4.13 shows the eigenvalues of the energy of the electron (in an energy region) for $\boldsymbol{k}$-points along two lines ΓB and ΓA of the BZ of a hypothetical crystal. We note that the allowed $\boldsymbol{k}$-points, according to (4.24), are distributed uniformly in $\boldsymbol{k}$-space and, when the dimensions of the crystal are

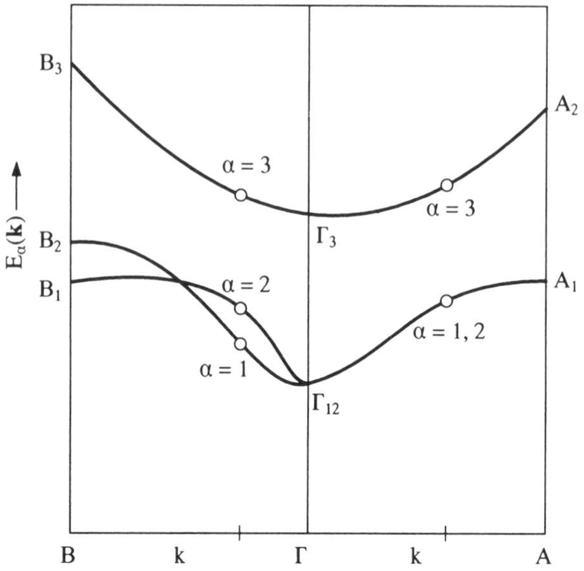

Figure 4.13. Energy bands along two lines ΓA and ΓB of the BZ of a hypothetical crystal. Γ denotes the centre of the BZ and A and B are two points on the surface of the BZ. ΓA is a symmetry line. The circles denote the energy levels for a certain point on ΓB and a certain point on ΓA

large enough ($L > 1000$ Å), lie so close to each other, that we are allowed, for the sake of clarity and convenience, to present $E_\alpha(\boldsymbol{k})$, in diagrams such as the one of Fig. 4.13, as functions defined over all $\boldsymbol{k}$. These appear as continuous functions of $\boldsymbol{k}$ because a small change in $\boldsymbol{k}$ leads to correspondingly small changes in the $E_\alpha(\boldsymbol{k})$, (see Section B.9). The energy levels $E_\alpha(\boldsymbol{k})$, of given α, corresponding to the N allowed $\boldsymbol{k}$-points in the BZ, constitute a band of energy levels: the energy band defined by the band-index α. To it corresponds the band of one-electron eigenstates denoted by $\psi_{\boldsymbol{k}\alpha}$.

Let us look at the dispersion curves (the name refers to a plot of E versus k along a given direction) shown in Fig. 4.13 more carefully. For every one of the (allowed) $\boldsymbol{k}$ points along the ΓB line we obtain three eigenstates $\psi_{\boldsymbol{k}\alpha}$, $\alpha = 1, 2, 3$, per spin, which correspond to different eigenvalues[8] $E_\alpha(\boldsymbol{k})$, $\alpha = 1, 2, 3$. The situation is different along the ΓA line. For a point on this line we obtain again three eigenstates $\psi_{\boldsymbol{k}\alpha}$, $\alpha = 1, 2, 3$, per spin; but in this instance we have

$$E_1(\boldsymbol{k}) = E_2(\boldsymbol{k}), \text{ for } \boldsymbol{k} \text{ on } \Gamma\text{A}$$

The dispersion curve $\Gamma_{12}A_1$ is two-fold degenerate (four-fold when spin is taken into account). We may say that the two bands $E_1(\boldsymbol{k})$ and $E_2(\boldsymbol{k})$ which extend, of course, over the whole of the BZ coincide along the ΓA line. This phenomenon is observed along special lines (so-called symmetry lines) of the BZ, which relate to the symmetry of the crystal (the group of symmetry operations, e.g., rotations about certain axes, reflections with respect to certain planes etc., which bring the crystal upon itself). At so-called symmetry points of the BZ, such as its centre, the degeneracy of an energy level may be even greater, as we shall see in the following sections. Of course, not all dispersion curves along a symmetry line will be degenerate, and there will be cases where none is; and the same applies to symmetry points.

We must remember that in every band ($\alpha = 1, 2, 3 \ldots$) which as we have already stated, extends over the whole of the BZ, there are N eigenstates, per spin, of the electron, corresponding to the N allowed $\boldsymbol{k}$-points in the BZ (as many as the number of primitive cells in the given crystal).

4.3.1 ENERGY BANDS OF COPPER

Let us now consider a real crystal: copper (Cu). Copper has a face-centred cubic lattice with one atom per primitive cell and, therefore, we may assume that the atomic nuclei are centred on the lattice points of (4.3). The corresponding reciprocal lattice (given by 4.5 and 4.15) is body-centred cubic and, therefore, copper has the BZ shown in Fig. 4.14. The reader

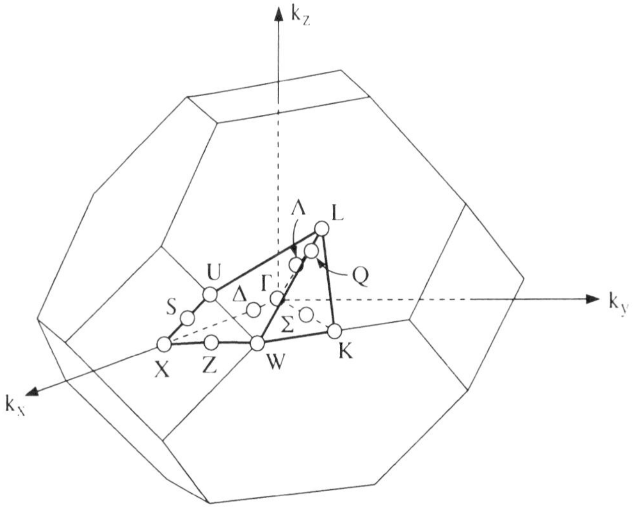

Figure 4.14. The Brillouin zone (BZ) of the face-centred cubic lattice

should familiarise himself with the points and lines of symmetry of this zone; these are denoted by capital letters in Fig. 4.14. The centre of the zone is denoted by Γ, the point where the k_x-axis meets the surface of the zone is denoted by X, the line along the k_x-axis from Γ to X is denoted by Δ, etc.

In Fig. 4.15 we present some of the energy bands of copper for $\boldsymbol{k}$ along various symmetry lines of the BZ shown in Fig. 4.14. We note that the line UX is equivalent to a continuation of the line ΓK beyond the Brillouin zone (see problem 4.4). The zero of energy in Fig. 4.15 has been chosen arbitrarily and has no obvious physical meaning. One can always determine experimentally (see problem 4.15) the value of E_F (the Fermi energy) relative to the vacuum level (the energy of an electron at rest on the vacuum side of a particular crystal surface). It turns out (see Section 4.13.1) that for most metal and semiconductor surfaces E_F lies a few eV below the vacuum level. Apart from the energy bands shown in Fig. 4.15 there are bands of higher energy (in the region from 0.3 Hartrees to infinity), and bands of lower energy ($E < -0.45$ Hartrees). The lower-energy bands are intimately connected, as we shall see, with the one-electron states in the inner shells of the atom of copper. We note, also, that in the case of copper the energy levels of the electron and the orbital parts of the corresponding eigenstates are the same for the two orientations of the spin (up and down), which implies that the potential (4.16) an electron sees in the crystal of copper is indeed independent of the spin.[9] At the absolute zero temperature ($T = 0$) all one-electron states with energy below E_F are occupied, and all states with energy above E_F are empty (see section 2.14). We shall see how E_F is determined in Section 4.5.

We know (see Table 2.5) that 18 out of the 29 electrons of the atom of copper are found in inner closed shells ($1s^2 2s^2 2p^6 3s^2 3p^6$) and the remaining 11 in outer shells ($3d^{10}4s$) with energies much higher than those of the inner shells. The inner shells are affected very little when the atoms bind together to form the crystal for the reasons stated in Section 2.15.3. Let us see what happens to the 1s atomic orbitals of the electron. We have N such orbitals: $\varphi_{1s}(\boldsymbol{r} - \boldsymbol{R}_n)$, $n = 1, 2, 3, \ldots, N$, one orbital for each of the N atoms that make up the crystal. We can approximate the corresponding states of the electron in the crystal by linear combinations of these atomic orbitals, as we have done in the construction of the molecular orbitals of Section 3.2. We seek linear combinations of the atomic orbitals which satisfy Bloch's theorem. We find N such combinations (Bloch waves): they correspond to the N allowed values of $\boldsymbol{k}$ in the BZ of the crystal and are written as follows

$$\psi_{\boldsymbol{k}\alpha}(\boldsymbol{r}, t) = \frac{A}{\sqrt{N}}\left\{\sum_{\boldsymbol{R}_n} \exp(\mathrm{i}\boldsymbol{k}\cdot\boldsymbol{R}_n)\varphi_\alpha(\boldsymbol{r} - \boldsymbol{R}_n)\right\}\exp(-\mathrm{i}E_\alpha(\boldsymbol{k})t/\hbar) \quad (4.27)$$

where $\alpha = 1s$; and A is a normalisation factor (about unity) which takes into account the fact that atomic orbitals centred on neighbour atoms overlap to

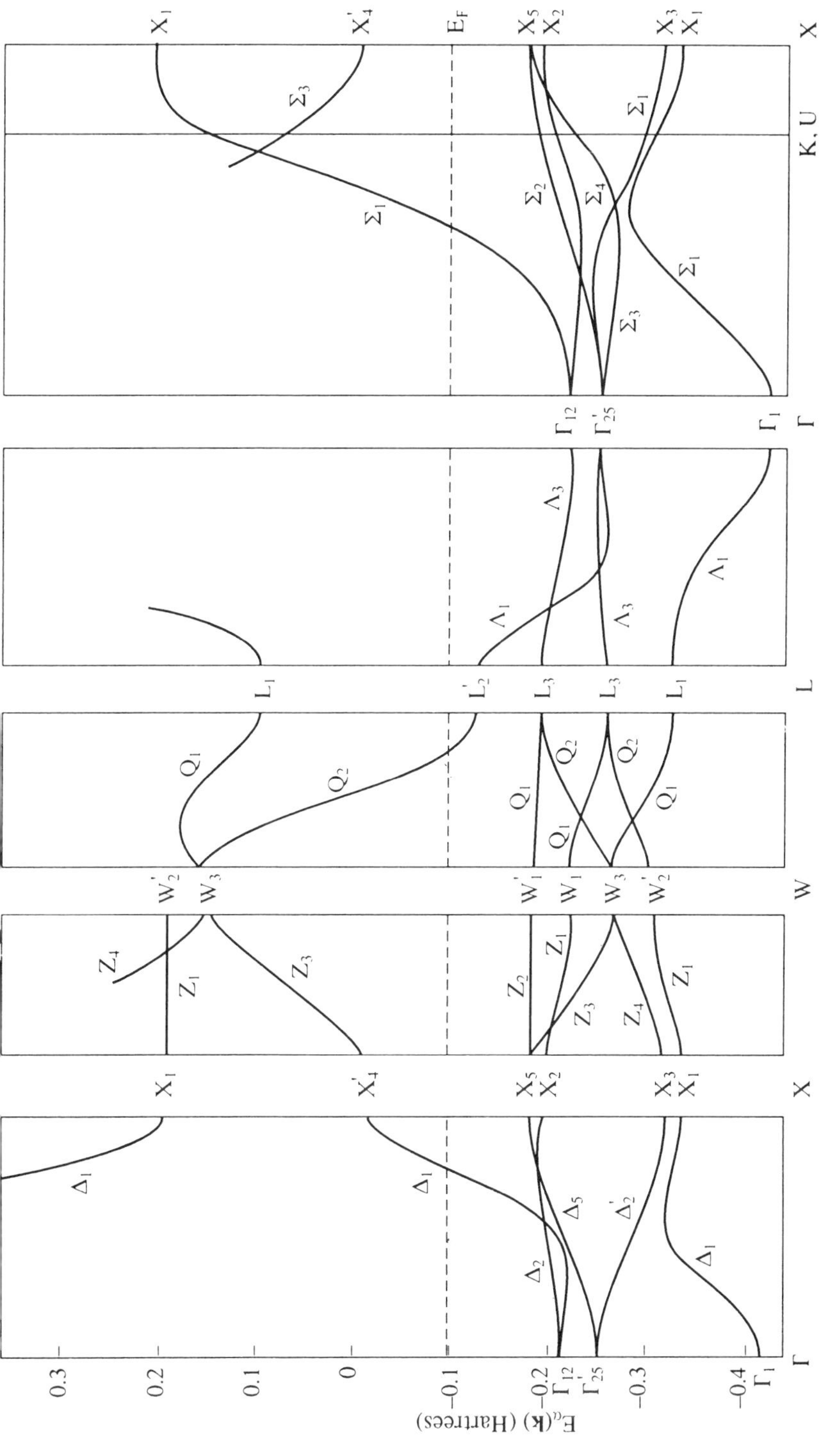

Figure 4.15. Energy bands of copper along various symmetry lines of the **BZ** (Fig. 4.14). [Calculated by B. Segall, *Physical Review*, **125** (1962) 109]

some degree (see problem 4.3). We can see that (4.27) is a Bloch wave, i.e. that it has the form (4.17), by rewriting it as follows

$$\psi_{k\alpha}(\boldsymbol{r}, t) = \exp(\mathrm{i}\boldsymbol{k}\cdot\boldsymbol{r})u_{k\alpha}(\boldsymbol{r})\exp(-\mathrm{i}E_\alpha(\boldsymbol{k})t/\hbar)$$

$$u_{k\alpha}(\boldsymbol{r}) \equiv \frac{A}{\sqrt{N}}\sum_{\boldsymbol{R}_n}\exp[-\mathrm{i}\boldsymbol{k}\cdot(\boldsymbol{r}-\boldsymbol{R}_n)]\varphi_\alpha(\boldsymbol{r}-\boldsymbol{R}_n) \qquad (4.27\mathrm{a})$$

and noting that $u_{k\alpha}$ satisfies (4.18).

The one-electron eigenstates described by (4.27) constitute a band of states: the band $\alpha = 1\mathrm{s}$. The corresponding energy eigenvalues are given by

$$E_\alpha(\boldsymbol{k}) = E_\alpha^{(0)} + \Delta E_\alpha + \Delta E_\alpha(\boldsymbol{k}) \qquad (4.28)$$

$E_\alpha^{(0)}$ is the energy of the electron in the corresponding state of the isolated atom (1s in the present case), ΔE_α is a relatively small correction ($|\Delta E_\alpha| \ll |E_\alpha^0|$) which derives from the change in the electronic cloud of the outer region of the atom caused by the formation of the crystal, and $\Delta E_\alpha(\boldsymbol{k})$, a correction which depends on $\boldsymbol{k}$, is a result of the possibility the electron has to tunnel from a state centred on a given atom to a state of the same energy centred on a neighbour atom, leading to the creation of the band of states (4.27), which extend over the entire crystal, and have energies spreading about the atomic level. We have here a phenomenon which is essentially the same to that we observed in the creation of the energy levels of an electron in the molecule H_2^+ (see Section 3.2, Fig. 3.5). It is worth remembering that tunnelling through a barrier does not by itself change the energy of the electron (see Section 1.8), so that, to begin with, only atomic states of the same energy interact via tunnelling; and this explains, for example, why the 1s state of a given atom interacts only with the 1s states on neighbour atoms and not, say, with the 2s states of these atoms. However, because of tunnelling the electron is found with some small probability between the atoms and this changes its energy. The above probability depends on $\boldsymbol{k}$ because of the form (4.27) of the wavefunction and, therefore, the energy changes by an amount $\Delta E_\alpha(\boldsymbol{k})$ which depends on $\boldsymbol{k}$. Evidently, the energy spread of the band will be larger the larger the tunnelling interaction between neighbour atoms, and this interaction will be larger when the overlap between the orbitals of neighbour atoms is larger. In the case under consideration ($\alpha = 1\mathrm{s}$, $\Delta E_\alpha(\boldsymbol{k})$ is very small because of the very small overlap of the 1s orbitals of neighbour atoms[10]. Therefore the energy spread (the bandwidth) of the 1s band is practically negligible. Finally, we can see that the electronic cloud around the nucleus at $\boldsymbol{R}_n$ due to the N states (4.27) is practically identical with that of the single atomic orbital $\varphi_{1\mathrm{s}}(\boldsymbol{r}-\boldsymbol{R}_n)$. We have (in the present case we can put $A = 1$ in (4.27)):

$$\sum_{\boldsymbol{k}}|\psi_{k\alpha}|^2 \underset{\boldsymbol{r}\simeq\boldsymbol{R}_n}{\Longrightarrow} \sum_{\boldsymbol{k}}\frac{1}{N}|\varphi_\alpha(\boldsymbol{r}-\boldsymbol{R}_n)|^2 = |\varphi_\alpha(\boldsymbol{r}-\boldsymbol{R}_n)|^2 \qquad (4.29)$$

where $\alpha = 1s$. The first step in the above equation is justified by the observation that in the region of $\boldsymbol{R}_n$ the wavefunctions $\varphi_{1s}(\boldsymbol{r} - \boldsymbol{R}_m)$, $\boldsymbol{R}_m \neq \boldsymbol{R}_n$ practically vanish. The second step is evident: the sum extends over the BZ of the crystal (N values of $\boldsymbol{k}$):

$$\sum_{\boldsymbol{k}} \frac{1}{N} |\varphi_\alpha(\boldsymbol{r} - \boldsymbol{R}_n)|^2 = |\varphi_\alpha(\boldsymbol{r} - \boldsymbol{R}_n)|^2 \sum_{\boldsymbol{k}} \frac{1}{N} = |\varphi_\alpha(\boldsymbol{r} - \boldsymbol{R}_n)|^2$$

Similar arguments apply to the bands of one-electron states which derive from the other inner shells of Cu (2s, 3s, 2p and 3p). We remember that in the isolated atom we have three linearly independent 2p states, and three linearly independent 3p-states, corresponding to the angular momentum quantum-numbers $(lm) = (1,1)$, $(1,0)$ and $(1,-1)$. Therefore, we obtain three bands of one-electron states in the crystal which are described to a good approximation by Bloch sums of 2p atomic orbitals, and three bands of states which are described to a good approximation by Bloch sums of 3p atomic orbitals.

The eigenstates of the three 2p bands have the form (4.27) where φ_α is, in general, a linear sum of the three 2p atomic orbitals. We have three linearly independent such sums, one for each band.[11] It is obvious that the spherical symmetry of the atomic potential field does not exist in the crystal. Therefore, the three bands of energy levels $E_\alpha(\boldsymbol{k})$, $\alpha = 2p$: 1, 2, 3 which derive from the triply degenerate E^0_{2p} atomic level are, in general, different: At a general point $\boldsymbol{k}$ of the BZ we obtain

$$\begin{gathered} E_{2p:\nu}(\boldsymbol{k}) = E^{(0)}_{2p} + \Delta E_{2p} + \Delta E_{2p:\nu}(\boldsymbol{k}) \\ \nu = 1, 2, 3 \\ \Delta E_{2p:1}(\boldsymbol{k}) \neq \Delta E_{2p:2}(\boldsymbol{k}) \neq \Delta E_{2p:3}(\boldsymbol{k}) \end{gathered} \tag{4.30}$$

The different terms in (4.30) can be understood in exactly the same way as the corresponding terms in (4.28). The 3p bands are described in the same manner. The correction terms (second and third terms in (4.28)) are very small in all the bands which derive from the inner shells of the atoms that make up the crystal. We often refer to the one-electron states in these bands as core states, and to the electrons occupying them as core electrons.

In every one of these bands there are N eigenstates of the electron (corresponding to the N allowed values of $\boldsymbol{k}$ in the BZ), where N is the number of primitive cells in the crystal. We have therefore a total of $18N$ (18 per atom, since we have one atom in the primitive cell of crystalline copper) one-electron states in the bands which derive from the inner shells of the atom: $2N$ states in the 1s bands (of spin up and spin down), $2N$ in the 2s bands, $2N$ in the 3s bands, $6N$ in the 2p bands and $6N$ in the 3p bands. In a crystal of copper containing N atoms we have $29N$ electrons (29 per atom). Under normal conditions (when the crystal is in thermodynamic equilibrium), $18N$ electrons are accommodated in the above mentioned core states (one electron in each state). As we have seen the properties of these electrons (energy and density of electronic cloud) do not differ essentially from their properties in the isolated atom. The core electrons do not contribute significantly to the binding of the crystal and do not affect significantly any of its properties (specific heat, conductivity etc.).

Let us now see what happens to the remaining $11N$ electrons (11 per atom) in the crystal of copper. These are accommodated in the bands shown in Fig. 4.15, occupying, at zero absolute temperature, all eigenstates with energy below E_F. These bands derive from the 3d and 4s one-electron states of atomic copper (see Table 2.5).

If we seek the corresponding states of the electron in the crystal as linear combinations (Bloch waves of the form 4.27) of 3d atomic orbitals, and 4s atomic orbitals respectively, as we have done for the energy bands deriving from the inner shells of the atom, we obtain for $\boldsymbol{k}$ points along the symmetry

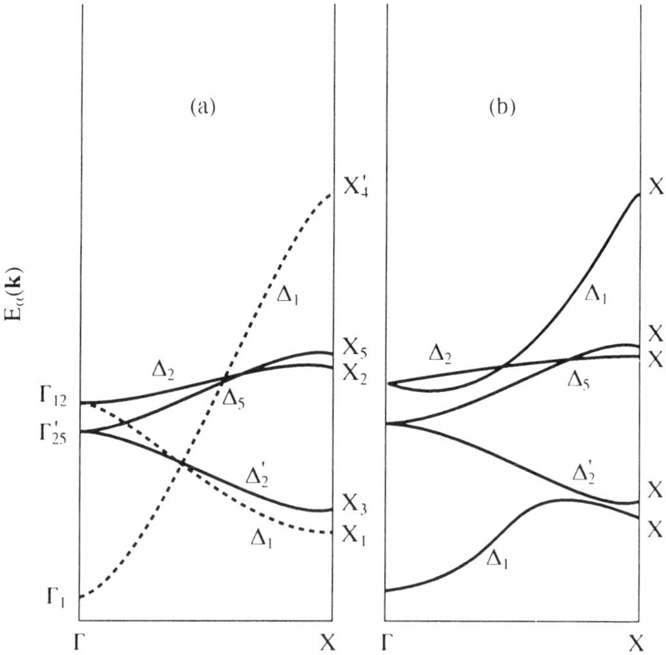

Figure 4.16. Energy bands of copper (schematic). (*a*) when the eigenstates of the electron (4.27) are Bloch sums of atomic orbitals (φ_α) of a single type (4s or 3d); (*b*) as in (*a*) but the atomic orbitals (φ_α) may be hybridizations of s and d orbitals

line ΓX the bands shown in Fig. 4.16a. (Similar results are obtained along the other lines of the BZ). The band represented by the broken line extending from the level Γ_1, to the level X_4' corresponds to eigenstates which are Bloch sums of atomic 4s orbitals: they are described by (4.27) with $\alpha = 4s$, and the corresponding energy levels are given by (4.28). In the present case, however, we expect a relatively large overlap between the 4s orbitals of neighbour atoms, because of the large radii (r_{max}) of these orbitals (see Table 2.6); we note (see also Section 2.15.3) that the atomic spheres (of the above radius) touch, or nearly touch, in the crystal. The considerable overlap between the 4s orbitals of neighbour atoms allows the electron to move relatively easily from atom to atom, and this leads to a relatively large variation of $\Delta E_\alpha(\boldsymbol{k})$ with $\boldsymbol{k}$. One can say that in the present case the electron can be found with considerable probability in the space between the atomic nuclei, and because this probability depends on the value of $\boldsymbol{k}$, the energy of the electron will vary appreciably with $\boldsymbol{k}$ leading to a relatively large bandwidth (a large spread of the 4s-band). The remaining bands in Fig. 4.16a correspond to eigenstates which are Bloch sums of atomic 3d orbitals: they are described by (4.27) where φ_α is an appropriate linear combination of the five atomic d-orbitals ($l = 2$, $m = -2, -1, 0, +1, +2$). We have five different (linearly independent) such combinations: five different φ_α and these lead, via (4.27), to five different d-bands (per spin) of one-electron states in the crystal. The five-fold degeneracy of the atomic d level (a consequence of the spherical symmetry of the atomic potential field) is lifted by the crystalline field and, therefore, at a general point $\boldsymbol{k}$ of the BZ the five electron states (belonging respectively to the five different d-bands) have a different energy. However, along symmetry lines, some degeneracy may remain. Along the ΓX line (see Fig. 4.16a) the dispersion curves belonging to the five d-bands are: Δ_1 (from the level Γ_{12} to the level X_1), Δ_2 (from Γ_{12} to X_2), Δ_2' (from Γ_{25}' to X_3), and the doubly degenerate Δ_5 (from Γ_{25}' to X_5). We note that at $\boldsymbol{k}$ points off the ΓX line the degeneracy of Δ_5 is lifted and the two bands, which coincide along the ΓX line, separate.

Finally, we note that at certain points of the BZ we obtain energy levels which are doubly or triply degenerate. We observe, for example, that at the point Γ (centre of the BZ) we obtain an energy level (Γ_1) which is not degenerate, one level (Γ_{12}) which is doubly degenerate, and one (Γ_{25}') which is triply degenerate. Naturally, when spin is taken into account, these levels become, respectively, two-fold, four-fold and six-fold degenerate.

It is obvious from Fig. 4.16a that the bandwidth (energy spread) of the d-bands is considerably smaller than that of the s-band. The term $\Delta E_\alpha(\boldsymbol{k})$ in (4.28) is, in general, smaller in the d bands, which implies that the movement of a d-electron (an electron in a state of a d-band) between atoms is not as easy as for an s-electron. This is easily understood if we observe (see Table 2.6) that the radius r_{max} of the atomic 3d-orbital of copper is much smaller than that of the 4s-orbital, which means that the overlap of 3d-orbitals centred on neighbour atoms of the crystal is much smaller than the corresponding overlap of 4s-orbitals.

An improved calculation of the one-electron states of copper leads to the results shown in Fig. 4.16b. The improvement comes from a better choice of the φ_α in (4.27). In the calculation of Fig. 4.16a, φ_α is either an 4s-orbital or a 3d-orbital; in the improved calculation φ_α may be a hybridization of the two (a linear combination of 4s and 3d atomic orbitals). The main difference between the two calculations concerns the two Δ_1-bands (broken lines in Fig. 4.16a) which are changed considerably by the opening of a gap between them. It turns out that the states with energy about the minimum of the higher-energy Δ_1-band (from Γ_{12} to X_4' in Fig. 4.16b) and the states about the maximum of the lower-energy Δ_1-band (from Γ_1 to X_1) are described by Bloch sums (4.27) of orbitals φ_α, which are (the φ_α) linear combinations of atomic orbitals of the s-type and the d-type, in approximately the same proportion.

The reader will easily recognise which of the dispersion curves shown in Fig. 4.15 belong to mainly 4s, mainly 3d, and to obviously hybridised bands, by comparison with the dispersion curves along the ΓX line.

At this stage we should mention that writing the one-electron orbitals of a crystal as linear combinations of atomic orbitals, as in (4.27), is not the best way to actually evaluate these orbitals and the corresponding energy levels. An apparently straightforward method of calculating the energy levels and corresponding states of an electron in a crystal, when the potential (4.16) is known (it must be calculated, in principle, self-consistently with the one-electron states; see Section 4.11.1), is described in Appendix B.9; and that is not the only alternative, or the best, method. (Information as to how the energy bands of copper, and those of other metals and semiconductors shown in this book, were calculated, can be found in the cited references.) However, once the energy bands and corresponding one-electron states have been evaluated (by some means or other) one finds that these agree with the picture we have described. The linear combination of atomic orbitals (LCAO) method which underlies the above picture provides us (and this is important) with an insight into the physical origin of the energy bands and corresponding one-electron states of the crystal, in much the same way as it did in relation to molecular orbitals (see Section 3.2).

Finally we note that in the energy-level diagram of Fig. 4.15 the energy bands of copper are given only along symmetry lines which lie within a small section of the BZ (Fig. 4.14). This section, known as the irreducible part of the BZ, covers 1/48 of the volume of the BZ. The 48 'symmetry operations' (rotations about an axis, reflections with respect to a plane, etc.) which keep the centre of a cube at the same point and retain its shape in space (bring the cube upon itself) bring the irreducible part to the other regions of the

BZ covering in this way its entire space. The bands (dispersion curves) along lines of the BZ which are obtained from a given line in the irreducible part of the BZ by the above mentioned symmetry operations are the same with the bands along the given line. For example, the dispersion curves along the line $-k_x$ are the same with those along the line $+k_x$. (The line $-x$ is the mirror reflection of the line $+x$ with respect to the yz plane which is a symmetry plane of the cubic lattice and also (and this is what ultimately matters) of the potential field the electron sees in the crystal ($U(-x, y, z) = U(x, y, z)$).) Similarly, the dispersion curves along the $+k_y$ line are the same with those along the $+k_x$ line, etc. And the argument applies not only for $\boldsymbol{k}$ on symmetry lines but for general points of the BZ as well, so that in order to know the energy bands everywhere in the BZ it is sufficient to know these bands in the irreducible part of the zone. The energy bands for $\boldsymbol{k}$ elsewhere in the BZ are obtained through the symmetry of the crystal. For a complete description of the energy band-structure we must know, of course, $E_\alpha(\boldsymbol{k})$ not only along the symmetry lines but everywhere in the irreducible part of the BZ. This is necessary for an accurate evaluation of the constant-energy surfaces, the density of states, and other quantities, which we shall introduce in the following sections. However, for a qualitative analysis of the electronic properties of a crystal, it is usually sufficient to know the dispersion of $E_\alpha(\boldsymbol{k})$ along the symmetry lines of the BZ, as in our example of Fig. 4.15. Because of the continuity of $E_\alpha(\boldsymbol{k})$ this gives us a fair picture of what the bands look like over the entire BZ.

4.3.2 ENERGY BANDS OF TUNGSTEN

Tungsten (W) has a body-centred cubic lattice with one atom per lattice site. Its reciprocal lattice is face-centred cubic and has the BZ shown in Fig. 4.17.

We know that the atom of tungsten has 74 electrons (see Table 2.7), 68 of which are in inner shells. In the crystal of tungsten these fill up low-energy bands of atomic-like states, as in copper, which do not contribute significantly to the binding of the crystal and to its metallic properties. The remaining six electrons of the atom tungsten are in outer shells 6s and 5d. These atomic states (of the 6s and 5d shells) develop, in the crystal of tungsten, into six bands of one-electron states corresponding to the energy bands shown in Fig. 4.18. We have, in fact, two same sets of these bands, one for spin up and one for spin down (tungsten, like copper, is not a magnetic material)[12]. The similarity of the energy bands of tungsten (shown in Fig. 4.18) with those of copper (shown in Fig. 4.15) is obvious. The reader will recognise bands of the d-type, of the s-type and hybridized ones in tungsten as easily as in the case of copper. The fact that the two metals have different space lattices is not important in this respect. At the absolute zero of temperature ($T = 0$) all eigenstates with energy below $E_{\rm F}$ are

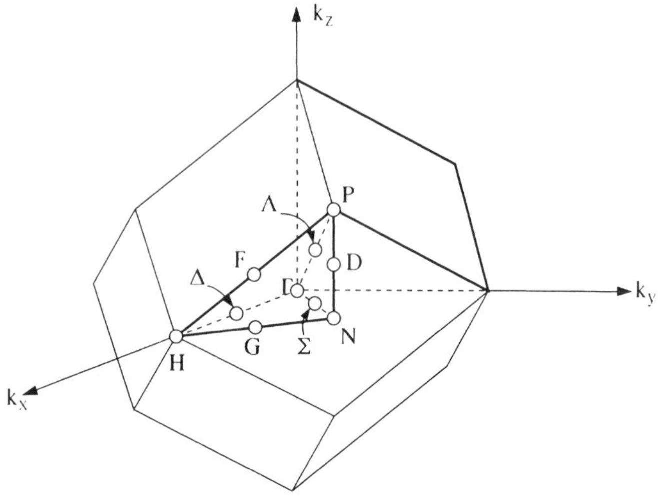

Figure 4.17. The Brillouin zone (BZ) of the body-centred cubic lattice

occupied and those above E_F are empty. Assuming that the crystal consists of N primitive cells, there must be $6N$ eigenstates of the electron with energy between zero and the Fermi level (6 eigenstates per atom of the crystal or, if you wish to say it like this, 3 eigenstates per spin per atom) to accommodate the $6N$ electrons from the outer shells of the tungsten atoms. This, in fact, is how E_F is determined (see Section 4.5), so that the above is true.

4.4 Surfaces of constant energy

From the energy-level diagrams of copper (Fig. 4.15) and tungsten (Fig. 4.18) which show $E_\alpha(\boldsymbol{k})$ along symmetry lines in the irreducible parts of their Brillouin zones, from the fact that in each case the same dispersion curves are obtained along symmetry lines in other regions of the respective zones, and knowing that $E_\alpha(\boldsymbol{k})$ vary continuously with $\boldsymbol{k}$ everywhere in BZ, we deduce that, in the above cases, and generally in crystals, there will be many one-electron eigenstates, corresponding to different values of $\boldsymbol{k}$, with the same energy E. The points $\boldsymbol{k}$ which satisfy the equation

$$E_\alpha(\boldsymbol{k}) = E \tag{4.31}$$

where α can be any band, define a so-called surface of constant energy in the $\boldsymbol{k}$-space of the BZ. This can, in principle, be a simple surface, e.g., a spherical surface, or a complex one consisting of different pieces that may or

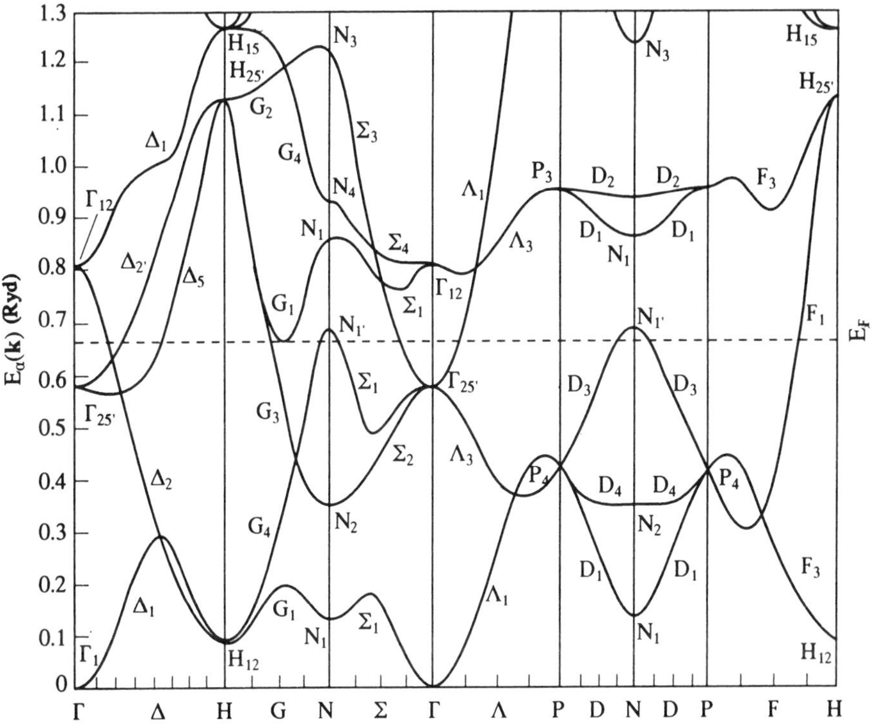

Figure 4.18. Energy bands of tungsten along various symmetry lines of the BZ (Fig. 4.17). The zero of the energy coincides with the minimum of the conduction band. $E_F = 0.675$ Ryd; 1 Rydberg (Ryd) = 0.5 Hartree = 13.6 eV. [Calculation by L. F. Mattheiss, Physical Review, 139 (1965) A 1893]

may not cross each other. The simplest form is obtained when $E(\boldsymbol{k})$ is given, as in the case of free electrons, by

$$E(\boldsymbol{k}) = \frac{\hbar^2 k^2}{2m^*} = \frac{\hbar^2}{2m^*}(k_x^2 + k_y^2 + k_z^2) \tag{4.32}$$

where m^* is some effective mass. The constant-energy surface of (4.32) corresponding to energy E is that of a sphere, centred at the origin of the coordinates ($\boldsymbol{k} = 0$), with radius equal to $(2m^*E)^{1/2}/\hbar$.

The calculation of the constant-energy surfaces of a real crystal is an elaborate one. We need to know the energy levels of the electron at many $\boldsymbol{k}$-points, not only along symmetry lines but at a sufficient number of other points as well (it is of course sufficient to consider points in the irreducible part of the BZ; the energy levels in the rest of the zone are obtained through symmetry; see Section 4.3.1). We can then determine numerically the points which satisfy (4.31) for any given value of the energy: the union of these points constitutes the constant-energy surface corresponding to this energy. We usually need to do the calculation for different values of the energy: $E = E_O$, $E_O + \Delta E$, $E_O + 2\Delta E$ and so on, covering the region of energy that is of interest to us. For example, in the case of tungsten (Fig. 4.18) we would like to know the constant-energy surfaces in the region from zero to about 1.2 Ryd at intervals of $\Delta E \simeq 0.01$ Ryd or so. We note that the shape of the constant-energy surface changes with the energy. It may be a simple one near the minimum of a band and a very complex one at higher energies. Of particular importance is the so-called Fermi surface of a metal: the constant-energy surface $E = E_F$. This surface can be determined experimentally and, therefore, we can check the theoretical calculation of this quantity and, indirectly, the calculated energy bands which lead to it.

The Fermi surface of tungsten as calculated by Mattheiss (loc. cit.) and other authors is shown in Fig. 4.19. Its correctness has been confirmed experimentally by a number of workers. We recognise a central piece which cuts into six spherical pieces, with centres on the $\pm k_x$, $\pm k_y$, $\pm k_z$ axes, which

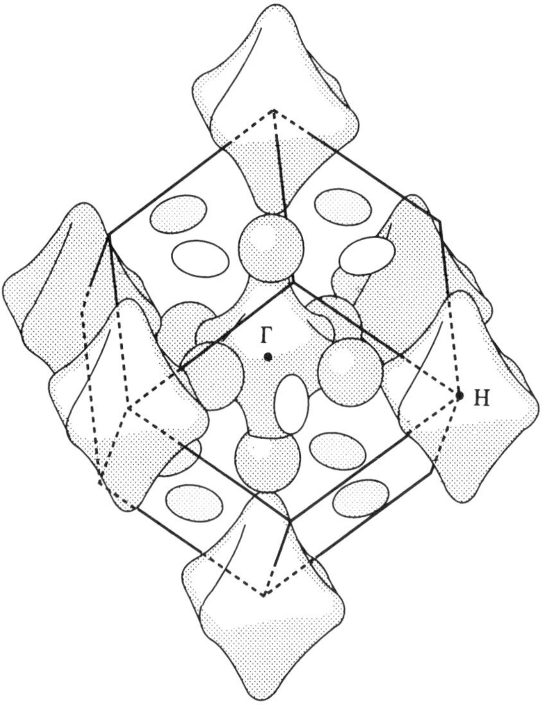

Figure 4.19. The Fermi surface of tungsten

are joined to the octahedra at the H points of the BZ, and finally we have ellipsoids centred on the N points and symmetric about the faces of the BZ. (The parts of the octahedra which lie outside the BZ are shown only for the sake of clarity.) We note that the Fermi surface (the same applies of course to every constant-energy surface) has the symmetry, cubic in the case of tungsten, of the crystal. The reader should also note the consistency of the above picture of the Fermi surface with the energy bands shown in Fig. 4.18.

We see that the Fermi surface of a metal can be a very complex surface indeed, and we are all the more satisfied with the fact that the theory is able to reproduce with very good accuracy this quantity.

4.5 Density of one-electron states in a crystal

We denote the density of the eigenstates of an electron, in a crystal, of spin up (down) by $\rho_\uparrow(E)$ ($\rho_\downarrow(E)$). For non-magnetic crystals we have: $\rho_\downarrow(E) = \rho_\uparrow(E) \equiv \rho(E)$. By definition, $\rho(E)\,\mathrm{d}E$ is the number of states, per spin, per unit volume of the crystal, with energy between E and $E + \mathrm{d}E$. We calculate $\rho(E)$ as follows. We draw the surface of constant energy E and that of constant energy $E + \Delta E$, and evaluate the $\boldsymbol{k}$-volume $\Delta V_k(E)$ between the two surfaces. We note that the constant-energy surfaces may consist of different pieces (as in Fig. 4.19) in which case $\Delta V_k(E)$ will consist of contributions from all the pieces. The number $\Delta N(E)$ of eigenstates of an electron in a crystal of volume V, per spin, with energy between E and $E + \Delta E$ is given by the ratio of the volume $\Delta V_k(E)$ to the volume $(2\pi)^3/V$ which gives, according to (4.24), the $\boldsymbol{k}$-volume per eigenstate (of given spin). We obtain

$$\Delta N(E) = \frac{V \Delta V_k(E)}{(2\pi)^3} = \frac{N V_o \Delta V_k(E)}{(2\pi)^3} \tag{4.33}$$

where V_o and N denote, respectively, the volume of the primitive cell and the number of primitive cells in the given crystal. By definition

$$\rho(E) = \frac{1}{V}\frac{\Delta N(E)}{\Delta E} \tag{4.34}$$

Therefore,

$$\rho(E) = \frac{1}{(2\pi)^3}\frac{\Delta V_k(E)}{\Delta E} \tag{4.35}$$

Multiplying the above with the volume per atom of the crystal (it equals V_o when there is only one atom in the primitive cell) we obtain the density of states, per spin, per atom; it is given by $V_o\rho(E)$. The density of states per atom is given by $2V_o\rho(E)$.

When the energy of the electron is given by (4.32) the $\boldsymbol{k}$-volume between the constant-energy surfaces corresponding to E and $E + \Delta E$ is given by

$$\Delta V_k(E) = 4\pi k^2 \Delta k \tag{4.36}$$

where Δk is obtained from the following relation

$$\Delta E = \frac{\hbar^2}{2m^*}[(k + \Delta k)^2 - k^2] \simeq \frac{\hbar^2}{m^*} k \Delta k \tag{4.37}$$

Substituting (4.36) and (4.37) in (4.35) and using (4.32) we obtain

$$\rho(E) = \frac{4\pi}{h^3}(m^*)^{3/2}\sqrt{2E} \tag{4.38}$$

which agrees with the density of states, per spin, of a free electron ($m^* = m$) given by (1.32).

The density of states of an electron in a real crystal, which one evaluates numerically in the manner we have described, is usually a more complicated function. In Fig. 4.20 we show, as an example, the density of one-electron

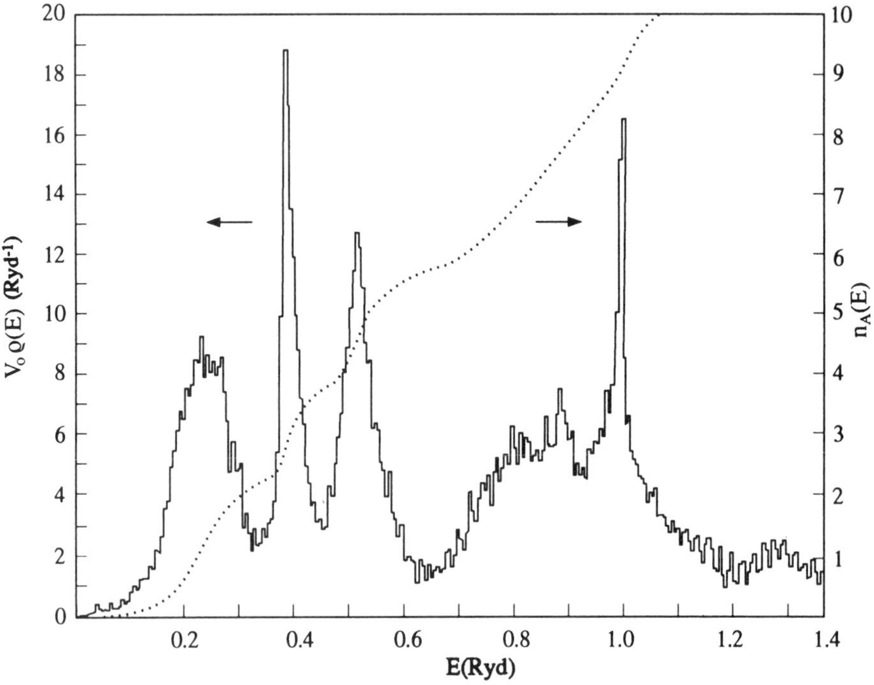

Figure 4.20. Density of one-electron states in tungsten (solid line). The dotted line gives the integrated density of states. [Mattheiss, loc. cit.]

states in tungsten, in the energy region of the bands shown in Fig. 4.18. We expect that $\rho(E)$ will be larger at those energies where $E_\alpha(\boldsymbol{k})$ change relatively little with $\boldsymbol{k}$, which means a larger $\boldsymbol{k}$-volume between the constant-energy surfaces of E and $E + \Delta E$. We expect, moreover, that these regions will coincide, more or less, with regions of energy where the dispersion of $E_\alpha(\boldsymbol{k})$ along the symmetry lines of the BZ has, at least for some of the energy bands, a very small slope ($\Delta E_\alpha/\Delta k \simeq 0$). One can see that, as far as tungsten is concerned, the maxima of $\rho(E)$ of Fig. 4.20 correspond, indeed, to regions of small slope of the dispersion curves of Fig. 4.18.

The dotted line in Fig. 4.20 shows the integrated density of one-lectron states, i.e. the number of eigenstates of the electron (per atom) in the bands of Fig. 4.18 with energy not grater than E. It is given by

$$n_A(E) \equiv 2V_o \int_0^E \rho(E')\,\mathrm{d}E' \qquad (4.39)$$

We note that, by construction, $n_A(E_\mathrm{F})$ gives us the number of occupied one-electron states, in the bands of Fig. 4.18, per atom of the crystal, when $T = 0$. In the case of tungsten we have

$$n_A(E_\mathrm{F}) = 6 \qquad (4.40)$$

since there are six electrons per atom accommodated in these bands (see Section 4.3.2). Equation (4.40) is, therefore, the equation by which E_F is determined. The graph (dotted line) of Fig. 4.20 tells us that the above equation is satisfied when $E_\mathrm{F} = 0.675$ Ryd. We now know how the Fermi level of tungsten, noted in the energy-level diagram of Fig. 4.18, has been determined. The Fermi level of copper, noted in Fig. 4.15, was determined in the same way, and in the same way one determines the Fermi level of every metal.

4.6 Metals, insulators, and semiconductors

The fact that the Fermi level cuts through the energy bands of copper (Fig. 4.15) and tungsten (Fig. 4.18) is significant. The same is true for every metal. It is a condition (as we shall see) for metallic behaviour (e.g., good conduction of electricity at any temperature including $T = 0$) that there must be, at any temperature, non-occupied one-electron states with energy in the immediate vicinity to that of occupied ones; and this can be true at $T = 0$ only when E_F cuts through one or more energy bands of the crystal. We refer to these energy bands as the conduction bands or, collectively, as the conduction band of the metal.

We shall describe the relation between the electronic structure of the crystal and macroscopic properties of the crystal, such as the specific heat and the conductivity of the crystal, in the following sections. At this stage we would like to present a few examples of energy bands of non-metals.

In Fig. 4.21 we present the valence and conduction bands (the terms are explained below) of germanium (Ge) for $\boldsymbol{k}$ along various symmetry lines of its BZ, which is that of an fcc lattice (Fig. 4.14). We remember, in relation to Fig. 4.14, that the line UX is equivalent to a continuation of the line ΓK beyond the Brillouin zone (see problem 4.4).

The unit cell of germanium is shown in Fig. 4.10; we remember that there are two atoms in its primitive cell (per lattice site). Out of the 32 electrons per atom of Ge, the 28 from the inner shells of the atom (see Table 2.5) are accommodated in corresponding bands, of atomic-like states, of lower energy (not shown in Fig. 4.21); these core states can be described in essentially the same way as the corresponding states of copper (see Section 4.3.1); they play no significant role in relation to the properties of crystalline Ge and we need not say anything more about them. The remaining four electrons per atom (from its outer 4s and 4p shells) are to be accommodated in the bands shown in Fig. 4.21.

The states corresponding to the energy bands shown in Fig. 4.21 are described by Bloch sums (4.27), but in the present case $\varphi_\alpha(\boldsymbol{r} - \boldsymbol{R}_n)$ itself is a combination of two hybridised (a mixture of 4s and 4p) atomic orbitals centred respectively on the two atoms of the nth primitive cell. We have one 4s and three 4p orbitals for each atom and we therefore obtain eight different (linearly independent) φ_α. Four of them, the so-called bonding orbitals, give rise to the four so-called valence bands of Ge, extending from about[13] -12 eV to 0 in Fig. 4.21. (We note that the energy levels Δ_5 (along the ΓX line) and the energy levels Λ_3 (along the ΓΛ line) are two-fold degenerate.) The remaining four φ_α (the so-called antibonding orbitals) give rise to the four so-called conduction bands of germanium, extending above the energy gap in Fig. 4.21. There is, of course, the spin-degeneracy which we have not taken into account in the above analysis. We obtain the same energy bands (and corresponding orbital wavefunctions) for the two orientations (up and down) of the spin.

We note that there are no eigenstates of the electron in the energy region from zero to about 0.7 eV. This is true not only for $\boldsymbol{k}$-points along the lines of the BZ represented in Fig. 4.21 but everywhere in the BZ. The region of energy between the minimum of the conduction band and the maximum of the valence band (we refer to the above mentioned valence bands and conduction bands, collectively, as the valence band and the conduction band respectively), where no electron states exist, is known as the energy gap.

We know that every band consists of N states, which correspond to the N points of the BZ defined by (4.24), where N denotes, as usual, the number of primitive cells in the given crystal. This means that there are $4N$ states of spin-up and $4N$ states of spin-down in the valence bands of germanium.

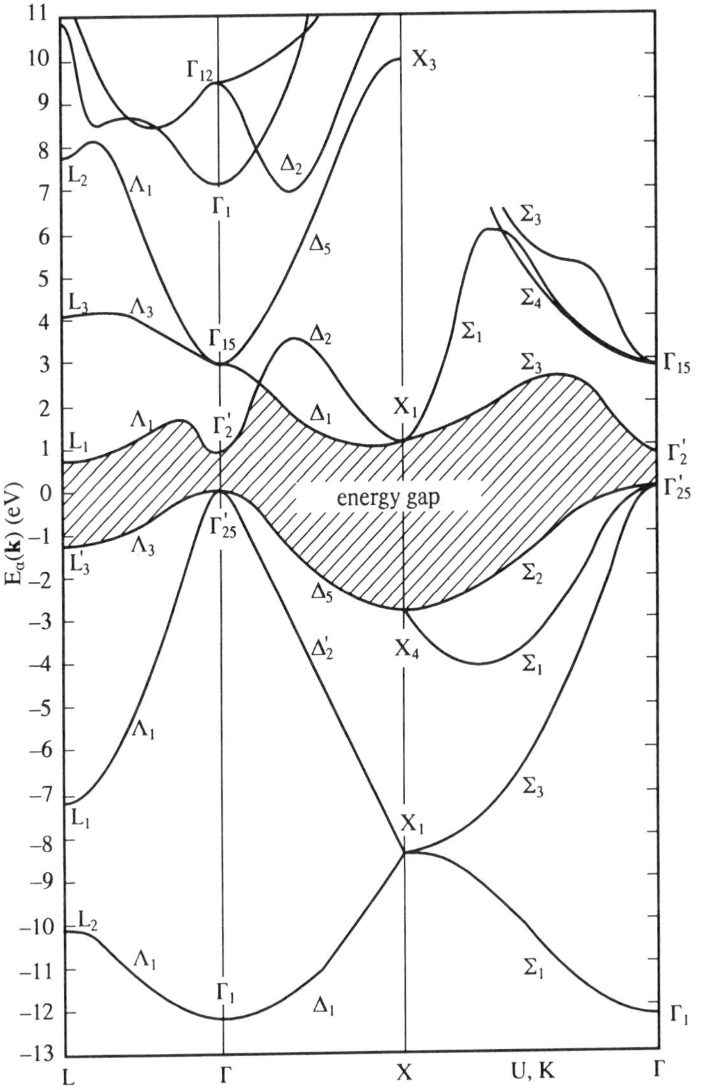

Figure 4.21. Energy bands of germanium. The zero of the energy coincides with the top of the valence band. (Calculation by F. Herman, R. L. Kortum and C. D. Kuglin, *Int. J. Quantum Chem.*, *1S* (1967) 533. Reprinted by permission of John Wiley & Sons Inc.)

Taken together they can accommodate $8N$ electrons, which is exactly the number of electrons that need to be accommodated (four electrons per atom times two atoms per primitive cell times N primitive cells in the crystal). Conclusion: at the absolute zero of temperature the valence bands of germanium are fully occupied and the conduction bands are entirely empty. Between the highest occupied energy level and the next empty level lies a gap of about 0.7 eV. In contrast to the situation in metals there are no empty electron states with energy in the immediate vicinity of that of occupied states. The Fermi level lies, in this case, somewhere near the middle of the energy gap (see Section 4.7).

Germanium is not the only crystal where an energy gap exists between occupied and empty electron states at $T = 0$. Figure 4.22 shows the valence and conduction bands, along typical symmetry lines of their Brillouin zones, of four different crystals: silicon (Si), gallium arsenide (GaAs), zinc sulphide (ZnS) and potassium chloride (KCl). In each case the zero of energy coincides with the top (maximum energy level) of the valence bands. At $T = 0$ the valence bands are fully occupied and the conduction bands entirely empty. We note that the energy gap, to be denoted by E_g, between the bottom (lowest energy level) of the conduction bands and the top of the valence bands differs from crystal to crystal. We have

	Ge	Si	GaAs	ZnS	KCl
E_g(eV)	0.67	1.11	1.43	3.58	8.2

We know that interatomic distances in the crystal change with temperature and, therefore, the energy bands will also change to some small degree with the temperature. For example, the energy gap of germanium changes from 0.744 eV near $T = 0$, to 0.67 eV at room temperature. The values listed above are obtained at room temperature.

The magnitude of the energy gap plays an important role in the determination of the thermal, electrical and optical properties of the crystal. Crystals with a relatively small energy gap are conventionally known as semiconductors and those with larger energy gaps as insulators. Ge, Si, and GaAs are, by this measure, semiconductors and KCl is an insulator. ZnS lies somewhere between between the two; whether we call it an insulator or a semiconductor is unimportant.

4.7 Specific heat

The specific heat is an important property of any material. A theoretical model of the structure of a material must account for the variation of its (internal) energy with the temperature. Let us, then, consider a crystal of volume V in thermodynamic equilibrium with its environment (this may be a

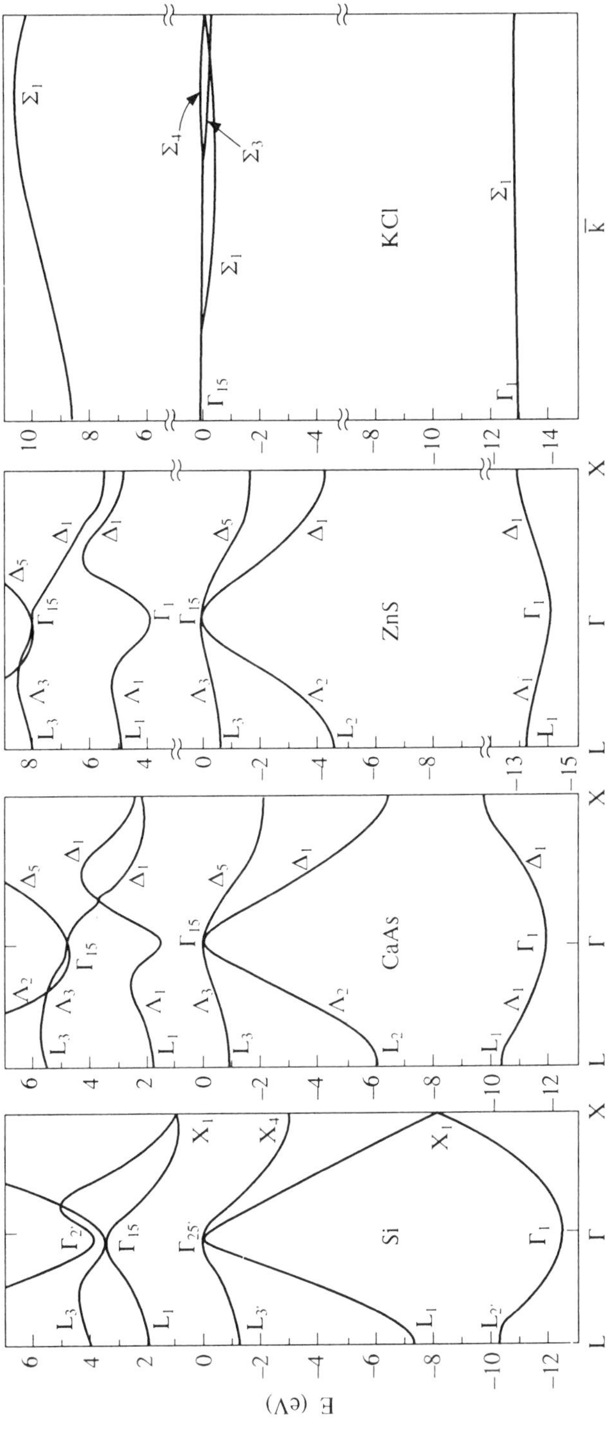

Figure 4.22. Energy bands of Si, GaAs, ZnS and KCl

gas of temperature T and pressure P which we can vary at will). At equilibrium the temperature of the crystal is the same as that of its environment, so that an increase ΔT of the latter leads to the same increase in the temperature of the former. We shall assume that this happens with the volume of the crystal staying the same (in practice we achieve this by changing the external pressure P by an appropriate amount). Now let $\Delta\varepsilon$ be the change in the energy of the crystal consequent to the increase ΔT of its temperature, under constant volume. The specific heat (under constant volume) of a material is defined by

$$C_V = \frac{\Delta\varepsilon}{\Delta T} \tag{4.41}$$

and, as a rule, it can be measured experimentally with great accuracy.

Let us now see what the theory we have described tells us about the specific heat of a metal. The probability $f(E)$ that an electron-eigenstate of energy E will be occupied at temperature T is given by the Fermi–Dirac distribution (2.79). One can show that the chemical potential μ of a metal is, at not so high temperatures ($T \leqslant 300\text{K}$), equal, to a very good approximation, with E_F (the Fermi level, evaluated at $T = 0$ in the manner we have described in Section 4.5). We obtain

$$f(E) = \frac{1}{\exp[(E - E_F)/k_B T] + 1} \tag{4.42}$$

In Fig. 4.23 we show the density of states per spin, $\rho(E)$, of a hypothetical metal, and the probability $f(E)$ that an eigenstate of energy E will be occupied when $T = 0$ and when $T > 0$. The zero of the energy in Fig. 4.23 coincides with the bottom of the conduction band[14]; the lower-energy bands (not shown) remain fully occupied and therefore do not affect the specific heat of the metal.

When $T = 0$, we have $f(E) = 1$ for $E < E_F$ and $f(E) = 0$ for $E > E_F$; and we note that the $f(E)$ for $T > 0$ differs from the $f(E)$ for $T = 0$ only over a small region of energy (its width about $k_B T$) about E_F. (This region has been exaggerated, for the sake of clarity, in Fig. 4.23. We note that when $T \simeq 300\text{K}$, $k_B T \simeq 1/40\,\text{eV}$ whereas, for a typical metal, $E_F \simeq 10\,\text{eV}$ when measured from the bottom of the conduction band.) We can say that the distribution of the electrons at the temperature $T > 0$ is obtained from that at $T = 0$ as follows: a considerable fraction of the electrons in eigenstates with energy between $E_F - k_B T$ (approximately) and E_F move up into eigenstates with energy between E_F and $E_F + k_B T$ (approximately). (The required energy passes into the crystal from the environment when the vibrating nuclei at the surface of the crystal collide with atoms of the environment; in turn the vibrating nuclei colliding with the electrons of the crystal pass part of this energy to the latter.) The number of electrons which

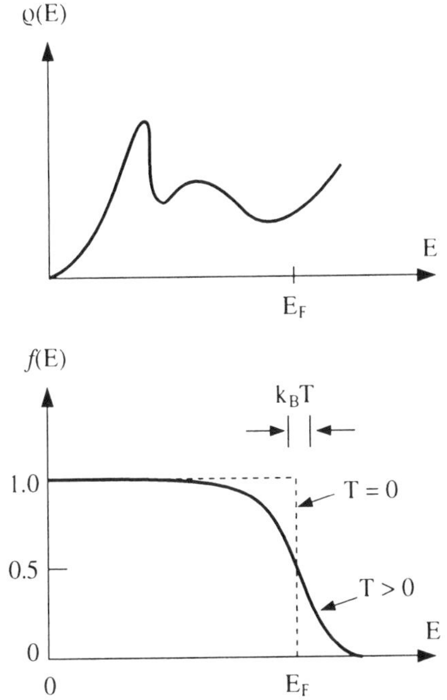

Figure 4.23. Schematic description of the density of one-electron eigenstates per spin, $\rho(E)$, of a metal; and the probability $f(E)$ that an eigenstate of energy E will be occupied when $T = 0$, and when $T > 0$. The change of $f(E)$ with temperature has been exaggerated; in reality $k_B T$ is about a thousand times smaller than E_F

gain energy in this way, when $k_B T \ll E_F$, is given approximately, by $2\rho(E_F)Vk_B T$, where $\rho(E_F)$ is the density of one-electron states per spin at the Fermi level. If we assume that every one of them receives, on the average, an amount of energy equal to $k_B T$, we conclude that the energy of the electrons of the metal, when the temperature of the latter is raised from 0 to T, is given by

$$\varepsilon_{el}(T) \simeq \varepsilon_{el}(0) + 2\rho(E_F)V(k_B T)^2 \tag{4.43}$$

from which we obtain

$$\Delta\varepsilon_{el} = \varepsilon_{el}(T + \Delta T) - \varepsilon_{el}(T) \simeq 4\rho(E_F)Vk_B^2 T\Delta T \tag{4.44}$$

where ΔT is assumed very small. Putting $\Delta\varepsilon = \Delta\varepsilon_{el}$ in (4.41) we find that the electronic contribution to the specific heat of the metal, at sufficiently low temperatures, is given by

$$C_{V,el} \simeq 4\rho(E_F)Vk_B^2 T \equiv c_1 T \tag{4.45}$$

which tells us that, at sufficiently low temperatures, the electronic specific heat (at constant volume) of a metal varies linearly with the temperature, and that the proportionality constant is determined by the density of the one-electron states at the Fermi level. This result, confirmed by experiment, has been one of the first major achievements of the quantum theory of solids, especially since a classical treatment of this quantity (based on the assumption that the conduction electrons of a metal behave like a Boltzmann gas of non-interacting particles) leads to a specific heat independent of temperature (see problem 2.14).

We know that the energy ε of a crystal is not only electronic. One must take into account the vibrational energy of the nuclei and its contribution to the specific heat of the metal. It turns out (see Section 4.12) that the latter, denoted by $C_{V,vb}$ varies in proportion to the third power of the temperature, when $T \to 0$. Therefore, at sufficiently low temperatures, the specific heat of the metal, at constant volume, is given by

$$C_V = C_{V,el} + C_{V,vb} = c_1 T + c_2 T^3 \tag{4.46a}$$

which can be rewritten as follows

$$C_V/T = c_1 + c_2 T^2 \tag{4.46b}$$

The constants c_1 and c_2 which characterise the electronic and vibrational contributions to the specific heat of the metal, respectively, are determined experimentally from a graphical representation of (4.46b) as shown in Fig. 4.24.

Let us now see what happens in a semiconductor. When $T = 0$, all eigenstates of the electron in the valence band of the semiconductor ($E \leqslant E_v$) are occupied ($f(E) = 1$ for $E \leqslant E_v$) and all the eigenstates in the conduction band ($E \geqslant E_c$) are empty ($f(E) = 0$ for $E \geqslant E_c$), as shown in Fig. 4.25.

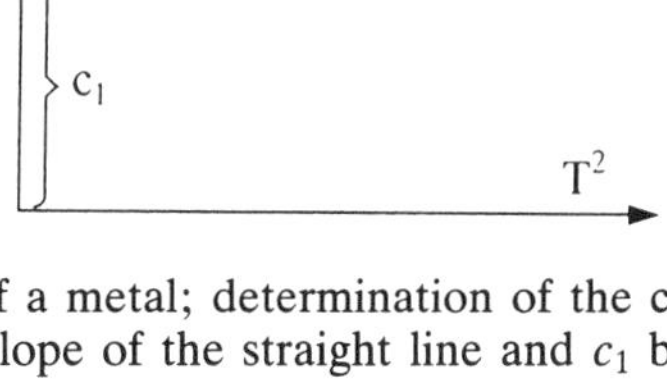

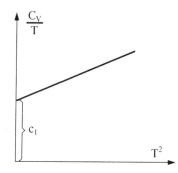

Figure 4.24. Specific heat of a metal; determination of the coefficients c_1 and c_2 of (4.46b). c_2 is given by the slope of the straight line and c_1 by its intercept with the C_V/T axis

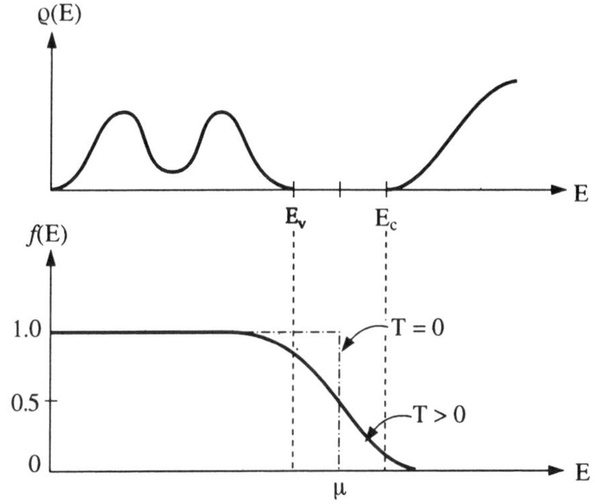

Figure 4.25. Schematic description of the density of one-electron eigenstates per spin, $\rho(E)$, of a semiconductor; and the probability $f(E)$ that an eigenstate of energy E will be occupied when $T = 0$, and when $T > 0$

When $T > 0$ a number (n) of electrons per unit volume move up from the valence band (from energy levels in the vicinity of the top of this band) into the conduction band (to levels in the vicinity of the bottom of this band), leaving a number (p) of empty states (so-called holes) per unit volume in the valence band. Obviously,

$$n = p \tag{4.47}$$

Moreover, one can see that (4.47) will be satisfied only if the chemical potential μ of the Fermi–Dirac distribution $f(E)$ lies somewhere near the middle of the energy gap (see Fig. 4.25). We write, approximately,

$$n \simeq 2\bar{\rho}(E_c)k_B T f(E_c) \tag{4.48}$$

where $\bar{\rho}(E_c)$ is an average value of the density of states per spin in the region of the conduction band minimum, and

$$f(E_c) = \frac{1}{1 + \exp[(E_c - \mu)/k_B T]} \simeq \exp[-(E_c - \mu)/k_B T] \tag{4.49}$$

given that $E_c - \mu \simeq E_g/2 \gg k_B T$. Substituting (4.49) into (4.48) we obtain

$$n \simeq 2\bar{\rho}(E_c)k_B T \exp[-(E_c - \mu)/k_B T] \tag{4.50}$$

On the other hand we have

$$p \simeq 2\bar{\rho}(E_v)k_B T[1 - f(E_v)] \tag{4.51}$$

where $\bar{\rho}(E_v)$ is an average value of the density of states per spin in the region of the valence band maximum, and

$$1 - f(E_v) = 1 - \frac{1}{1 + \exp[(E_v - \mu)/k_B T]} \simeq \exp[-(\mu - E_v)/k_B T]$$

given that $(\mu - E_v) \simeq E_g/2 \gg k_B T$. Substituting the above equation into (4.51) we obtain

$$p \simeq 2\bar{\rho}(E_v) k_B T \exp[-(\mu - E_v)/k_B T] \tag{4.52}$$

Substituting (4.50) and (4.52) in (4.47), and assuming that $\bar{\rho}(E_v) \simeq \bar{\rho}(E_c)$, we obtain

$$\mu \simeq (E_c + E_v)/2 \simeq E_v + E_g/2 \tag{4.53}$$

which confirms our initial assumption as to the position of the chemical potential. Finally, substituting (4.53) into (4.50) we obtain[15]

$$n = p \simeq 2\bar{\rho}(E_c) k_B T \exp[-E_g/(2k_B T)] \tag{4.54}$$

Clearly, the electronic energy of the crystal at temperature T differs from that at $T = 0$ in proportion to the number (n) of electrons transferred from the valence band to the conduction band. Since this varies, essentially, as $\exp(-E_g/2k_B T)$ with the temperature, so does the electronic contribution to the specific heat. We have

$$C_{V,el} \propto \exp[-E_g/(2k_B T)] \tag{4.55}$$

The rapid decrease to zero of the specific heat as $T \rightarrow 0$, implied by (4.55), clearly distinguishes a semiconductor from a metal, whose electronic specific heat goes to zero proportionally to the temperature ((4.45)).

We referred to semiconductors but the same formulae apply to insulators; however, the large values of E_g met in insulators lead to vanishingly small values of the electronic specific heat at ordinary temperatures.

4.8 Electron transport—conductivity of metals

The Bloch wavefunctions (4.17) which describe the eigenstates of an electron in an infinite crystal extend, like the plane waves of free particles, over all space. We have seen in Sections 1.7.2 and 1.8 that, for certain purposes, a free particle is better described by a wavepacket (a superposition of plane waves of approximately the same momentum and energy) which does not extend that much about its moving centre. We can construct a wavepacket of Bloch waves in the same way we have constructed a wavepacket of plane waves. We find that

$$\Phi_{k\alpha}(\boldsymbol{r}, t) = \sum_{\boldsymbol{k}'} A(\boldsymbol{k}') \psi_{\boldsymbol{k}'\alpha}(\boldsymbol{r}, t) \tag{4.56}$$

where $\psi_{k\alpha}$ are Bloch waves (4.17) and

$$A(\boldsymbol{k}') = \text{const}\exp[-\tfrac{1}{2}b^2(\boldsymbol{k}' - \boldsymbol{k})^2], \tag{4.57}$$

has a corresponding probability-density $W_{k\alpha}(\boldsymbol{r}, t) = |\Phi_{k\alpha}(\boldsymbol{r}, t)|^2$ which decays rapidly to zero outside a limited region (a few units cells, provided the parameter b in (4.57) is sufficiently small) about its moving centre at $\boldsymbol{v}(\boldsymbol{k})t$, as shown schematically in Fig. 4.26. The component of the transport velocity $\boldsymbol{v}(\boldsymbol{k})$ parallel to a given axis is determined by the slope of the dispersion curve $E_\alpha(\boldsymbol{k})$, at the point $\boldsymbol{k}$, along this axis. For example, the x-component of the transport velocity is given by

$$v_x(\boldsymbol{k}) = \frac{1}{\hbar}\frac{\partial E_\alpha}{\partial k_x} \equiv \frac{1}{\hbar}\frac{E_\alpha(k_x + \mathrm{d}k_x, k_y, k_z) - E_\alpha(k_x, k_y, k_z)}{\mathrm{d}k_x} \tag{4.58}$$

The y and z components of $\boldsymbol{v}(\boldsymbol{k})$ are calculated in similar manner.

The mean energy of an electron described by (4.56) equals $E_\alpha(\boldsymbol{k})$ and the spread about this mean is relatively small, provided the parameter b in (4.57) is sufficiently large. The choice of b must be balanced between this requirement and that for space confinement which demands that b must be as small as possible. Assuming that this is possible (and it is for many purposes), one can think of the electron described by $\Phi_{k\alpha}$ of (4.56) as a nearly classical particle of energy $E_\alpha(\boldsymbol{k})$ moving in an infinite crystal with velocity $\boldsymbol{v}(\boldsymbol{k})$ given by (4.58). It is reasonable to assume that, at equilibrium, the probability of the state $\Phi_{k\alpha}$ being occupied is the same with the

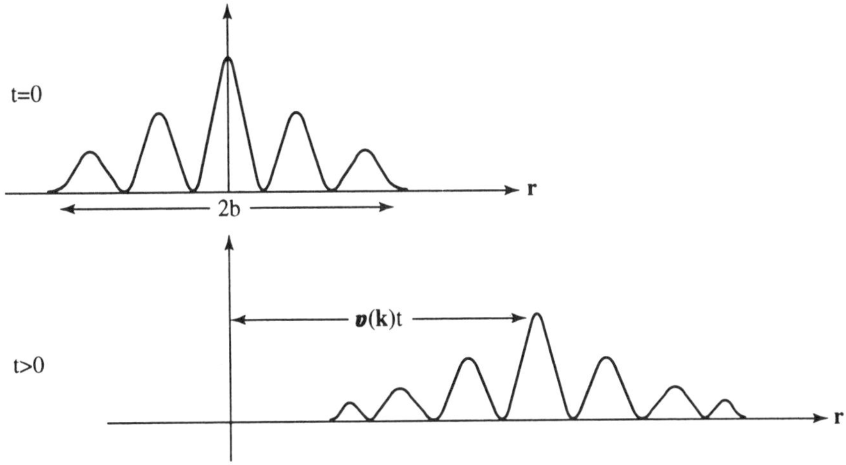

Figure 4.26. Probability-density of a Bloch-wavepacket (4.56) along the direction of propagation at two different moments ($t = 0$, and $t > 0$). The probability-density is, of course, confined in every direction

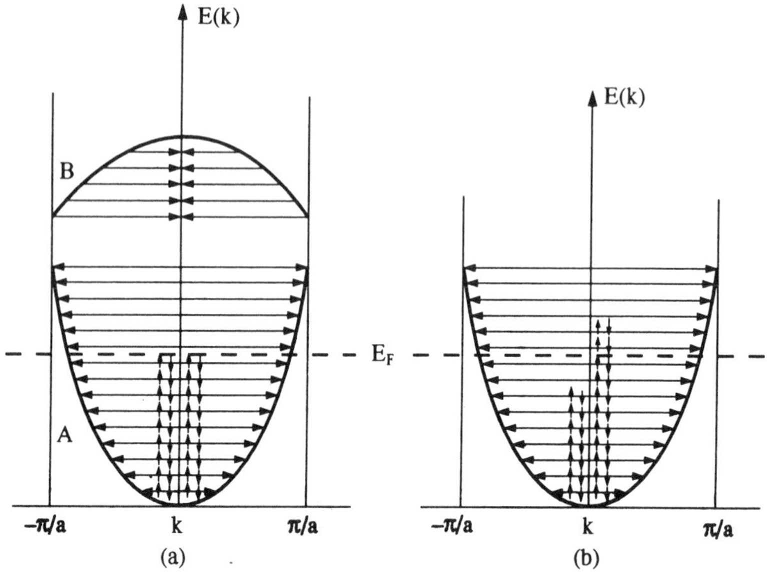

Figure 4.27. (*a*) For every electron travelling to the 'right' there is one travelling to the 'left' so that the total current equals zero (equilibrium situation described by a Fermi-Dirac distribution (4.42)). (*b*) More electrons travel to the 'right' than to the 'left' and, therefore, the total current is different from zero. The velocity of the electron is defined by the horizontal arrows. The vertical arrow ↑ (↓) signifies the existence of an electron with spin up (down) at the given energy level. In drawing the above figure we have assumed that $T = 0$, but our arguments apply for $T > 0$ as well

probability of the eigenstate $\psi_{k\alpha}$ being occupied and to evaluate this probability from the Fermi–Dirac distribution $f(E)$ putting $E = E_\alpha(\boldsymbol{k})$. Now, because $f(E)$ depends only on the energy of the electron (see (4.42)), when the state $\Phi_{\boldsymbol{k}\alpha}$ is occupied, the state $\Phi_{-\boldsymbol{k}\alpha}$ (which always exists for symmetry reasons and has the same energy $E_\alpha(-\boldsymbol{k}) = E_\alpha(\boldsymbol{k})$) is occupied too. And one can easily see that the velocity of the electron in the state $\Phi_{-\boldsymbol{k}\alpha}$ is opposite to that of the electron in the state $\Phi_{\boldsymbol{k}\alpha}$, so that the net flow of electrons (the total current) vanishes at equilibrium. The situation is described schematically in Fig. 4.27(a). Band A represents the conduction band, along a direction in $\boldsymbol{k}$-space, of a hypothetical metal; we assume that each energy level (denoted by a horizontal line) is doubly degenerate (because of the spin of the electron). The presence of an electron of spin up (down) at a given level is indicated by a vertical arrow ↑(↓). An electron occupying a state of A with $k > 0$ travels to the right ($\partial E_\mathrm{A}/\partial k > 0$ when $k > 0$) and an electron occupying a state with $k < 0$ travels to the left ($\partial E_\mathrm{A}/\partial k < 0$ when $k < 0$). (We remember (see (4.58)) that the direction of the velocity of the electron (indicated by horizontal arrows in Fig. 4.27) is determined by the slope $\partial E/\partial k$ of the dispersion curve and not according to whether k is positive or negative. An electron in band B, for example, with $k > 0$ travels to the left.) Fig. 4.27(a) tells us that, at equilibrium, for every electron travelling in one direction there is another travelling with the same speed in the opposite direction so that the total (macroscopic) current equals zero. In the same way we can argue that in a metallic wire forming a closed circuit (Fig. 4.28) a flow of electrons in one direction is balanced, at equilibrium, by a flow of electrons in the opposite direction.

The application of an electric field parallel to the axis of the wire produces, as we know, a net electric current given by (Ohm's law):

$$\boldsymbol{j} = \sigma \boldsymbol{F} \tag{4.59a}$$

$$i = jA \tag{4.59b}$$

where $\boldsymbol{j}$ is the current density (charge passing per unit time through a unit area normal to the direction of $\boldsymbol{j}$), $\boldsymbol{F}$ is the applied electric field, and σ is a constant, known as the conductivity of the metal, which depends on the properties of the metal and the temperature. With $\boldsymbol{F}$ along the axis of the wire and $\boldsymbol{j}$ having the same direction as $\boldsymbol{F}$, the current i (charge passing through a cross-section of the wire per unit time) is given by (4.59b), where A is the area of the above cross-section. In this section our aim will be to understand the conductivity of metals in relation to their electronic structure.

It is obvious that the application of an electric field destroys the balance between 'left' and 'right' currents depicted in Fig. 4.27(a). The equilibrium distribution (Fermi–Dirac) of this figure is replaced by a distribution more like that of Fig. 4.27(b). The probability that an electron will occupy the eigenstate $\psi_{\boldsymbol{k}\alpha}$, and by extension $\Phi_{\boldsymbol{k}\alpha}$, depends, now, not only on the energy $E_\alpha(\boldsymbol{k})$ of the electron but also on its wavevector $\boldsymbol{k}$ and therefore, indirectly, on its transport velocity $\boldsymbol{v}(\boldsymbol{k})$. Our task will be to relate this non-equilibrium

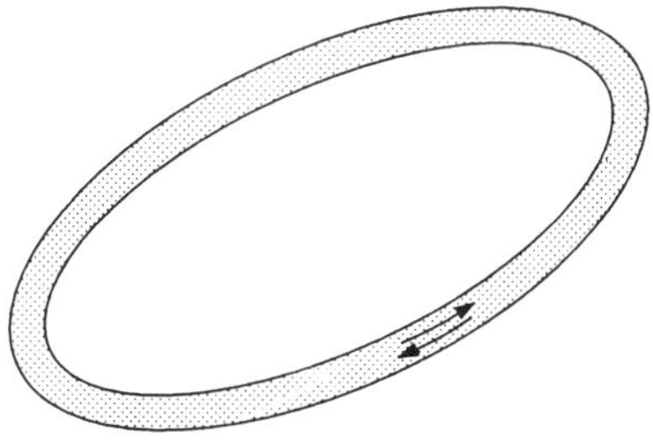

Figure 4.28. In a closed circuit (as in a wire) at equilibrium, the flow of electrons in one direction is balanced by a flow of electrons in the opposite direction

distribution (and the current density that results from it) to the applied electric field.

It can be shown that the electric field causes an electron to change its wavevector as follows (see problem 4.10)

$$\delta \boldsymbol{k} = (\delta k_x, 0, 0)$$
$$\delta k_x = -(eF/\hbar)\delta t \tag{4.60}$$

where we have assumed (without loss of generality) that the electric field is applied along the x-direction)[16]. We read (4.60) as follows: In the time interval δt the electron moves out of the state $\Phi_{\boldsymbol{k}\alpha}$ and into the state $\Phi_{\boldsymbol{k}+\delta\boldsymbol{k},\alpha}$; and, of course, its energy changes by an amount $\delta E = E_\alpha(\boldsymbol{k} + \delta\boldsymbol{k}) - E_\alpha(\boldsymbol{k})$. It follows from (4.60) that the distribution of the electrons after the application of the electric field, to be denoted by $g(\boldsymbol{k})$, is obtained from the equilibrium distribution $f(\boldsymbol{k}) \equiv f(E_\alpha(\boldsymbol{k}))$ by shifting the wavevector of every electron by an amount $\delta\boldsymbol{k}$; we have

$$g(\boldsymbol{k}) \simeq f(\boldsymbol{k} - \delta\boldsymbol{k}) = f(k_x - \delta k_x, k_y, k_z) \tag{4.61}$$

The distribution of Fig. 4.27(b) is obtained from the Fermi–Dirac distribution of Fig. 4.72(a) in the above manner.

It follows from (4.60) that the electric field by itself would lead to a continuous change of $\boldsymbol{k}$ and, therefore, to an ever-changing distribution of the electrons in $\boldsymbol{k}$-space, which is obviously in contradiction with Ohm's law (4.59), which tells us that a constant electric field induces a constant current which implies a stationary (non-equilibrium but constant) distribution of the electrons.

In order to proceed with our description of the conduction mechanism it will be necessary to introduce, at this stage, the so-called defects of the crystal. A defect may be anything which destroys the perfect periodicity of the crystal. Here we shall mention only two kinds of defects (which matter in relation to conductivity):

(i) lattice vibrations
(ii) impurities, by which we mean a small concentration of foreign atoms in an otherwise pure crystal.

The one-electron states we have described (Bloch waves or wavepackets) are obtained only in a perfect crystal, i.e. one without any impurities and with the nuclei frozen at their equilibrium positions. A foreign atom or for that matter a missing atom, a so-called vacancy in the crystal structure, modifies the potential field around it, and so does a vibrating nucleus. The change $\Delta U(\boldsymbol{r})$ in the potential field in the vicinity of the defect acts as a scatterer. It scatters, with certain probability, an otherwise uninhibited electron out of its initial state, say $\Phi_{\boldsymbol{k}\alpha}$, into another state $\Phi_{\boldsymbol{k}'\alpha'}$, as illustrated schematically in Fig. 4.29. The phenomenon is analogous to the scattering of a particle by a potential barrier (see Section 2.3.2), except that in the present case the incident and scattered waves are not plane waves but Bloch waves. We note, also, that when the electron scatters off a vibrating nucleus it may loose to it or gain from it energy equal to $k_B T$ or so.

It is important to realise that collisions between electrons and crystal defects (impurities if such exist and vibrating nuclei) go on all the time, even when the metal is at thermal equilibrium. If we could follow a particular electron as it went through the crystal, we would see that its motion consisted of periods of free travelling between successive collisions, as ilustrated schematically in Fig. 4.30. The state of the electron between collisions, denoted by $\boldsymbol{k}\alpha$ in the figure, is described by a wavepacket $\Phi_{\boldsymbol{k}\alpha}$ in the manner we have described at the beginning of this section. Moreover, at equilibrium, the distribution of the electrons among the one-electron states of the crystal does not change; it is always given by the Fermi–Dirac formula (4.42). When an electron vacates a certain state after a collision, another occupies the emptied state within a very short time as a result of some other collision, so that on average that state is occupied with a probability determined by (4.42). Finally, we convince ourselves that only electrons with energy within a small region about E_F are scattered. A scattering event can take place only when there is an empty state, for the electron to scatter into, with energy which does not differ from its initial energy by more than $k_B T$ or so (this being the amount of energy that can be gained or lost during

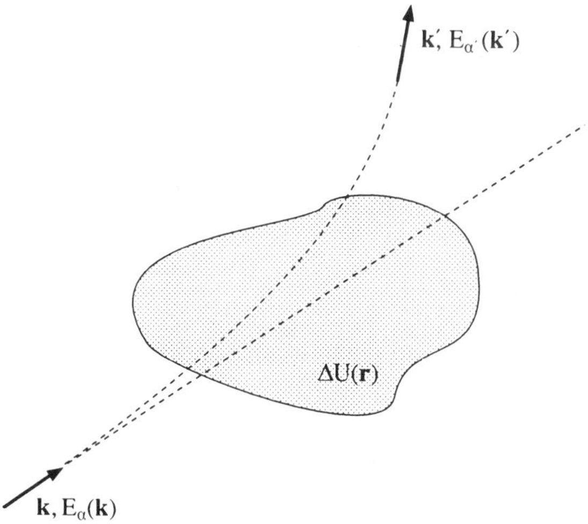

Figure 4.29. Scattering of an electron by a 'defect' $\Delta U(\boldsymbol{r})$

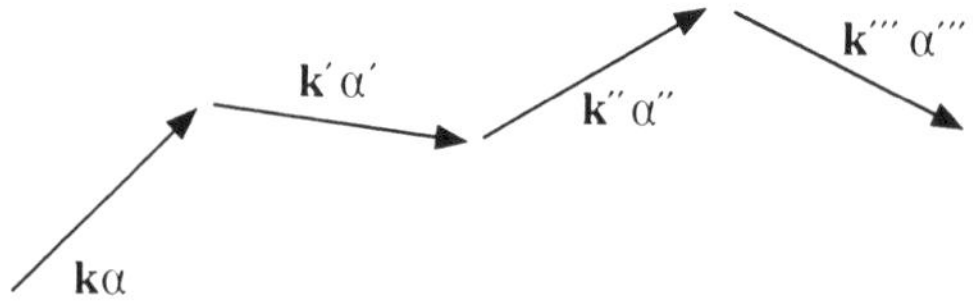

Figure 4.30. Periods of free travelling of an electron between collisions with crystal defects

a collision with a vibrating nucleus). It is obvious (see Fig. 4.23) that only electrons with energy $E \gtrsim E_{\mathrm{F}} - k_{\mathrm{B}}T$ satisfy the above condition.

Let us denote the average time between two successive collisions of an electron by 2τ. The smaller the scattering probability of the electron is, the larger the value of τ. In a perfect crystal $\tau = \infty$. Let us see why τ is an important quantity. Imagine a distribution of electrons in a metal which is different, but not much, from the equilibrium one. This might be due to an applied electric field. We know that when the cause (the electric field in our example) for the deviation from equilibrium is removed, the metal returns to equilibrium (a Fermi–Dirac distribution) and, obviously, this comes about by collisions of the electrons with defects of one kind or the other (there is no other way). The time required for the return to equilibrium cannot be less than the average time 2τ between two successive collisions of an electron. We find that the two are, in fact, practically the same. This need not surprise us if we consider that many electrons collide with defects at the same time (independently of each other). We shall, in what follows, refer to τ as the relaxation time (of the electrons).

We can now return to our discussion of conductivity.[17] We claim that the stationary distribution of the electrons after the application of the electric field will be given by (4.61) with

$$\delta \boldsymbol{k} = (\delta k_x, 0, 0)$$
$$\delta k_x \simeq -(eF/\hbar)\tau \tag{4.62}$$

which is obtained from (4.60) by putting $\delta t = \tau$. We can justify (4.62) as follows. According to (4.60), between collisions ($0 < \delta t < 2\tau$) δk_x increases, on the average, linearly with the time from zero to $-(eF/\hbar)2\tau$ when a collision brings it back to zero. Therefore the time-averaged value of this quantity is given by (4.62). Now it turns out that this value is quite small; when δk_x is small we can approximate (4.61) by

$$g(\boldsymbol{k}) = f(\boldsymbol{k}) - \frac{\partial f}{\partial k_x}\delta k_x \tag{4.63}$$

which we can rewrite as

$$g(\boldsymbol{k}) = f(E(\boldsymbol{k})) - \frac{\mathrm{d}f}{\mathrm{d}E}\frac{\partial E}{\partial k_x}\delta k_x \tag{4.64}$$

Substituting (4.62) into the above equation and noting (see (4.58)) that $\partial E/\partial k_x = \hbar v_x(\boldsymbol{k})$, where v_x is the x-component of the transport velocity of the electron, we finally obtain the following formula for the stationary distribution of the electrons

$$g(\boldsymbol{k}) = f(E(\boldsymbol{k})) + \frac{\mathrm{d}f}{\mathrm{d}E}v_x(\boldsymbol{k})e\tau F \tag{4.65}$$

The current density in the direction of the applied field (x-direction) is given by[18]

$$j = -2e\sum_{\boldsymbol{k}} v_x(\boldsymbol{k})g(\boldsymbol{k}) \tag{4.66}$$

The sum over $\boldsymbol{k}$ (extending over all BZ) includes all states of the conduction band. The factor of two takes into account spin degeneracy. The contribution to (4.66) from the first term of (4.65) describes the current density at equilibrium and, therefore, vanishes. We then obtain

$$j = 2e^2\tau\left(\sum_{\boldsymbol{k}}\left(-\frac{\mathrm{d}f}{\mathrm{d}E}\right)v_x^2(\boldsymbol{k})\right)F \tag{4.67}$$

We observe (see Fig. 4.23) that $(\mathrm{d}f/\mathrm{d}E)$ is different from zero only within a narrow region about E_F, and that over this region $\rho(E)$ and $\overline{v_x^2(E)}$ (defined below) do not vary appreciably and can therefore be replaced in the evaluation of the sum (4.67) by their respective values at $E = E_\mathrm{F}$. We obtain

$$\sum_{\boldsymbol{k}}\left(-\frac{\mathrm{d}f}{\mathrm{d}E}\right)v_x^2(\boldsymbol{k}) = \rho(E_\mathrm{F})\overline{v_x^2(E_\mathrm{F})} \tag{4.68}$$

where $\rho(E)$ denotes, as usual, the density of states of the electron per spin, and $\overline{v_x^2(E)}$ is the average value of $v_x^2(\boldsymbol{k})$ taken over all states of energy E. For crystals with cubic symmetry we can write

$$\overline{v_x^2(E_\mathrm{F})} = \tfrac{1}{3}\overline{(v_x^2 + v_y^2 + v_z^2)} = \tfrac{1}{3}\overline{v^2(E_\mathrm{F})} \tag{4.69}$$

Substituting (4.68) and (4.69) into (4.67) we obtain

$$j = \tfrac{2}{3}e^2\overline{v^2(E_\mathrm{F})}\rho(E_\mathrm{F})\tau F \tag{4.70}$$

The above formula gives the current density in the direction of the electric field (x-direction). One can easily show that the current densities in the y

and z directions vanish. It follows from (4.70) that the conductivity of the metal, defined by (4.59), is given by

$$\sigma = \tfrac{2}{3}e^2\overline{v^2(E_F)}\rho(E_F)\tau \tag{4.71}$$

The fact that the conductivity is proportional to $\rho(E_F)$ reminds us that metallic conduction is possible because there are empty electron energy levels in the immediate vicinity of occupied ones (only then is $\rho(E_F)$ different from zero). Only then is the redistribution of electrons that leads to a macroscopic current possible under the action of a weak electric field. (The applied electric field is, usually, weak in the sense that the resultant force on the electron is very much smaller than that which derives from the crystal potential (4.16).) Similarly, the presence of $v(E_F)$ in (4.71) reminds us that metallic conduction is due to carriers (electrons represented by Bloch wavepackets) moving freely, with a velocity given by (4.58), through the crystal. When the one-electron states of a solid are localised in space, as happens in some non-crystalline materials, the conductivity vanishes, when $T \to 0$, even though $\rho(E_F)$ may be different from zero, because the localisation of the wavefunction implies a zero velocity of the electron (see Section 4.18.3). The existence of free carriers and the availability of empty energy levels next to occupied ones are the basic requirements for metallic, or metallic-like, conduction.

Finally the presence of the relaxation time τ in (4.71) shows the importance of the vibrating nuclei and other crystal defects in limiting conduction. At sufficiently high temperatures ($T \gtrsim 300$K) τ is determined almost entirely by collisions of the electron with the vibrating nuclei. The probability ($\propto 1/\tau$) that an electron will be scattered by a vibrating nucleus in a unit of time is proportional to the scattering cross-section of a vibrating nucleus, and this is proportional to the square of the amplitude of the vibration which increases linearly with the temperature. Therefore we obtain

$$\sigma \propto \tau \propto 1/T \tag{4.72}$$

which means that the resistivity $\rho \equiv 1/\sigma$ of a metal must increase linearly with the temperature; and so it does (see Fig. 4.31).

At low temperatures the classical picture of a nucleus vibrating independently of all others, which underlies the derivation of (4.72) ceases to be valid (see text following (4.110) in Section 4.12). One finds that, as $T \to 0$ the resistivity goes to zero like $\rho \propto T^5$, assuming that the metal is pure (has no impurities). If there are impurities, the resistivity of the metal, at low temperatures, will be determined by their concentration as shown in Fig. 4.32. We see from this figure that in the presence of impurities, the resistivity becomes independent of temperature (as $T \to 0$), its magnitude determined by the concentration of impurities.

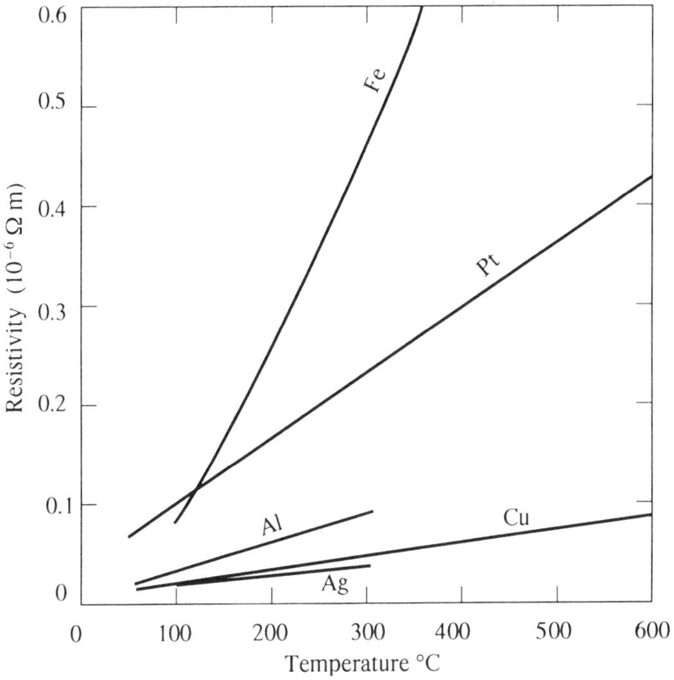

Figure 4.31. Variation of the resistivity of the metals Fe, Pt, Al, Cu, Ag with temperature (experimental). Note that the temperature is given on a Celcius scale. (K. J. Pascoe, *Properties of Materials for Electrical Engineers*, (New York: Wiley) 1973. Reprinted by permission of John Wiley & Sons Ltd)

4.8.1 MEAN FREE PATH

Equation (4.71) makes it possible to determine the relaxation time τ from conductivity measurements; σ is measured and $\overline{v^2(E_F)}$ and $\rho(E_F)$ are obtained from calculated dispersion curves $E_\alpha(\boldsymbol{k})$, such as those of Figs. 4.15 and 4.18. Substitution of the above in (4.71) then gives τ. We find that at normal temperatures ($T \simeq 300$K) $\tau \simeq 10^{-14}$ s. Remembering that 2τ gives the average time interval between two collisions of an electron (with energy $E \simeq E_F$), we can estimate the so-called mean free path of an electron at $E = E_F$ (this is the average distance an electron travels between two successive collisions) from the following formula

$$L \simeq 2\tau(\overline{v^2(E_F)})^{1/2} \tag{4.73}$$

The mean free path, like τ, varies from metal to metal and with temperature (see (4.72)); it equals, in most cases, a few hundred angstrom when $T \simeq 300$K.

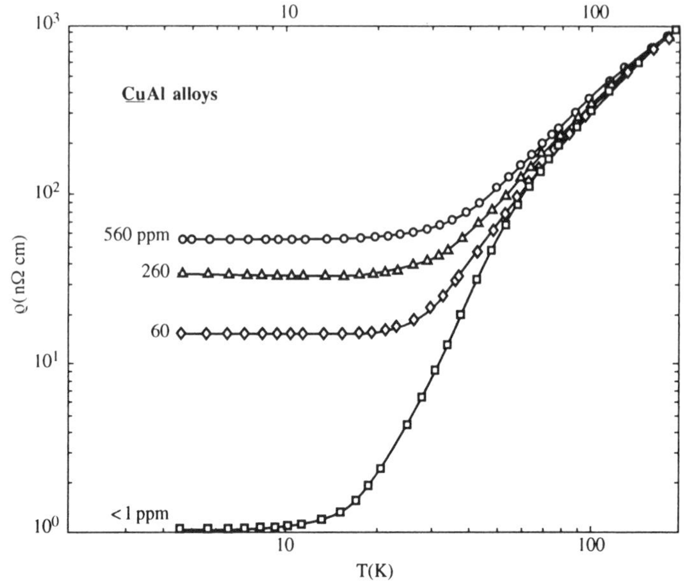

Figure 4.32. Resistivity of copper containing impurities (aluminium atoms) at low temperatures. The different curves correspond to different impurity-concentrations, measured in parts per million (ppm). $1\ n\Omega = 10^{-9}\ \Omega$. [G. Apostolopoulos, private communication]

4.9 The free-electron model of metals

In this model, proposed by Sommerfeld in 1928, the conduction-band electrons are treated as free particles: their energy is given by (4.32) with $m^* = m$ (the electron mass) and with the bottom of the conduction band taken as the zero of the energy; the corresponding eigenstates are described by ordinary plane waves: $\psi_{\boldsymbol{k}} = (V)^{-1/2} \exp(\mathrm{i}\boldsymbol{k} \cdot \boldsymbol{r} - \mathrm{i}E(\boldsymbol{k})t/\hbar)$, where V denotes as usual the volume of the crystal and $\boldsymbol{k}$ is given by (4.24). The density of states, per spin, is given by (4.38) with $m^* = m$.

The number of free (conduction-band electrons) in the Sommerfeld model is usually treated as an adjustable (empirical) parameter. At the absolute zero of temperature they occupy, of course, all eigenstates of the conduction band with energy up to the Fermi level. The latter, when the density of states is given by (4.38), can be obtained analytically:

$$E_{\mathrm{F}} = \frac{\hbar^2 k_{\mathrm{F}}^2}{2m} = \frac{\hbar^2}{2m}(3\pi^2 n)^{2/3} \tag{4.74}$$

where n denotes the number of free electrons per unit volume. Substituting

(4.74) in (4.38), one obtains

$$\rho(E_F) = \frac{m}{2\pi^2\hbar^2}(3\pi^2 n)^{1/3} \tag{4.75}$$

The conductivity of the free-electron metal can now be calculated from (4.71): $v(E_F) = \hbar k_F/m$, where k_F is the magnitude of the wavevector at $E = E_F$ (given by (4.74)) and $\rho(E_F)$ is given by (4.75); we obtain

$$\sigma = \frac{ne^2\tau}{m} \tag{4.76}$$

The above formula is extensively used even today. It is valid provided $n \simeq (2/3)m\overline{v^2(E_F)}\rho(E_F)$ where $\overline{v^2(E_F)}$ and $\rho(E_F)$ are the properly calculated values of these quantities for the metal under consideration. In earlier times, as we have already mentioned, n was treated as an empirical parameter, and there of course lies the difference between formulae (4.71) and (4.76). The first contains only one empirical parameter (τ), the second has two such parameters (n and τ). In a complete theory one would like to be able also to calculate τ from first principles.

The model of Sommerfeld describes reasonably well not only the conductivity of a metal but other quantities such as the specific heat, etc., and it has been used successfully in the description of electron emission from metal surfaces as well (see Section 4.13.3). In other cases it fails. For example, it cannot, as a rule, reproduce the Fermi surface of a metal which can be, as we have seen in Section 4.4, a very complex surface. The optical properties of a metal cannot be adequately described by a free-electron model either. In this respect, Sommerfeld's model is not different from many other empirical models which are proposed at times in order to explain certain experimental data. The empirical model will explain these data, may be a few others, but it is unlikely that it will have a general validity.

4.10 Conductivity of semiconductors

It is obvious that the conductivity of an intrinsic (pure) semiconductor at zero ($T = 0$) temperature is zero. One of the basic requirements for conduction, namely that there must be empty energy levels of the electron next to occupied ones, is not satisfied. At higher temperatures one finds (see Section 4.7) a small number of electrons per unit volume (denoted by n) in the conduction band and a number of empty states (holes) per unit volume (denoted by p) in the valence band ($p = n$ in a pure semiconductor). Next to the occupied energy levels of the conduction band there are, obviously, empty ones and, similarly, next to the empty levels of the valence band are occupied ones; and in every case the corresponding states are of the Bloch type (free carriers with a transport velocity given by (4.58)). Therefore, the conditions for metallic-like conduction are satisfied when $T > 0$ and, obviously, the conductivity will be larger, the larger the value of n and p. We write (by analogy to (4.76)).

$$\sigma = e\mu_p p + e\mu_n n \tag{4.77}$$

μ_p (the so-called mobility of holes) and μ_n (the mobility of electrons) are coefficients to be determined. μ_n is in general different from μ_p; we find, for example, that for silicon at $T \simeq 300\text{K}$, $\mu_n = 1350\,\text{cm}^2/\text{Vs}$ and $\mu_p = 480\,\text{cm}^2/\text{Vs}$; μ_p and μ_n depend respectively on the average velocity of the holes in the valence band (which is the same as that of the electrons that were occupying them) and on the average velocity of the electrons in the conduction band, and these average velocities depend on the temperature (unlike the situation in metals, where at any reasonable temperature electrons at the Fermi level dominate); μ_n and μ_p depend also on the relaxation times of electrons and holes respectively, which are determined, as in metals, by collisions with impurities and the vibrating lattice. A careful calculation shows (see problem 4.12) that, in the absence of impurities, and at sufficiently high temperatures ($T \gtrsim 300\text{K}$), when the vibrating nuclei can be treated classically (as in the derivation of (4.72)), μ_n and μ_p vary with temperature like $\mu_n, \mu_p \propto T^{-3/2}$. On the other hand n and p vary with the temperature (see Problem 4.8) like $n = p \propto T^{3/2} \exp(-E_g/2k_B T)$. Therefore the conductivity, given by (4.77), varies with temperature lke $\sigma \propto \exp(-E_g/2k_B T)$. The variation of σ with T is shown schematically in Fig. 4.33. Evidently, for given temperature, the conductivity gets smaller exponentially as the energy gap increases. For an insulator, the gap is so large that the conductivity is vanishingly small at any reasonable temperature (it is worth remembering that any crystalline structure will break down above a certain temperature).

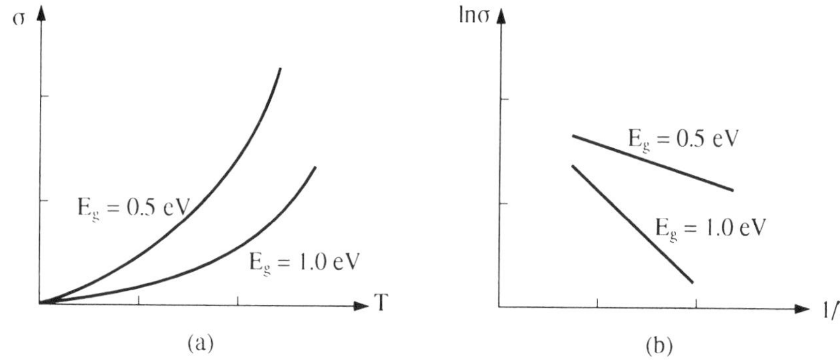

Figure 4.33. The conductivity of a semiconductor as a function of temperature (schematic). The slope of the straight line in (b) equals $-E_g/2k_B$

4.10.1 DOPED SEMICONDUCTORS

Substitution of a small fraction ($< 10^{-4}$) of the atoms of a semiconductor by foreign atoms (impurities) of an appropriate electronic structure can change dramatically the conductivity of the semiconductor. We take germanium as an example; silicon, gallium arsenide and other semiconductors behave in a similar manner. We know (see Section 4.6) that the four valence electrons per germanium atom (those occupying the outer shells of the free atom) fill up the valence bands (shown in Fig. 4.21) of the germanium crystal. Let us, now, assume that a small number of Ge atoms (e.g., 10^{16} atoms per cm^3) have been replaced with arsenic (As) atoms. The As atom has 33 electrons (see Table 2.5), 28 of which are in inner shells and five are in outer shells. The inner shells of the atom are affected very little by the crystal environment and will not concern us here. Four out of the five valence electrons of As are accommodated in the valence bands of the Ge crystal which are not affected significantly by the doping of the crystal as long as the concentration of impurities is relatively low. (We remember that there are approximately 10^{23} atoms per cm^3 of the crystal; impurity concentrations up to 10^{19} atoms per cm^3 will not change significantly the energy bands of the host crystal.) A question arises as to what happens to the fifth electron of As. In the free As atom the 'fifth' electron is attracted to a positive ion ($\overset{+}{\text{As}}$) made up of the atomic nucleus and the other electrons of the atom (see (2.90)). When the atom is placed in the crystal, this attraction is reduced considerably by the polarisation of the electronic cloud of the crystal in between the electron and the ion (see Fig. 4.34); we obtain (instead of (2.90))

$$U \simeq -\left(\frac{1}{4\pi\epsilon_0\epsilon'}\right)\frac{e^2}{r} \tag{4.78}$$

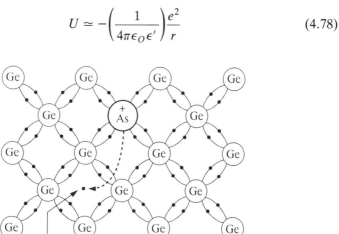

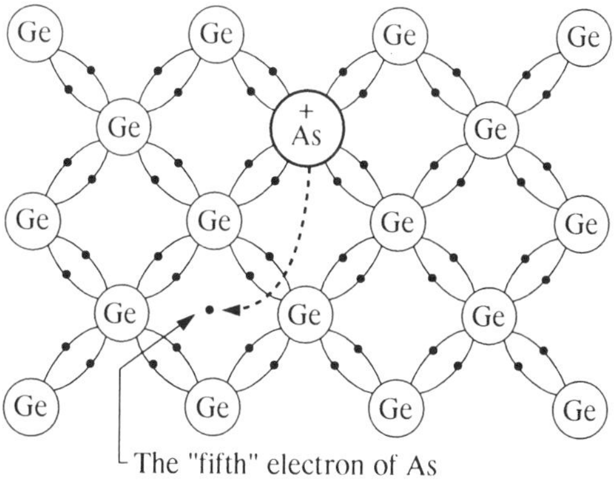

Figure 4.34. Donor-impurity (As) in a germanium crystal (schematic description)

The dielectric constant ϵ' is a measure of the polarisability of the electronic cloud of the host material ($\epsilon' = \epsilon(\omega = 0)$ of Section 4.15; for germanium $\epsilon' \simeq 16$). In the free atom the interaction (2.90) can keep the electron in orbits[19] close to the atomic nucleus; in the crystal the much weaker interaction (4.78) can only sustain an orbit of the 'fifth' electron of As about its nucleus of much larger radius, containing many germanium atoms. It is worth noting that it is this fact (of many Ge atoms in between the $\overset{+}{\mathrm{As}}$ and the 'fifth' electron of As) which justifies eventually the use of ϵ' in (4.78). Quantum-mechanically one finds a localised one-electron state, loosely bound to the $\overset{+}{\mathrm{As}}$ ion, with energy E_d a bit below the bottom E_c of the conduction band (see Fig. 4.35). One can obtain a rough estimate of $E_d - E_c$ using (2.12) with $Z = 1$, replacing ϵ_O by $\epsilon_O\epsilon'$ and m (the mass of the electron) by an effective mass m^* which takes into account the fact that the electron moves not in vacuum but in the germanium crystal; E_c corresponds to the zero energy of vacuum and $(E_c - E_d)$ to the ionisation energy of hydrogen (given by the negative of (2.12) with $Z = 1$ and $n = 1$). The effective mass of the electron is obtained from the $E(\boldsymbol{k})$ of the conduction band at $E \simeq E_c$. We assume, for simplicity, that we have a single band with its minimum at $\boldsymbol{k} = 0$, and that there $E(\boldsymbol{k})$ can be written as $E(\boldsymbol{k}) = E_c + \hbar^2k^2/2m^*$. The velocity of the electron, is, then, given, in accordance with (4.58), by $\boldsymbol{v} = \hbar\boldsymbol{k}/m^*$. We can identify $\hbar\boldsymbol{k}$ with an effective momentum of the electron (as in (4.60)) in which case m^* may be identified with the effective mass of the electron. If the conduction band consists not of one band but of a number of bands, as is usually the case, with their minima at $E \simeq E_c$, or if the $E(\boldsymbol{k})$ about a minimum is not spherically

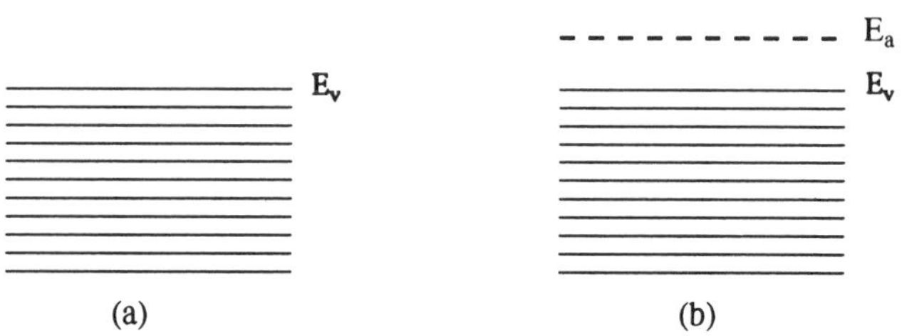

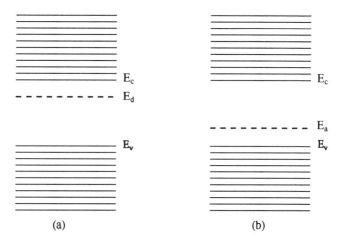

Figure 4.35. One-electron levels in a doped semiconductor. (*a*): *n*-type, (*b*): *p*-type. The energy differences $(E_c - E_d)$ and $(E_a - E_v)$ have been exaggerated

symmetric, one must use an average value of m^* (for germanium $m^* \simeq 0.14m$; see problem 4.9). In this way one obtains $E_d - E_c \simeq -13.6 (m^*/m)/(\epsilon')^2$ eV. For germanium this gives $E_c - E_d \simeq 0.007$ eV in rough agreement with the experimental value of 0.012 eV.

The electron state of the above energy, loosely bound to the impurity centre, accommodates, at $T = 0$, the 'fifth' valence electron of the impurity atom (arsenic in our example). The electron occupying this so-called impurity state may have spin up or spin down. We emphasise, however, that we cannot have a second electron (of different spin) in the above impurity orbital. The potential field about $\overset{+}{\mathrm{As}}$ is far too weak to hold two mutually repelling electrons. So we shall assume that for each impurity atom there is one impurity energy level with a corresponding orbital which can accommodate a single electron with spin up or spin down. We have N_d such orbitals (impurity states) per unit volume, where N_d is the number of impurity atoms per unit volume. In Fig. 4.35(a) the impurity energy levels E_d are indicated by a broken line, which should remind us that the corresponding states are localised about their respective atoms. The mean velocity of an electron in a localised state vanishes and, therefore, such electrons do not contribute directly to conduction. However, they play an important role in the determination of the conductivity of the semiconductor at finite temperatures; when $k_B T \gtrsim E_c - E_d$ the electrons are excited from the impurity states to the conduction band thereby contributing to metallic-like conduction.[20] In Fig. 4.36 we show n, the number of electrons per unit volume in the conduction band of an n-type semiconductor as a function of the temperature. (A semiconductor is of the n-type (also called donor-type) if the impurities donate electrons to the conduction band at finite temperatures. There are other atoms with five valence electrons, such as phosphorous (P) and antimony (Sb), which as dopants behave in the same way as As.) We understand Fig. 4.36 as follows. At $T = 0$ there are no electrons in the conduction band; as T increases (region A of the diagram) electrons are excited from the impurity states (of the E_d levels of Fig. 4.35(a)) into the conduction band:

$$n = N_d \exp(-(E_c - E_d)/2k_B T) \tag{4.79}$$

The conductivity in region A is due, almost entirely, to electrons (p, the number of holes per unit volume in the valence band, is practically zero) and therefore it is given by $\sigma = e\mu_n n$. When the temperature is further increased (region B in Fig. 4.36) all impurities are ionised:

$$n = N_d \tag{4.80}$$

and the conductivity is given by $\sigma = e\mu_n N_d$. Finally, at higher temperatures (region Γ in Fig. 4.36) electrons are excited into the conduction band from

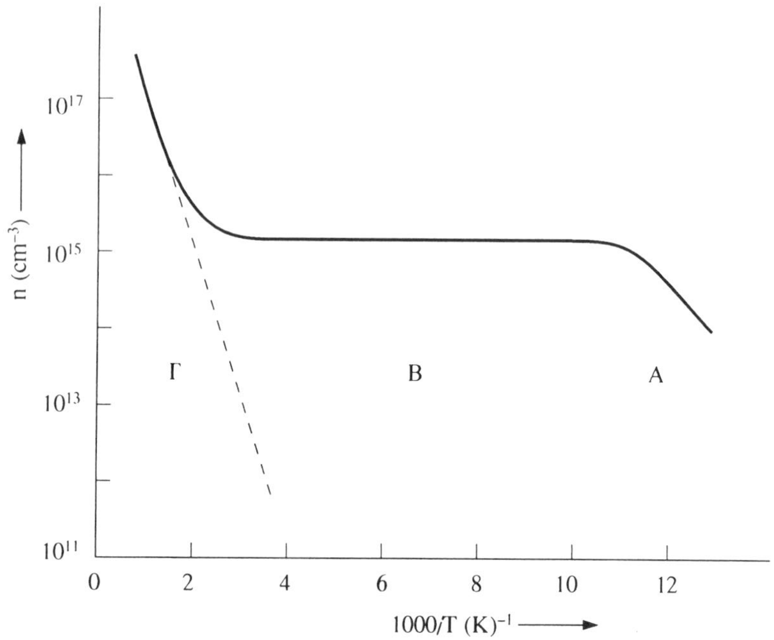

Figure 4.36. n vs. T^{-1} in an n-type semiconductor ($N_d \simeq 10^{15}\,\mathrm{cm}^{-3}$)

the valence band as in an intrinsic semiconductor (see Section 4.7) and we obtain

$$N_\mathrm{d} \ll n \simeq p \propto \exp(-E_\mathrm{g}/2k_\mathrm{B}T) \qquad (4.81)$$

The conductivity in this region is given by (4.77).

Besides n-type semiconductors, today's technology has at its disposal p-type semiconductors. We obtain a p-type Ge crystal (the same applies to Si, GaAs, etc.) by replacing a small percentage of the Ge atoms by trivalent impurities (atoms, such as boron (B) and aluminium (Al), with three electrons in their outer shells). In this case the ground state of the semiconductor (its state at $T = 0$) can be described as follows. We have fully occupied valence bands (essentially those of germanium) and empty localised states about the impurity atoms with energy E_a close to the top of the valence band ($E_\mathrm{a} - E_\mathrm{v} \simeq 0.01\,\mathrm{eV}$, see Fig. 4.35(b)). These, so-called acceptor impurity states, have very much the same properties as the electron states of donor impurities. We have N_a of them per unit volume (as many as the number of impurity atoms per unit volume). What happens is that a number of extended one-electron states (one per impurity atom) are squeezed out of the valence band transformed by the process into the localised acceptor states described above. By their presence these states determine the concentration of holes in the valence band in the same way that donor states determine the concentration of electrons in the conduction band. We obtain p as a function of T from Fig. 4.36 substituting p for n and N_a for N_d. At $T = 0$, the acceptor impurity states are empty and $p = 0$. As the temperature increases (region A of the diagram) electrons are excited into the impurity states (of the E_a levels of Fig. 4.35(b)) from the valence band creating a corresponding number of holes:

$$p \simeq N_a \exp(-(E_a - E_v)/2k_B T) \tag{4.82}$$

The conductivity in region A is due, almost entirely, to holes and therefore it is given by $\sigma = e\mu_p p$. When the temperature is further increased (region B in Fig. 4.36) the impurity states are filled up:

$$p = N_a \tag{4.83}$$

and the conductivity is given by $\sigma = e\mu_p N_a$. At higher temperatures (region Γ in Fig. 4.36) electrons are excited into the conduction band and the conductivity is due equally to electrons and holes; it is given by (4.77), with

$$N_a \ll n \simeq p \propto \exp(-E_g/2k_B T) \tag{4.84}$$

The variation of the conductivity with temperature is similar to that of n and p (shown in Fig. 4.36) but not exactly the same, because of the dependence of μ_p and μ_n on the temperature. We have already noted (in Section 4.10) that at sufficiently high temperatures a mobility determined by collisions of the carriers (electrons and holes) with the vibrating lattice will vary with temperature as $T^{-3/2}$. In the case of doped semiconductors, however, scattering by the ionised impurities is bound to play an important role especially at not-so-high temperatures; it turns out that a mobility limited by impurity scattering varies with the temperature as $T^{3/2}$. (As the temperature decreases the average electron (hole) becomes slower and has a better chance of being scattered by the impurity, leading to a lower mobility.) The above discussion suggests that at relatively low temperatures (when impurity scattering is dominant) the mobility will increase with the temperature as $T^{3/2}$ and at the higher temperatures (when lattice scattering is important) will decrease with temperature as $T^{-3/2}$; at intermediate temperatures (when both mechanisms are important) the mobility reaches a maximum as it changes smoothly from one behaviour to the other. Of course, as the concentration of impurities increases, the effect of impurity scattering on the mobility will be felt at higher temperatures. For example, μ_n of intrinsic silicon at 300K is 1350 cm^2/Vs. With a donor doping concentration of $10^{17}\ \text{cm}^{-3}$ this is reduced (at $T = 300\text{K}$) to $\mu_n = 700\ \text{cm}^2/\text{Vs}$ because of impurity scattering.

In the above discussion we have assumed that the semiconductor contains either donor or acceptor impurities. But it may have both. A predominance of donors ($N_d \gg N_a$) will make the material n-type. In this case, however, a donor atom can loose its electron to an empty acceptor state and, therefore, in the intermediate region of temperatures (region B in Fig. 4.36) there will be no holes (the acceptor states being occupied with electrons from the donor states), and the number of electrons in the conduction band will be given by the difference ($N_d - N_a$) of the two concentrations. This process is called compensation. By this process it is possible to begin with an n-type material and add acceptors until $N_a = N_d$ and then no donated electrons remain in the conduction band. With further acceptor doping the semiconductor becomes p-type with a hole concentration of ($N_a - N_d$).

Many devices of modern electronics are based on semiconductors: p–n diodes, photodiodes, transistors and many other. A description of these devices lies beyond the scope of the present book. One could say, however, that the many and varied applications of semiconductors depend, to a large degree, on our ability to determine, by doping, the conductivity of the semiconductor, and also the type of carrier (electrons and/or holes) of the electric current. It is worth noting (see Fig. 4.31) that the resistivity of metals is of the order of 10^{-6} Ωm and does not vary much from metal to metal; while in semiconductors, the resistivity may have a value between 10^{-4} and 10^7 Ωm depending on the value of E_g and on the presence or otherwise of different impurities.

4.11 The self-consistent field in crystals

The description we have given, in the preceding sections, of the electronic properties of crystals is based on the assumption that every electron is moving in a mean potential field $U(\boldsymbol{r})$ due to the nuclei and all other electrons of the crystal (see Section 4.3). We emphasised the periodic nature of $U(\boldsymbol{r})$ but we have not said much about the way one calculates this potential field. We think it worthwhile to describe the basic rules of such a calculation. The way to proceed is essentially the same as that followed in the self-consistent calculation of the potential an electron sees in a many-electron atom (Sections 2.15 and 2.15.1). In the present case we write

$$U_{\uparrow(\downarrow)}(\boldsymbol{r}) = U_c(\boldsymbol{r}) + U_{ex\uparrow(\downarrow)}(\boldsymbol{r}) \tag{4.85}$$

We assume that the potential $U_\uparrow(\boldsymbol{r})$ seen by an electron of spin up may be different from the potential $U_\downarrow(\boldsymbol{r})$ seen by an electron of spin down. In non-magnetic crystals, such as Cu, W, Ge, etc., $U_\uparrow(\boldsymbol{r}) = U_\downarrow(\boldsymbol{r})$; but in the so-called ferromagnetic metals, such as nickel (Ni) and iron (Fe), $U_\uparrow(\boldsymbol{r}) \neq U_\downarrow(\boldsymbol{r})$ below a critical temperature[21]. The first term of (4.85) describes the electrostatic interaction of an electron with the nuclei and with the electronic cloud (due to all electrons) of the crystal. We write (by an obvious extension of (2.83))

$$U_c(\boldsymbol{r}) = -\frac{e^2}{4\pi\epsilon_O}\sum_j \frac{Z_j}{|\boldsymbol{r} - \boldsymbol{R}_j|} + \frac{e^2}{4\pi\epsilon_O}\int \frac{n(\boldsymbol{r}')\,\mathrm{d}V'}{|\boldsymbol{r} - \boldsymbol{r}'|} \tag{4.86}$$

The first term of (4.86) gives the interaction of the electron with the stationary nuclei; eZ_j is the charge of the nucleus stationed at $\boldsymbol{R}_j$, and the sum extends over all the nuclei of the crystal. In crystals such as Cu, Ge, etc., which consist of same atoms $Z_j = Z$. In crystals, such as NaCl, GaAs, etc., which consist of atoms of more than one kind the value of Z_j is defined by the nucleus at $\boldsymbol{R}_j$.[22] The second term of (4.86) gives the interaction of the electron with the electronic cloud of the crystal (the integral extends over the whole crystal). The density of the electronic cloud

$$n(\boldsymbol{r}) = n_\downarrow(\boldsymbol{r}) + n_\uparrow(\boldsymbol{r}) \tag{4.87}$$

is defined as in Section (2.15); the two terms of (4.87) give the contributions to $n(\boldsymbol{r})$ from electrons of spin up and spin down respectively. The second term of (4.86) includes an 'interaction energy of the electron with its own electronic cloud' which does not exist in reality and must therefore be cancelled out. We have dealt with the same problem (in Section 2.15) in relation to the one-electron potential of a many-electron atom. We can apply the same arguments to the present case. The second term of (4.85), which is evaluated in the same way as (2.88), cancels out the above self-interaction energy of the electron. In the present case $n_\uparrow(\boldsymbol{r})$ may be different from $n_\downarrow(\boldsymbol{r})$ and, therefore, the quantity $n(\boldsymbol{r})/2$ in (2.88) is replaced by $n_\uparrow(\boldsymbol{r})$ for spin up and by $n_\downarrow(\boldsymbol{r})$ for spin down. We write

$$U_{\mathrm{ex}\uparrow(\downarrow)}(\boldsymbol{r}) = -\frac{3\alpha e^2}{4\pi\epsilon_O}\left[\frac{3}{4\pi}n_{\uparrow(\downarrow)}(\boldsymbol{r})\right]^{1/3} \tag{4.88}$$

In a crystal of atoms A the value of the parameter α has the same value as in the corresponding expression (2.88) of the atomic potential. When the nuclei are arranged periodically in space, $n_{\uparrow(\downarrow)}(\boldsymbol{r})$ and the potential field defined by equations (4.85)–(4.88) are periodic functions in accordance with (4.16); but the distribution of $n_{\uparrow(\downarrow)}(\boldsymbol{r})$ and the corresponding variation of $U_{\uparrow(\downarrow)}(\boldsymbol{r})$ within a primitive cell may, in general, be quite complicated. Usually, in order to simplify matters, we approximate the actual potential field by a more symmetric one. This can be done, for example, as follows[23]. We draw a sphere about each nucleus; the spheres centred on neighbouring nuclei may touch but do not overlap each other; in this way the space of a primitive cell (a Wigner–Seitz cell) is separated into a central region (within the sphere) and an outer region (between the sphere and the boundary of the Wigner–Seitz cell). Within the sphere we replace $n_{\uparrow(\downarrow)}(\boldsymbol{r})$ by a spherically symmetric distribution $n_{\uparrow(\downarrow)}(r)$ as we have done in the atomic case (see Fig. 2.21); in the outer region of the cell we replace $n(\boldsymbol{r})$ by its volume-average over that region. In this way (4.88) is replaced by a spherically symmetric quantity inside the sphere and by a constant outside the sphere. In the same way we approximate the Coulomb potential (4.86) by a spherically symmetric one inside the above sphere and by its volume average outside this sphere. A simplified form of the potential field obtained in this manner, or in some similar manner, allows one to proceed with the solution of the problem which consists in finding the eigenvalues and corresponding eigenstates (Bloch waves) of an electron in this field. When, in what follows, we refer to equations (4.85)–(4.88), we shall assume that where necessary the relevant quantities have been approximated as described above.

At this stage we must observe that the density of the electronic cloud is obtained, as in the case of the many-electron atom, from the occupied one-electron states and, therefore, the two must be calculated self-consistently as in the atomic case (Section 2.15.1). We shall describe the basic steps of such a calculation in relation to the electronic structure of non-magnetic and ferromagnetic metals; the electronic structure of a semiconductor or an insulator can be calculated in more or less the same way.

4.11.1 NON-MAGNETIC METALS

In this case we have

$$n_\uparrow(\boldsymbol{r}) = n_\downarrow(\boldsymbol{r}) = \tfrac{1}{2}n(\boldsymbol{r}) \tag{4.89}$$

$$U_\uparrow(\boldsymbol{r}) = U_\downarrow(\boldsymbol{r}) = U(\boldsymbol{r}) \tag{4.90}$$

The potential field $U(\boldsymbol{r})$ is obtained from equations (4.85)–(4.88) if we know $n(\boldsymbol{r})$. In many cases we obtain a reasonably good approximation of $n(\boldsymbol{r})$ from a superposition of atomic densities; we write

$$n(\boldsymbol{r}) \simeq n_0(\boldsymbol{r}) \equiv \sum_j n_{\mathrm{A}}(|\boldsymbol{r} - \boldsymbol{R}_j|) \tag{4.91}$$

where $n_{\mathrm{A}}(|\boldsymbol{r} - \boldsymbol{R}_j|)$ is the density of the electronic cloud of a free atom A stationed at $\boldsymbol{R}_j$ (we assume that the metal consists of same atoms A). Let $U_0(\boldsymbol{r})$ be the potential field obtained, through equations (85)–(88), from $n_0(\boldsymbol{r})$. To the field $U_0(\boldsymbol{r})$ corresponds a set of one-electron energy levels $E_\alpha^{(0)}(\boldsymbol{k})$ and corresponding eigenfunctions (Bloch waves) $\psi_{\boldsymbol{k}\alpha}^{(0)}$, the same for spin up and spin down, in accordance to what has been said in Sections 4.3 and 4.3.1. At the absolute zero of temperature, all states with energy below E_{F} are occupied and those with energy above E_{F} are empty (E_{F} is determined as described in Section 4.5). We obtain the density of the electronic cloud from the occupied states as follows

$$n_1(\boldsymbol{r}) = 2\sum_{\boldsymbol{k}\alpha} |\psi^{(0)}_{\boldsymbol{k}\alpha}(\boldsymbol{r})|^2 f(E^{(0)}_\alpha(\boldsymbol{k})) \tag{4.92}$$

where $f(E)$ is the Fermi–Dirac distribution (4.42); at $T = 0$, $f(E^{(0)}_\alpha(\boldsymbol{k})) = 1$, if $E^{(0)}_\alpha(\boldsymbol{k}) < E_\mathrm{F}$ and $f(E^{(0)}_\alpha(\boldsymbol{k})) = 0$, if $E^{(0)}_\alpha(\boldsymbol{k}) > E_\mathrm{F}$. The sum over α extends over all bands and the sum over $\boldsymbol{k}$ includes all the allowed (by (4.24)) points of the BZ. The multiplicative factor (2) takes account of the spin degeneracy.

If $n_1(\boldsymbol{r})$ is different from $n_0(\boldsymbol{r})$ we repeat the calculation. Substituting $n_1(\boldsymbol{r})$ in (4.85)–(4.88) we obtain the next approximation to the potential field: $U_1(\boldsymbol{r})$. Obviously, $U_1(\boldsymbol{r}) \neq U_0(\boldsymbol{r})$, if $n_1(\boldsymbol{r}) \neq n_0(\boldsymbol{r})$. We then find the energy eigenvalues $E^{(1)}_\alpha(\boldsymbol{k})$ and corresponding eigenfunctions $\psi^{(1)}_{\boldsymbol{k}\alpha}$ corresponding to $U_1(\boldsymbol{r})$; and from the occupied states we obtain the next approximation to the density of the electronic cloud: $n_2(\boldsymbol{r})$. If $n_2(\boldsymbol{r}) \neq n_1(\boldsymbol{r})$ we repeat the calculation once again; and so on until convergence, by which we mean that:

$$n_i(\boldsymbol{r}) = n_{i-1}(\boldsymbol{r}) \text{ and } U_i(\boldsymbol{r}) = U_{i-1}(\boldsymbol{r}) \tag{4.93}$$

The above convergence secures the self-consistency of the potential field and of the density of the electronic cloud. The one-electron energy levels $E_\alpha(\boldsymbol{k})$ and corresponding eigenfunctions $\psi_{\boldsymbol{k}\alpha}$ obtained from the self-consistent potential $U(\boldsymbol{r}) \equiv U_i(\boldsymbol{r})$ are registered as the energy levels and corresponding eigenfunctions of an electron in the given metal.

4.11.2 FERROMAGNETIC METALS

In this case the ground state of the metal (the state of least total energy) is obtained with unequal distributions of spin up and spin down electrons: $n_\uparrow(\boldsymbol{r}) \neq n_\downarrow(\boldsymbol{r})$. The difference between the number of electrons, per atom, with spin up and the number of electrons with spin down multiplied with the magnetic moment μ_B of an electron (defined by (2.34)) gives the magnetisation (magnetic moment per atom) of the metal. Let us see how $n_\uparrow(\boldsymbol{r})$ (corresponding to the majority of the electrons) and $n_\downarrow(\boldsymbol{r})$ (corresponding to the minority of the electrons) are calculated for a ferromagnetic metal such as iron or nickel. The potential field (4.85) an electron sees in a ferromagnetic metal depends on its spin, because the exchange term of this field given by (4.88) is different for spin up and spin down when $n_\uparrow(\boldsymbol{r}) \neq n_\downarrow(\boldsymbol{r})$. This means that the energy levels and the corresponding eigenfunctions (Bloch waves) will be different for the two spin orientations. In Fig. 4.37 we present the conduction bands of nickel for spin up and spin down as calculated by Moruzzi, Janak and Williams. The space lattice of

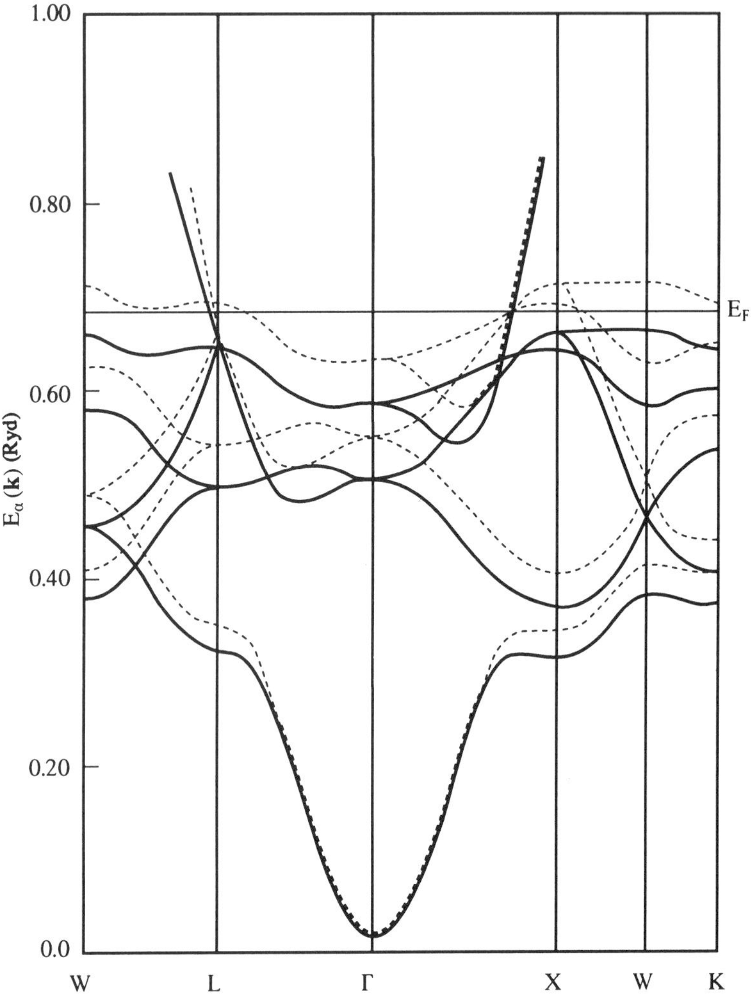

Figure 4.37. Energy bands of ferromagnetic nickel along various symmetry lines of the BZ (Fig. 4.14). Solid lines: spin up (majority electrons); broken lines: spin down (minority electrons). (Redrawn with permission from L. Moruzzi, J. F. Janak and A. R. Williams, *Calculated Electronic Properties of Metals*, (New York: Pergamon) 1978)

nickel is fcc (there is one atom per unit cell); and its BZ is that of Fig. 4.14; the electrons (18 per atom) from the inner shells of the atoms occupy lower-energy bands (not shown in Fig. 4.37) and do not affect the magnetisation of the metal. The electrons (ten per atom according to Table 2.5) from the outer shells of the atoms occupy, at $T = 0$, all the states in the conduction bands shown in Fig. 4.37 with energy below E_F. All states with energy above E_F are empty. We see that the minority bands (spin down) are shifted upward in energy relative to the majority bands (spin up); it can be seen that while certain majority bands lie entirely below E_F, the corresponding minority bands lie partly above E_F. Consequently there are more states of spin up occupied than are states of spin down, which tells us that nickel has a spontaneous magnetisation at $T = 0$.

The densities of one-electron states per atom, denoted by $V_o\rho_\uparrow(E)$ and $V_o\rho_\downarrow(E)$, and the corresponding integrated densities per atom, denoted by $n_{A\uparrow}(E)$ and $n_{A\downarrow}(E)$ respectively (we remember that $n_{A\uparrow(\downarrow)}(E)$ gives the number of one-electron states of spin up (down) in the conduction bands with energy below E), are shown in Fig. 4.38. The number of conduction-band electrons with spin up, per atom, is given by $n_{A\uparrow}(E_F)$ and the corresponding quantity for spin down by $n_{A\downarrow}(E_F)$. The difference between the two times μ_B gives the magnetisation of nickel at $T = 0$. It turns out to be $0.59\mu_B$ per atom, in good agreement with the experimental value ($0.6\mu_B$ per atom)[24]. It is worth noting that $n_{A\uparrow}(E_F) + n_{A\downarrow}(E_F) = 10$ as it should be, given that there are 10 electrons per atom in the conduction bands of nickel.

The quantities $n_\uparrow(\boldsymbol{r})$ and $n_\downarrow(\boldsymbol{r})$, which define, through equations (4.85)–(4.88) the mean potential fields $U_\uparrow(\boldsymbol{r})$ and $U_\downarrow(\boldsymbol{r})$ seen by an electron of spin up and spin down respectively, are calculated self-consistently as in the case of non-magnetic metals, except that the calculation is twice as lengthy, since the potential and the corresponding orbitals of the electron are different for the two spin orientations. We have (by analogy to (4.92))

$$\begin{aligned} n_\uparrow(\boldsymbol{r}) &= \sum_{\boldsymbol{k}\alpha} |\psi_{\boldsymbol{k}\alpha\uparrow}(\boldsymbol{r})|^2 f(E_{\alpha\uparrow}(\boldsymbol{k})) \\ n_\downarrow(\boldsymbol{r}) &= \sum_{\boldsymbol{k}\alpha} |\psi_{\boldsymbol{k}\alpha\downarrow}(\boldsymbol{r})|^2 f(E_{\alpha\downarrow}(\boldsymbol{k})) \end{aligned} \tag{4.94}$$

where $\psi_{\boldsymbol{k}\alpha\uparrow(\downarrow)}$ is the orbital part of an electron state of spin up (down) and energy $E_{\alpha\uparrow(\downarrow)}(\boldsymbol{k})$. The requirement of self-consistency means that $n_\uparrow$ and $n_\downarrow$ of (4.94) must be the same with the densities which led to the $U_\uparrow$ and $U_\downarrow$ which led in turn to the $\psi_{\boldsymbol{k}\alpha\uparrow}$ and $\psi_{\boldsymbol{k}\alpha\downarrow}$ appearing in (4.94). This is achieved by repeating the cycle of calculation of the above quantities, in the manner we have described in the previous section, until convergence.

It is worth noting that, if we begin our calculation with $n_\uparrow^{(0)}(\boldsymbol{r}) \neq n_\downarrow^{(0)}(\boldsymbol{r})$ and the metal under consideration is non-magnetic, we shall obtain, after convergence has been achieved, $n_\uparrow(\boldsymbol{r}) = n_\downarrow(\boldsymbol{r})$ as expected. On the other hand, if we begin with $n_\uparrow^{(0)}(\boldsymbol{r}) = n_\downarrow^{(0)}(\boldsymbol{r})$ we shall always obtain a non-magnetic final result $n_\uparrow = n_\downarrow$ even if the metal is magnetic (!) Once a symmetry has been imposed on our potential field we cannot escape from it

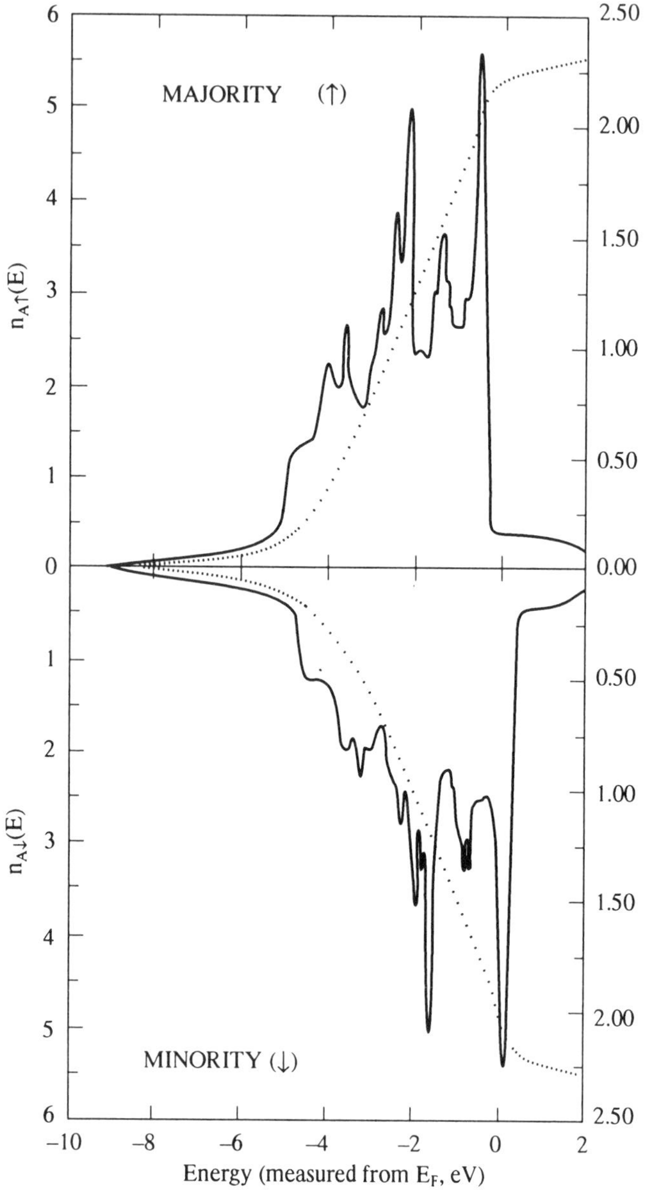

Figure 4.38. Densities of one-electron states per atom of ferromagnetic nickel. The dotted line gives the integrated densities of states $n_{A\uparrow(\downarrow)}(E)$ [Redrawn with permission from L. Moruzzi, *et al.* loc. cit.]

by a self-consistency cycle. The non-magnetic solution of a ferromagnetic metal does not correspond to the ground state (the state of least total energy) of the metal. It describes, approximately, the paramagnetic phase of it, which however does not exist below the Curie temperature.

4.11.3 TOTAL ENERGY OF THE METAL

The total energy of the metal, in its ground state, with the nuclei motionless at their equilibrium positions (4.4), is given, in the mean-field approximation, by the following expression (obtained by an obvious extension of (2.97))

$$\begin{aligned}
\varepsilon^0 = {} & \sum_{k\alpha}\{E_{\alpha\uparrow}(\boldsymbol{k})f(E_{\alpha\uparrow}(\boldsymbol{k})) + E_{\alpha\downarrow}(\boldsymbol{k})f(E_{\alpha\downarrow}(\boldsymbol{k}))\} \\
& - \int\{U_\uparrow(\boldsymbol{r})n_\uparrow(\boldsymbol{r}) + U_\downarrow(\boldsymbol{r})n_\downarrow(\boldsymbol{r})\}\,\mathrm{d}V - \frac{Ze^2}{4\pi\epsilon_O}\sum_i\int\frac{n(\boldsymbol{r})\,\mathrm{d}V}{|\boldsymbol{R}_i - \boldsymbol{r}|} \\
& + \frac{e^2}{8\pi\epsilon_O}\iint\frac{n(\boldsymbol{r})n(\boldsymbol{r}')}{|\boldsymbol{r} - \boldsymbol{r}'|}\,\mathrm{d}V\,\mathrm{d}V' \\
& - \frac{9}{4}\alpha\left(\frac{e^2}{4\pi\epsilon_O}\right)\int\left\{\left(\frac{3}{4\pi}n_\uparrow(\boldsymbol{r})\right)^{1/3}n_\uparrow(\boldsymbol{r}) + \left(\frac{3}{4\pi}n_\downarrow(\boldsymbol{r})\right)^{1/3}n_\downarrow(\boldsymbol{r})\right\}\mathrm{d}V \\
& + \frac{(Ze)^2}{8\pi\epsilon_O}\sum_i\sum_{j\neq i}\frac{1}{|\boldsymbol{R}_i - \boldsymbol{R}_j|}
\end{aligned} \tag{4.95}$$

The first two terms of (4.95), taken together, give the kinetic energy of the electrons; the third term gives the potential energy of attraction between the electrons and the nuclei (*Ze* is the charge of each nucleus); the fourth term gives the potential energy of repulsion between the electrons; the fifth term, the so-called exchange energy, relates to the interaction of the electrons with their Fermi-holes; and the last term gives the repulsion energy between the stationary nuclei. It is understood that the one-electron energy levels and electronic densities appearing in (4.95) must be calculated self-consistently.

We must emphasise, however, that even the best of mean-field calculations will not reproduce accurately the total energy of the metal. For that, one must take into account, as in the case of atoms (see Section 2.15.5) the correlation energy (the residual interaction between the electrons of the metal which the mean-field calculation does not take into account). The negative of the total energy of the metal represents the energy required to remove all constituents (electrons and nuclei) to infinite distance from each other at rest, and it is a very large quantity indeed; the correlation energy is, as in the atomic case, a very small fraction of the total energy, and in this respect ε^0 is a very good approximation to the total energy of the metal. However, ε^0 is not of much interest by itself. What we should like to know, for example, is the so-called cohesive energy of the metal, which we define to be the total energy of the atom minus the total energy of the metal (per atom), which equals the energy required (per atom) to dissociate the metal into its constituent atoms. In this respect, correlation energy is not negligible and must be taken into account. Now it turns out that one can calculate, approximately, the correlation energy of a metal, or for that matter of any many-electron system (atom, molecule, solid), by introducing into the mean field seen by an electron a correlation hole, by which we mean a suppression of the electronic cloud within a region about the electron, in addition to that produced by our familiar Fermi hole. We shall not be concerned with the details of the method which is known in the literature as the density functional method. Moruzzi, Janak and Williams[25] using this method calculated the total energies (including correlation) of the atoms with atomic numbers between 19 and 48 and those of the corresponding metals; they tell us that although the calculated total energies as such are not accurate (for either metals or atoms), their differences are, due to a systematic cancellation of errors, a good approximation to the cohesive energies of these metals. The calculated values of the cohesive energies of these metals varying between 0.10 and 0.50 Rydbergs per atom are in good agreement with the experimental values of these quantities.

We note that as far as the correlation energy of an atom is concerned there are more reliable ways of calculating it, and the corresponding total energy of the atom, than the above mentioned density functional method. However, it would not do to calculate the energy of the atom by one method (though more accurate) and the energy of the metal by the density functional method, for then the systematic cancellation of errors, referred to above, will not occur and the result for the cohesive energy will be worse rather than better.

One may inquire, at this stage, as to the nature of the chemical bond in a metal. We can answer this question by looking at $n(\boldsymbol{r})$, i.e. at the distribution of the electronic cloud in space. We find an increase in the electronic cloud in between the nuclei of the metal which suggests a mechanism of bonding very similar to that of the covalent bond of a homonuclear diatomic molecule (see Section 3.4). In metals the close packing of atoms blurrs to some degree the angular variation of this distribution about any given atom. In the more open structures of the semiconductors germanium and silicon, one can see more clearly the concentration of electronic cloud in 'bonds' between an atom and its four neighbours (see Fig. 4.10). Finally, in the so-called ionic crystals one has the analogue of an ionic bond. In NaCl, for example, there is an excess of about one electron about each Cl and a deficiency of about one electron about each Na, the crystal being held together by the electrostatic forces between the oppositely charged ions.

In conclusion we may say that (4.95) gives a good estimate of the total energy of a metal (and the same is true for a semiconductor or insulator), but for the evaluation of the cohesive energy of a crystal one must take into account, by some means or another, the electron correlation energy. The same applies if one wishes to estimate the stability of a certain structure in relation to some other; in order to know, for example, why a certain material crystallises as fcc, or bcc and not as some other space lattice. Similarly one may wish to know how the total energy varies with the lattice constant in order to estimate the compressibility of the material, etc. In such calculations the correlation energy must be taken into account.

We shall close this section with a brief discussion of the total energy (at $T = 0$) of a ferromagnetic metal. Ferromagnetism implies that (4.95) has a lower value when $n_{\uparrow}(\boldsymbol{r}) \neq n_{\downarrow}(\boldsymbol{r})$. Taking into account correlation does not change the above result. That spin-polarisation lowers the ground-state energy of a metal should not surprise us; we know that spin-polarisation lowers the ground-state energy of an atom (see Hund's rule; Section 2.15.2) and the physical reason behind that fact applies to a metal as well. Spin-polarisation implies that any two electrons of the metal are more likely to have the same spin than otherwise, and to keep apart because of this (due to the Pauli principle); and this reduces the average Coulomb repulsion energy between the electrons. However, the ground state of a metal is not always spin-polarised; if after a set of would-be majority bands (see Fig. 4.37) are complete with electrons, one has to raise the Fermi level way up in order to accommodate the remaining electrons, the increase in the so-called band energy (first term of (4.95)) will outweigh the above-mentioned decrease in the energy due to exchange correlation (the tendency of electrons of the same spin to stay away from each other), and a ferromagnetic ground state will not be sustained. Only when most of the conduction bands of the metal are rather narrow (as are the d-bands of nickel shown in Fig. 4.37), may spin-polarisation be sustained.

Finally, we note that, in the above discussion we have assumed that the nuclei of the crystal are stationary. In reality, the nuclei vibrate and there is a minimum of energy (at $T = 0$) associated with their vibration, and that must be added to the total electronic energy to obtain the total energy of the crystal.

4.12 Lattice vibrations

These can be described in essentially the same way as the vibrations of a polyatomic molecule (see Section 3.6). The basic assumption is the same, namely that the displacement of the atomic nuclei from their equilibrium positions is sufficiently small, so that the harmonic approximation remains valid; and then we proceed in the same manner to find the classical normal modes of vibration (coupled harmonic oscillations of the nuclei of the crystal) and, finally, we obtain the quantum-mechanical oscillators corresponding to these normal modes in exactly the same way as we have done for the molecular vibrations.

The displacement from its equilibrium position[26] of the nth atom of the ith primitive cell associated with a certain normal mode of vibration ($\boldsymbol{k}\alpha$) can be written as in (3.32):

$$X_{ni}^{(\boldsymbol{k}\alpha)}(t) = X_{oni}^{(\boldsymbol{k}\alpha)} \exp(\mathrm{i}\omega_\alpha(\boldsymbol{k})t), \quad Y_{ni}^{(\boldsymbol{k}\alpha)}(t) = Y_{oni}^{(\boldsymbol{k}\alpha)} \exp(\mathrm{i}\omega_\alpha(\boldsymbol{k})t),$$
$$Z_{ni}^{(\boldsymbol{k}\alpha)}(t) = Z_{oni}^{(\boldsymbol{k}\alpha)} \exp(\mathrm{i}\omega_\alpha(\boldsymbol{k})t) \tag{4.96a}$$

In the present case, however, the periodicity of the crystal allows us to write

$$X_{oni}^{(\boldsymbol{k}\alpha)} = \exp(\mathrm{i}\boldsymbol{k}\cdot\boldsymbol{R}_i)X_{on}^{(\boldsymbol{k}\alpha)}, \quad Y_{oni}^{(\boldsymbol{k}\alpha)} = \exp(\mathrm{i}\boldsymbol{k}\cdot\boldsymbol{R}_i)Y_{on}^{(\boldsymbol{k}\alpha)},$$
$$Z_{oni}^{(\boldsymbol{k}\alpha)} = \exp(\mathrm{i}\boldsymbol{k}\cdot\boldsymbol{R}_i)Z_{on}^{(\boldsymbol{k}\alpha)} \tag{4.96b}$$

Where $\boldsymbol{R}_i$ is the lattice vector (4.4) associated with the ith primitive cell of the crystal, and $\boldsymbol{k}$ is a wavevector within the BZ of the reciprocal lattice of the crystal defined in Section 4.2. We can see (4.96b) as the application of Bloch's theorem (Section 4.3) to the description of lattice vibrations. Equation (4.96b) tells us that in every cell the atom on site n moves in the same direction with the same amplitude; only the phase changes from cell to cell. We note that, as in the case of polyatomic molecules (see (3.33)), the actual dispacement of the atom (a real quantity) is given by a linear sum of the real and imaginary parts of (4.96a).

We can argue, as in the description of electron states, that values of $\boldsymbol{k}$ differing by a reciprocal lattice vector (4.5) do not lead to different modes (4.96), and that therefore we need only consider values of $\boldsymbol{k}$ within the BZ. We remember, also, that Bloch's theorem disregards the finite extension of the real crystal, which must be taken into account, in order to obtain the correct number of normal modes in any given frequency interval. We do this by imposing periodic boundary conditions on (4.96), as for the electron states, and this, naturally, leads to the same set (4.24) of allowed values of $\boldsymbol{k}$, namely

$$\boldsymbol{k} = (k_x, k_y, k_z) = \frac{2\pi}{L}(n_x, n_y, n_z) \tag{4.97}$$
$$n_x, n_y, n_z = 0, \pm 1, \pm 2, \pm 3, \ldots$$

where L^3 is the volume of the crystal; we know, from Section 4.3, that there are N such points in the BZ, where N is the number of primitive cells in the crystal. We find that the replacement of the finite crystal by an infinite one (with imposition of periodic boundary conditions) gives a very good description of the vibrations of the nuclei in the interior (bulk) of the crystal; it fails, as we should expect, to describe correctly the vibration of the nuclei within a region of 10 Å or so from the surface of the crystal.

One can show that for given $\boldsymbol{k}$ (any one of the allowed points (4.97) in the BZ) one obtains 3ν normal modes, where ν is the number of atoms in a primitive cell. The corresponding normal frequencies are denoted by $\omega_\alpha(\boldsymbol{k})$, $\alpha = 1, 2, \ldots, 3\nu$; these and the corresponding displacement vectors defined by (4.96) are obtained by solving the classical equations of motion of the nuclei.[27]

Because the allowed points (4.97) are distributed uniformly and densely in BZ and because a small change in $\boldsymbol{k}$ produces a small change in $\omega_\alpha(\boldsymbol{k})$ we can, for the sake of simplicity, present $\omega_\alpha(\boldsymbol{k})$ as continuous functions of $\boldsymbol{k}$. In Fig. 4.39 we show the normal frequency bands $\omega_\alpha(\boldsymbol{k})$, $\alpha = \mathrm{T}_1, \mathrm{T}_2, \mathrm{L}$ along various symmetry lines of the BZ (Fig. 4.14) of a typical fcc crystal (lead) with one atom per primitive cell. We note, to begin with, that the three bands extend over the whole BZ, and since there are N normal modes in each band (corresponding to the N allowed points (4.97)) we have in all $3N$ normal modes, i.e. three modes per atomic nucleus, as expected.

We observe that for $\boldsymbol{k}$ along any given direction (defined by angles θ, φ as in Fig. 2.1) every one of the above bands varies linearly with k near $k = 0$ (the centre Γ of the BZ). We may write

$$\omega_\alpha(k) = c_\alpha(\theta, \varphi)k \tag{4.98}$$

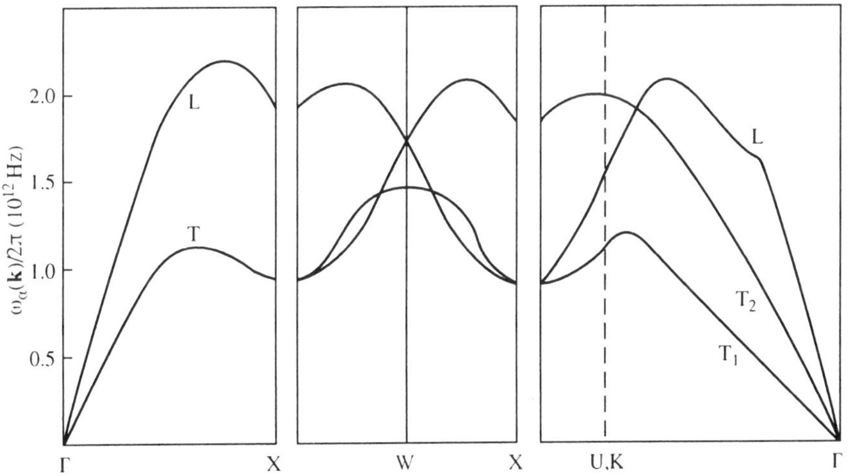

Figure 4.39. Normal frequency-bands of the vibrating lattice, along various symmetry lines of the BZ (Fig. 4.14), of lead. (Redrawn from B. N. Brockhouse, T. Arase, G. Caglioti, K. R. Rao and A. D. B. Woods, *Physical Review*, **128** (1962) 1099)

for k along the direction (θ, φ) and small. These long-wavelength vibrational waves are acoustic waves and $c_\alpha(\theta, \varphi)$ is the sound velocity in the given crystal for the given direction.

We note that one can build wave packets from vibrational Bloch waves (4.96) of approximately the same wavevector in the same way as we have done for electrons (see Section 4.8); the transport velocity in the direction (θ, φ) of such a wavepacket, in the long-wavelength limit $(\lambda = 2\pi/k \to \infty)$, will be given by $\partial\omega_\alpha/\partial k = c_\alpha(\theta, \varphi)$, according to (4.98), assuming that only waves from a single band α contribute to the wavepacket. The fact that $c_\alpha(\theta, \varphi)$ does not depend on k means that there will be no dispersion of the kind depicted in Figs. 1.13 and 4.26. However, the variation of $\omega_\alpha(k)$ with k, as k increases, becomes more complicated, especially near the zone boundary as can be seen from Fig. 4.39, and there of course dispersion does occur.

Finally, we note that along certain symmetry lines, frequency bands may become degenerate. For example, the two bands $\alpha = T_1, T_2$ (see Fig. 4.39) of a cubic crystal, coincide along the ΓX line. In the corresponding modes of vibration, the nuclei move normal to the direction of $\boldsymbol{k}$ (x-direction); and by symmetry, if the nuclei move in, say, the y-direction in the $\alpha = T_1$ mode, then in the $\alpha = T_2$ mode they move in the z-direction. We refer to the above two modes of vibration as transverse modes. In the third mode, the so-called longitudinal mode ($\alpha = L$ in Fig. 4.39), the nuclei vibrate in the direction of $\boldsymbol{k}$ (x-direction). For $\boldsymbol{k}$-points off the ΓX symmetry line the degenerate band splits into two separate frequency bands. We continue to refer to the corresponding modes as the transverse modes and to the third mode as the longitudinal mode, although this is somewhat misleading, since the actual vibration of the atoms need not be strictly along or normal to $\boldsymbol{k}$.

In Fig. 4.40 we show schematically the normal-frequency bands $\omega_\alpha(\boldsymbol{k})$, $\alpha = 1, 2, \ldots, 6$ of a crystal with two atoms per primitive cell. The three lower-frequency bands correspond to the so-called acoustic modes of vibration and the three higher-frequency bands to the so-called optical modes of vibration. In crystals, such as NaCl, with large electric dipole moments per unit volume, the optical modes couple with electromagnetic waves of the right frequency, often so strongly, leading in some instances to modes of vibration which cannot be disentangled from the associated electromagnetic field.

In many applications one needs only to know the density of normal modes, to be denoted by $\mathcal{D}(\omega)$. By definition $\mathcal{D}(\omega)\,d\omega$ is the number of normal modes of vibration, per unit volume of the crystal, with frequency between ω and $\omega + d\omega$. Once the frequency bands $\omega_\alpha(\boldsymbol{k})$ have been obtained, one can proceed with the calculation of $\mathcal{D}(\omega)$ in exactly the same way as with the calculation of the density of one-electron states described in Section 4.5. We need not say anything more in relation to the general

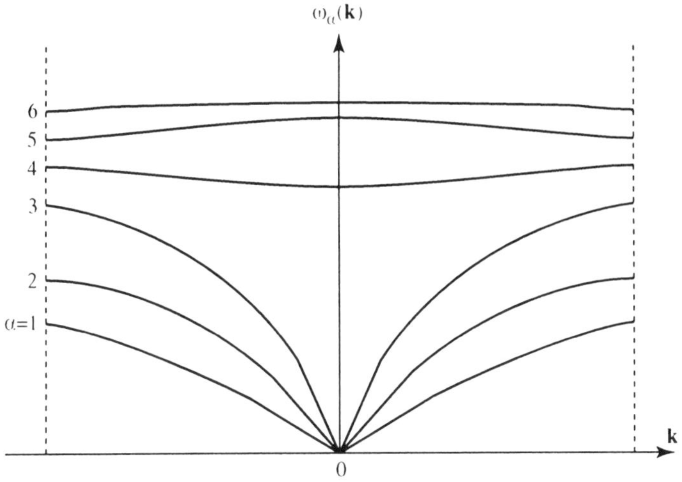

Figure 4.40. Vibrational frequencies $\omega_\alpha(\boldsymbol{k})$, $\alpha = 1, 2, 3$ (acoustic modes) and $\omega_\alpha(\boldsymbol{k})$, $\alpha = 4, 5, 6$ (optical modes) of a hypothetical crystall with two atoms per primitive cell, for $\boldsymbol{k}$-points along a general line of the BZ

problem of calculating $\mathcal{D}(\omega)$; but it is worthwhile to obtain $\mathcal{D}(\omega)$ for the so-called Debye model of the lattice vibrations of a crystal, for which the calculation can be carried out analytically.

In the Debye model one assumes that there are three acoustic-like bands described by

$$\omega_\alpha(k) = c_s k = c_s(k_x^2 + k_y^2 + k_z^2)^{1/2}, \quad \alpha = 1, 2, 3 \tag{4.99}$$

where c_s is an average velocity of sound, and $\boldsymbol{k}$ takes all values (4.97) within a sphere of radius k_D, such that

$$3[(4\pi k_D^3/3)/(8\pi^3/L^3)] = 3N\nu$$

We recall that L^3 denotes the volume of the crystal, N the number of primitive cells in it, and ν the number of atoms per primitive cell. The quantity in the square brackets is the number of normal modes per band in the Debye sphere (according to (4.97)), and multiplied by three (the number of bands) gives the total number of normal modes in the Debye model of the lattice, which must equal $3N\nu$ (the number of normal modes in the actual crystal). Introducing v (the volume per atom) and solving the above equation with respect to k_D we obtain

$$k_D = (6\pi^2/v)^{1/3} \tag{4.100}$$

We define a Debye wavelength by

$$\lambda_D \equiv 2\pi/k_D \tag{4.101}$$

and a Debye cut-off frequency by

$$\omega_{\mathrm{D}} \equiv c_{\mathrm{s}} k_{\mathrm{D}} \tag{4.102}$$

Now, in the Debye model the number of normal modes with frequency between ω and $\omega + \mathrm{d}\omega$, equals three times the number of $\boldsymbol{k}$-points (4.97) with magnitude between k and $k + \mathrm{d}k$, where $k = \omega/c_{\mathrm{s}}$ and $\mathrm{d}k = \mathrm{d}\omega/c_{\mathrm{s}}$. This is easily calculated and one finds

$$\begin{aligned}\mathscr{D}(\omega)\,\mathrm{d}\omega &= \frac{3\omega^2\,\mathrm{d}\omega}{2\pi^2 c_{\mathrm{s}}^3}, \quad \omega < \omega_{\mathrm{D}} \\ &= 0, \quad \omega > \omega_{\mathrm{D}}\end{aligned} \tag{4.103}$$

The quantisation of the normal modes of vibration of the crystal is effected in the same way as that of the vibrational modes of a polyatomic molecule. To each mode, of frequency $\omega_\alpha(\boldsymbol{k})$, corresponds a quantum-mechanical oscillator with energy eigenvalues $\hbar\omega_\alpha(\boldsymbol{k})(v + \frac{1}{2})$ where $v = 0, 1, 2, 3, \ldots$ as usual. The average energy of the oscillator, at thermal equilibrium, is given in accordance with (3.39) by

$$\overline{E_\alpha(\boldsymbol{k})} = \frac{\hbar\omega_\alpha(\boldsymbol{k})}{2} + \frac{\hbar\omega_\alpha(\boldsymbol{k})}{\exp\left[\hbar\omega_\alpha(\boldsymbol{k})/k_{\mathrm{B}}T\right] - 1} \tag{4.104}$$

The physical meaning of the two terms of (4.104) is the same as for the corresponding terms of (3.39); and, as then, we can think of

$$n(\boldsymbol{k}\alpha) = \frac{1}{\exp\left[\hbar\omega_\alpha(\boldsymbol{k})/k_{\mathrm{B}}T\right] - 1} \tag{4.105}$$

as the average number of vibration quanta (we call them phonons) of the $\boldsymbol{k}\alpha$ type (mode of vibration). Phonons are obviously bosons; the distribution (4.105), like the corresponding one (3.40) for polyatomic molecules, is the Bose–Einstein distribution (2.72) with $\mu = 0$.

In evaluating the temperature variation of the energy associated with the lattice vibrations of the crystal, we can disregard the first term of (4.104), in which case the vibrational energy of the crystal can be written

$$\varepsilon_{\mathrm{vb}} = \sum_{\boldsymbol{k}\alpha} n(\boldsymbol{k}\alpha)\hbar\omega_\alpha(\boldsymbol{k}) = V\int \left(\frac{\hbar\omega}{\exp(\hbar\omega/k_{\mathrm{B}}T) - 1}\right)\mathscr{D}(\omega)\,\mathrm{d}\omega \tag{4.106}$$

and the corresponding contribution to the specific heat of the crystal becomes

$$C_{\mathrm{V,vb}} = \frac{\mathrm{d}\varepsilon_{\mathrm{vb}}}{\mathrm{d}T} = k_{\mathrm{B}}V\int \frac{(\hbar\omega/k_{\mathrm{B}}T)^2 \exp(\hbar\omega/k_{\mathrm{B}}T)}{\left[\exp(\hbar\omega/k_{\mathrm{B}}T) - 1\right]^2}\mathscr{D}(\omega)\,\mathrm{d}\omega \tag{4.107}$$

The integration in (4.107) can be performed analytically when $\mathcal{D}(\omega)$ is given by (4.103) and then we obtain

$$C_{V,vb} = N\nu k_B (T/\Theta_D)^3 I(\Theta_D/T) \tag{4.108}$$

where $N\nu$ is the total number of atoms in the crystal; Θ_D is the so-called Debye temperature defined by

$$k_B\Theta_D \equiv \hbar\omega_D \tag{4.109}$$

and $I(x)$ is defined by

$$I(x) = 9\int_0^x \frac{z^4 e^z}{(e^z - 1)^2}\,dz$$

The variation of $C_{V,vb}$ with temperature obtained from (4.108) is shown graphically in Fig. 4.41. At high temperatures (when $T \gg \Theta_D$) we obtain

$$C_{V,vb} = 3N\nu k_B \tag{4.110}$$

which is what we would have obtained if we had modelled each nucleus by three classical linear harmonic oscillators (along the x, y and z directions) vibrating independently from each other and from all other oscillators in the crystal. At low temperatures (when $T \ll \Theta_D$) we obtain

$$C_{V,vb} \simeq \frac{12\pi^4}{5} N\nu k_B (T/\Theta_D)^3 \tag{4.111}$$

in accordance with our earlier statement of this result (in Section 4.7).

The Debye law (4.108) describes very well $C_{V,vb}$ for most crystals, and the Debye temperature is tabulated as a physical parameter of the solid. $k_B\Theta_D$ represents the energy of the maximum quantum that can be absorbed or emitted by a lattice mode of vibration. We conclude this section with a short list of Θ_D values:

	Al	Fe	Ni	Cu	As	Se
$\Theta_D(K)$	390	420	375	330	224	135

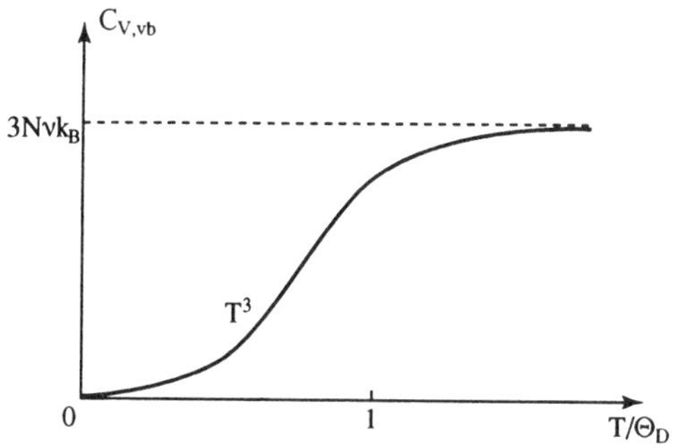

Figure 4.41. The Debye law of specific heat

4.13 Surfaces

4.13.1 GENERAL PROPERTIES

In discussing surface properties, it is necessary to refer to crystallographic planes. A crystallographic plane is defined by the length of its intercepts on the crystal axes (the three edges of the unit shell shown in Fig. 4.7), measured from the origin of coordinates. The intercepts are expressed in terms of the dimensions of the unit cell, which define the unit lengths along the three axes. The reciprocals of these intercepts reduced to the smallest three integers having the same ratio are known as Miller indices. We note that while the intercepts of parallel planes are different, their ratio is constant; so that a particular set of Miller indices defines a set of parallel planes.

Examples of Miller indices of planes are shown in Fig. 4.42. The shaded plane in (1) has intercepts 1, ∞, ∞ and therefore its Miller indices are (1 0 0); the shaded plane in (2) has intercepts 1, 1, ∞ and Miller indices (1 1 0); the one in (3) has intercepts ∞, 1, 1 and indices (0 1 1); the one in (4) has intercepts 1, 1, 1 and indices (1, 1, 1); the one in (5) has intercepts -1, 1, 1 and indices ($\bar{1}$ 1 1) where the line above the number indicates a negative sign, and finally the shaded plane in (6) has intercepts 1, 1, $\frac{1}{2}$ and Miller indices (1 1 2).

Parentheses around a set of Miller indices, as in Fig. 4.42, denote a single

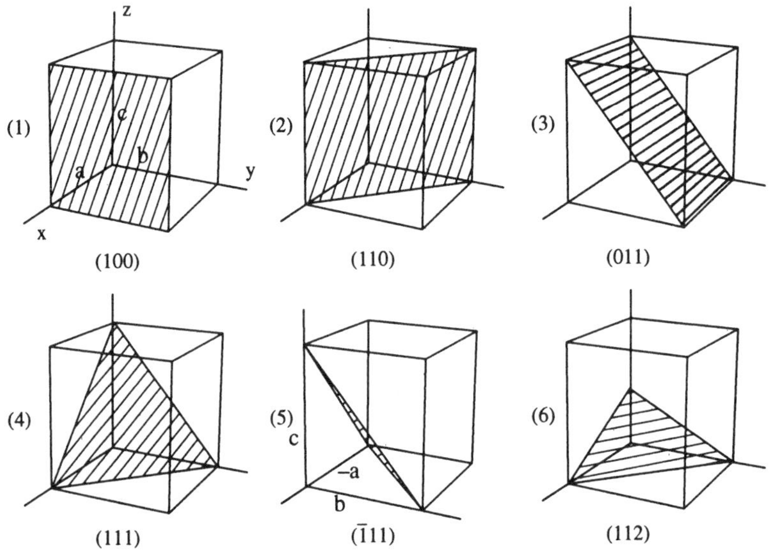

Figure 4.42. Examples of Miller indices of planes

set of parallel planes. In many space lattices, different sets of planes are equivalent to one another. For example, in a cubic lattice the planes (1 1 0), (1 0 1), (0 1 1), $(1\,\bar{1}\,0)$, $(1\,0\,\bar{1})$ and $(0\,1\,\bar{1})$ are equivalent to one another; and we note that reversing the signs of all indices merely denotes another parallel plane, e.g., $(\bar{1}\,\bar{1}\,0)$ is parallel to (1 1 0).

We can view a given space lattice as a sequence of parallel planes. These could be the (1 0 0) planes, or the (1 1 0) planes or any other set of parallel planes. As an example, we show in Fig. 4.43 the lattice sites of a body-centred cubic (bcc) lattice as a sequence of (1 1 0) planes. In this figure the circles represent the lattice sites of a given plane (which we take as the xy plane) and the squares, those of the next plane along the positive z direction. The sites on the xy plane (circles in Fig. 4.43) constitute a two-dimensional (2D) lattice:

$$\boldsymbol{R}_n^{(2)} = n_1\boldsymbol{a}_1 + n_2\boldsymbol{a}_2 \tag{4.112}$$

where $\boldsymbol{a}_1$ and $\boldsymbol{a}_2$ are two primitive vectors in the xy plane, and n_1 and n_2 take all integer values between $-\infty$ and $+\infty$. The sites of the second plane (squares in Fig. 4.43) are obtained from (4.112) by a simple translation described by a primitive vector $\boldsymbol{a}_3$. In an infinite lattice (extending from $z = -\infty$ to $z = +\infty$) the sites of the ith plane ($i = 0, \pm1, \pm2, \pm3, \ldots$) are obtained from those of the $(i-1)$th plane by the same primitive translation $\boldsymbol{a}_3$. In our example

$$\boldsymbol{a}_1 = \frac{a}{\sqrt{2}}\hat{x} + \frac{a}{2}\hat{y},\ \boldsymbol{a}_2 = \frac{a}{\sqrt{2}}\hat{x} - \frac{a}{2}\hat{y} \tag{4.113a}$$

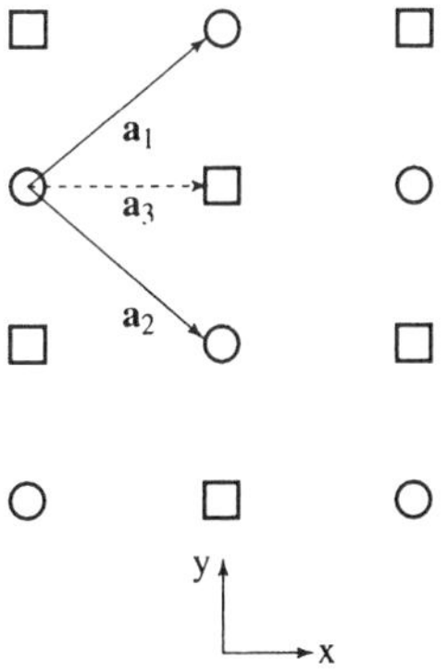

Figure 4.43. The lattice sites of two successive (1 1 0) planes of a (bcc) lattice. The circles represent the sites of a given (1 1 0) plane (taken as the xy plane) and the squares those of the next (1 1 0) plane along the positive z direction. The i_{th} plane along the z direction is obtained from the $(i-1)_{\text{th}}$ plane by the primitive translation $\boldsymbol{a}_3$. We note that the sites of the i_{th} plane have the same projection on the xy plane as those of the $(i-2)_{\text{th}}$ plane

and

$$\boldsymbol{a}_3 = \frac{a}{\sqrt{2}}(\hat{x} + \hat{z}) \tag{4.113b}$$

where a is the lattice constant. We note that by construction the interplane distance is equal to the z-component of $\boldsymbol{a}_3$.

Obviously, $\boldsymbol{a}_1$, $\boldsymbol{a}_2$, and $\boldsymbol{a}_3$ constitute a set of primitive vectors of the three-dimensional (3D) lattice

$$\boldsymbol{R}_n^{(3)} = n_1\boldsymbol{a}_1 + n_2\boldsymbol{a}_2 + n_3\boldsymbol{a}_3 \tag{4.114}$$

where n_1, n_2, n_3 take all integer values from $-\infty$ to $+\infty$. Formulae (4.114) and (4.113) provide an alternative, exactly equivalent to that given by (4.1) and (4.2), description of the bcc lattice. One can easily check, for example, that the primitive cell defined by $\boldsymbol{a}_1$, $\boldsymbol{a}_2$ and $\boldsymbol{a}_3$ has volume $a^3/2$, as expected.

We can, as we have already noted, describe the bcc lattice as a sequence of any set of parallel planes. The procedure is the same as for the (1 1 0) planes. In each case one defines two primitive vectors $\boldsymbol{a}_1$ and $\boldsymbol{a}_2$ in the xy plane (which coincides with one of the planes under consideration) and a 2D lattice (4.112) corresponding to them. The above 2D lattice taken together with $\boldsymbol{a}_3$ (a primitive vector which takes us from a lattice point in a given plane to one in the next plane along the z direction) provides an equivalent description of the bcc lattice. And we can do the same for the fcc lattice or any other space lattice.

When there is only one atom per lattice site, we can assume that the atoms are centred on the lattice sites and in this case we can imagine the crystal as a stack of planes of atoms (atomic layers) parallel to a given crystallographic plane. The case of two, or more than two, atoms per lattice site is not very different; we obtain two or more planes of atoms for each plane of lattice points. In what follows we may assume for the sake of simplicity that there is only one atom per lattice site.

Let us then consider a crystal as a stack of 2D atomic layers, parallel to a given crystallographic plane, extending from $z = -\infty$ to $z = +\infty$ (the z-axis being normal to the crystallographic plane); and let us, then, remove all atomic layers on the positive side of the z axis; the result is a semi-infinite crystal extending from $z = -\infty$ to $z \simeq 0$. The resulting surface (metal-vacuum interface at $z \simeq 0$) bears the name of the corresponding crystallographic plane; we call the surface corresponding to the (1 0 0) crystallographic plane the (1 0 0) surface; the one corresponding to the (1 1 0) plane we call the (1 1 0) surface, etc..

The first question one has to answer is whether the equilibrium positions of the atomic nuclei in the surface region are the same as in the corresponding infinite crystal. The spatial arrangement of the nuclei in the surface region can be determined experimentally; a powerful technique for this purpose is that of low-energy electron diffraction (LEED); a beam of electrons with well defined energy and momentum is directed on the surface; a small fraction of the electrons are elastically scattered backward into certain directions (directions of diffraction) which define the 2D periodicity of the surface atomic layers; and moreover the variation with energy of the intensity of the diffracted beams allows one to determine, with very good accuracy in many cases, the equilibrium positions of the nuclei in the surface region. We find that the arrangement of the atomic nuclei at most metal surfaces is not very different from that in the bulk of the metal; in many cases, one observes a small (less than 10%) increase or decrease in the interlayer distance near the surface, mainly between the top surface layer and the one beneath it; this difference rapidly diminishes and practically vanishes two or three layers in from the surface. Semiconductor surfaces are less stable and in some cases considerable reconstruction has been observed. The surface layers (say the top two or three atomic layers) usually retain a 2D periodicity but the surface unit cell is larger than that of the unreconstructed surface. This may happen, for example, because the atomic nuclei in every other row of the top atomic plane move away from their original positions by a specified amount. A discussion of surface reconstruction lies beyond the scope of the present book. In what follows we shall assume that the surface under consideration has not reconstructed; i.e. that the positions of the nuclei in the surface region are the same as in the corresponding infinite crystal. What we have to say applies not only to metal surfaces, for which this is a reasonably good model, but also, at a semi-quantitative level, to semiconductor surfaces as well.

The potential field an electron sees near the surface of a crystal is of course different from that in the interior of the crystal, even for an unreconstructed surface. In Fig. 4.44 we show, schematically, the variation of the potential field in a metal which extends from $z = -\infty$ to $z \simeq 0$. In this instance the metal–vacuum interface (the plane $z = 0$) is taken a short distance above the top surface atomic plane. The solid line represents the average value $\bar{U}(z)$ of the potential field $U(\boldsymbol{r})$ over the plane $z =$ constant. We see that on the vacuum side of the metal–vacuum interface ($z > 0$) the potential field increases monotonically and reaches at some distance from the interface, of the order of a few Å, a constant value $E_F + \varphi$, where E_F denotes as usual the Fermi level of the metal and φ is the so-called work function of the electron for the given surface. We note that for $z \gtrsim 2$ Å, U becomes independent of x and y, and therefore $U = U(z) = \bar{U}(z)$ for $z \gtrsim 2$ Å. It turns out that asymptotically $U(z)$ is given by the so-called image law of classical electrostatics, i.e.

$$U(z) \simeq E_F + \varphi - \frac{e^2}{16\pi\epsilon_0 z} \tag{4.115}$$

when z is greater than 2 Å or so.

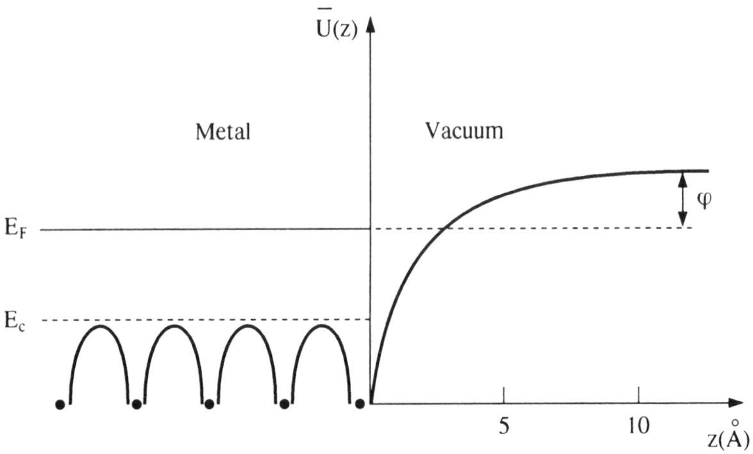

Figure 4.44. Potential barrier at the surface of a metal. The metal extends from $z = -\infty$ to $z \simeq 0$. The solid circles denote the positions of the atomic planes parallel to a given surface. The metal-vacuum interface (the plane $z = 0$) is taken a short distance above the surface atomic plane. E_c denotes the bottom of the conduction band

It follows from Fig. 4.44 that an electron cannot escape from the metal if its energy is less that $E_F + \varphi$. And since an electron cannot have energy greater than E_F (when $T = 0$), we can identify φ with the least energy that must be supplied to the metal to get an electron out, at the absolute zero of temperature. We should mention that different surfaces, have different work functions: $\varphi = \varphi(hkl)$, where h, k, l are the Miller indices of the corresponding crystallographic plane. It turns out that φ equals a few eV (3–5 eV) for most metal surfaces and, as a rule, it increases almost linearly with the density of surface atoms. The work function is greater for close-packed planes (the area of the unit surface cell defined by the primitive vectors $\boldsymbol{a}_1$ and $\boldsymbol{a}_2$ is smaller for these planes and, therefore, the corresponding density of surface atoms is larger); the work function is smaller for relatively open surface (the area of the unit surface cell is larger for these planes and therefore the corresponding density of surface atoms is comparatively small). To demonstrate the above we show in Fig. 4.45 the work function of different surfaces (so-called faces) of tungsten. The circles and squares denote experimental values of the work function determined by a variety of experimental methods. $A_0(hkl)$ is the area of the unit cell of the (hkl) surface in units of a^2 ($a = 3.16$ Å is the lattice constant of tungsten). We see that $\varphi(hkl)$ increases almost linearly with $A_o^{-1}(hkl)$ which implies an almost linear increase with the density of surface atoms.

Similar observations can be made about semiconductor surfaces. The

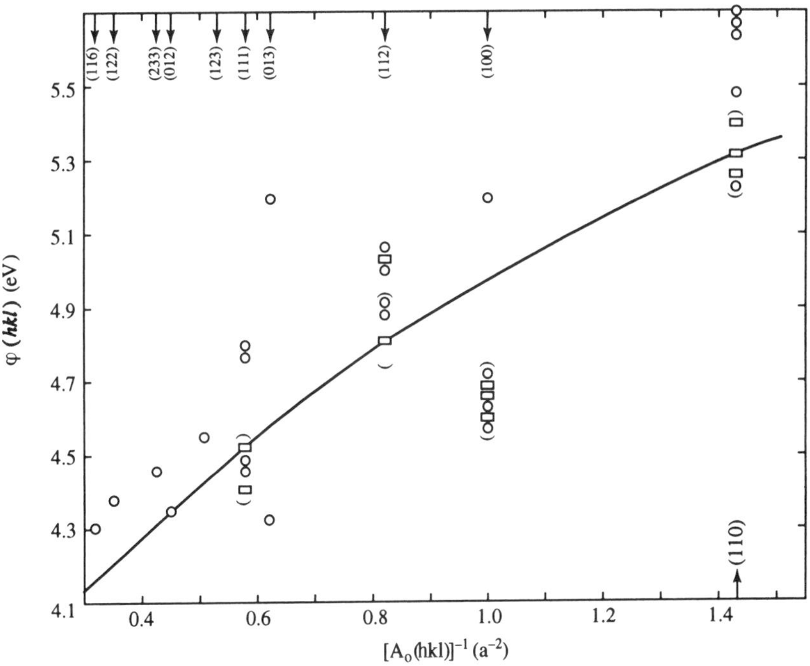

Figure 4.45. Work function of different faces of tungsten. The circles and squares denote experimental values of the work function determined by a variety of experimental methods. $A_o(hkl)$ is the area of the surface unit cell (a = 3.16 Å)

energy of an electron at rest on the vacuum side of the interface (at infinity) lies a few eV above E_c (the bottom of the conduction band of the semiconductor), and varies from surface to surface as in metals.

4.13.2 ONE-ELECTRON STATES OF A SEMI-INFINITE CRYSTAL—SURFACE STATES

We define the 2D reciprocal lattice $\{\boldsymbol{g}\}$ corresponding to a 2D space lattice (4.112) as follows (the procedure is analogous to that of Section 4.2):

$$\boldsymbol{g} = m_1 \boldsymbol{b}_1^{(2)} + m_2 \boldsymbol{b}_2^{(2)} \tag{4.116}$$

where m_1 and m_2 take all integer values between $-\infty$ and $+\infty$, and $\boldsymbol{b}_1^{(2)}$ and $\boldsymbol{b}_2^{(2)}$ are 2D primitive vectors in reciprocal space, defined as follows:
$\boldsymbol{b}_1^{(2)}$ is normal to $\boldsymbol{a}_2$ and has magnitude $b_1^{(2)} = 2\pi/(a_1 \cos\theta_1)$, where θ_1 is the angle between $\boldsymbol{b}_1^{(2)}$ and $\boldsymbol{a}_1$.
$\boldsymbol{b}_2^{(2)}$ is normal to $\boldsymbol{a}_1$ and has magnitude $b_2^{(2)} = 2\pi/(a_2 \cos\theta_2)$, where θ_2 is the angle between $\boldsymbol{b}^{(2)}$ and $\boldsymbol{a}_2$.

As an example we show in Fig. 4.46 the reciprocal vectors $\boldsymbol{b}_1^{(2)}$ and $\boldsymbol{b}_2^{(2)}$ for the (1 1 0) plane of a bcc lattice (defined by (4.112) and (4.113a) and shown in Fig. 4.43).

The first Brillouin zone, or surface Brillouin zone (SBZ), as it is most commonly known, corresponding to a given crystallographic plane is a primitive unit cell of (4.116) defined as follows. We draw the $\boldsymbol{g}$ vectors from the origin (a site of the 2D reciprocal lattice) to all neighbour lattice sites. Then we draw straight lines through the midpoints and normally to these vectors. The area (of reciprocal space) enclosed by these lines constitutes, by the way it was constructed, a primitive cell of the 2D reciprocal lattice and has in addition the symmetry of the crystallographic plane under consideration. The SBZ of the bcc (1 1 0) plane is shown in Fig. 4.46 and it is obvious that it has the rectangular symmetry (see Fig. 4.43) of this plane.

We note the following property of the $\boldsymbol{g}$-vectors (analogous to (4.9)):

$$\exp(\mathrm{i}\boldsymbol{g}\cdot\boldsymbol{R}_n^{(2)}) = 1 \tag{4.117}$$

from which follows that the functions $\exp(\mathrm{i}\boldsymbol{g}\cdot\boldsymbol{r}_\parallel)$, where $\boldsymbol{r}_\parallel \equiv (x, y)$, are periodic functions in the xy plane:

$$\exp[\mathrm{i}\boldsymbol{g}\cdot(\boldsymbol{r}_\parallel + \boldsymbol{R}_n^{(2)})] = \exp(\mathrm{i}\boldsymbol{g}\cdot\boldsymbol{r}_\parallel) \tag{4.118}$$

which is the analogue of (4.11) in two dimensions.

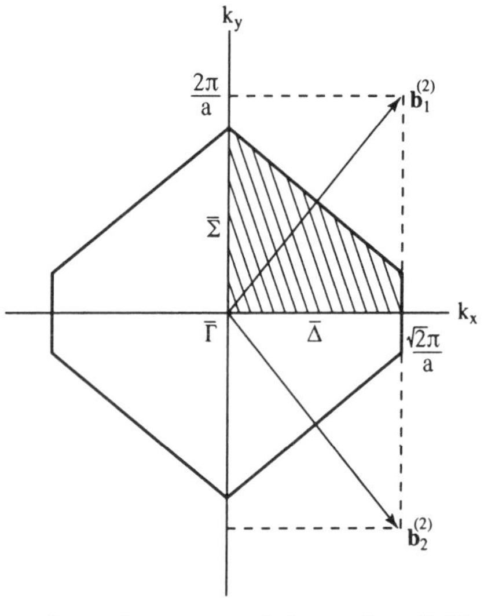

Figure 4.46. Primitive reciprocal vectors and the surface Brillouin zone (SBZ) of the (110) plane of a bcc lattice

The 3D reciprocal lattice of the space lattice defined by the primitive vectors $\boldsymbol{a}_1$, $\boldsymbol{a}_2$, $\boldsymbol{a}_3$ is given by

$$\boldsymbol{G} = m_1\boldsymbol{b}_1^{(3)} + m_2\boldsymbol{b}_2^{(3)} + m_3\boldsymbol{b}_3^{(3)} \tag{4.119}$$

where m_1, m_2, m_3 take all integer values from $-\infty$ to $+\infty$ and $\boldsymbol{b}_1^{(3)}$, $\boldsymbol{b}_2^{(3)}$, $\boldsymbol{b}_3^{(3)}$ are primitive vectors (in 3D reciprocal space) constructed in terms of $\boldsymbol{a}_1$, $\boldsymbol{a}_2$, $\boldsymbol{a}_3$ as explained in Section 4.2 (in the present case $\boldsymbol{t}_1$, $\boldsymbol{t}_2$, $\boldsymbol{t}_3$ are replaced by $\boldsymbol{a}_1$, $\boldsymbol{a}_2$, $\boldsymbol{a}_3$ where $\boldsymbol{a}_1$ and $\boldsymbol{a}_2$ are primitive vectors parallel to the plane under consideration). The reciprocal lattice (4.119) is, of course, exactly the same with $\{\boldsymbol{K}_\mathrm{h}\}$ defined in Section 4.2; the different choice of primitive vectors is a matter of convenience. We see immediately (from (4.6)) that $\boldsymbol{b}_3^{(3)}$ is normal to the crystallographic plane under consideration (the xy plane). It is also easy to show that the projections of $\boldsymbol{b}_1^{(3)}$ and $\boldsymbol{b}_2^{(3)}$ on the xy plane coincide with $\boldsymbol{b}_1^{(2)}$ and $\boldsymbol{b}_2^{(2)}$ respectively, which means that (4.116) is the projection of (4.119) on the xy plane. We also note that the primitive cell of the 2D reciprocal lattice defined by the parallelogram which is defined by $\boldsymbol{b}_1^{(2)}$ and $\boldsymbol{b}_2^{(2)}$ coincides with the projection on the xy plane of the primitive cell of the 3D reciprocal lattice defined by the parallelepiped which is defined by $\boldsymbol{b}_1^{(3)}$, $\boldsymbol{b}_2^{(3)}$ and $\boldsymbol{b}_3^{(3)}$. A more suitable (for our purposes) primitive cell of the 3D reciprocal lattice is the primitive cell whose cross-section with the xy plane is the SBZ of the plane under consideration and, normally to that plane, extends from $k_z = -|\boldsymbol{b}_3^{(3)}|/2$ to $k_z = |\boldsymbol{b}_3^{(3)}|/2$. In other words, we choose the so-called reduced $\boldsymbol{k}$-zone (see Section 4.3) as follows:

$$\boldsymbol{k} = (\boldsymbol{k}_\parallel, k_z);\ \boldsymbol{k}_\parallel \equiv (k_x, k_y) \text{ in the SBZ}, \ -|\boldsymbol{b}_3^{(3)}|/2 \leqslant k_z < |\boldsymbol{b}_3^{(3)}|/2 \tag{4.120}$$

We must emphasise that the above zone is exactly equivalent to the Brillouin zone (BZ) introduced in Section 4.2. By which we mean that to every point in the zone (4.120) corresponds a point in the BZ in that the difference of the two is a vector of the reciprocal lattice (4.119) and, similarly, to every point in the BZ corresponds in the above sense a point in the zone (4.120). The BZ has the full symmetry of the 3D space lattice, while (4.120) has a lesser symmetry appropriate to the crystallographic plane under consideration.

Let us now consider an infinite crystal extending from $z = -\infty$ to $z = +\infty$, the xy plane being parallel to a crystallographic surface (a crystallographic plane) of the crystal. We shall further assume, for the sake of simplicity, that the plane under consideration is a mirror plane of symmetry of the given crystal (the (1 1 0) and (1 0 0) planes of a bcc crystal with one atom per lattice site are examples of such planes). We know (see Section 4.3) that the eigenstates of an electron in the periodic field of the infinite crystal are Bloch waves (4.17) characterised by a wavevector $\boldsymbol{k} = (\boldsymbol{k}_\parallel, k_z)$ in the reduced $\boldsymbol{k}$-zone and a band index α; we rewrite (4.17) as follows

$$\psi_{k_z,\boldsymbol{k}_\parallel\alpha}(\boldsymbol{r}, t) = A\exp(\mathrm{i}\boldsymbol{k}\cdot\boldsymbol{r})u_{k_z,\boldsymbol{k}_\parallel\alpha}(\boldsymbol{r})\exp[-\mathrm{i}E_\alpha(k_z, \boldsymbol{k}_\parallel)t/\hbar] \tag{4.121}$$

to indicate that we have chosen (4.120) as the reduced $\boldsymbol{k}$-zone instead of the BZ. We choose to present the energy bands $E_\alpha(k_z, \boldsymbol{k}_\parallel)$ as in Fig. 4.47. For given $\boldsymbol{k}_\parallel$ (in the SBZ of the plane under consideration) we give the energy bands (dispersion curves) as functions of k_z (the component of the reduced wavevector normal to the plane under consideration). In Fig. 4.47 we do that for $\boldsymbol{k}_\parallel = 0$ (the centre of the SBZ); if, for example, we were dealing with the $(1\bar{1}0)$ plane of a bcc crystal (such as tungsten), the SBZ would be that of Fig. 4.46 and $\boldsymbol{k}_\parallel = 0$ would be its centre denoted by $\bar{\Gamma}$. We can obtain the energy bands, in the same way, for every $\boldsymbol{k}_\parallel$ in the SBZ. However, because of the symmetry of the crystal we need only do that for points $\boldsymbol{k}_\parallel$ within the irreducible part of the SBZ (the shaded area in the example of Fig. 4.46). The energy bands at points outside the irreducible part of the SBZ are the same with those at points in the irreducible part of the SBZ related to them through a symmetry operation. For example, the dispersion curves at a point $(-k_x, 0)$ on the negative k_x axis in the SBZ of Fig. 4.46 will be the same with those at the point $(k_x, 0)$ on the positive k_x axis, because the operation $x, y \to -x, y$ is a symmetry operation of the given lattice and of the corresponding potential field (we assume throughout that we are dealing with a crystal, such as tungsten, with one atom per lattice site); similarly the dispersion curves at a point $(0, -k_y)$ on the negative k_y axis will

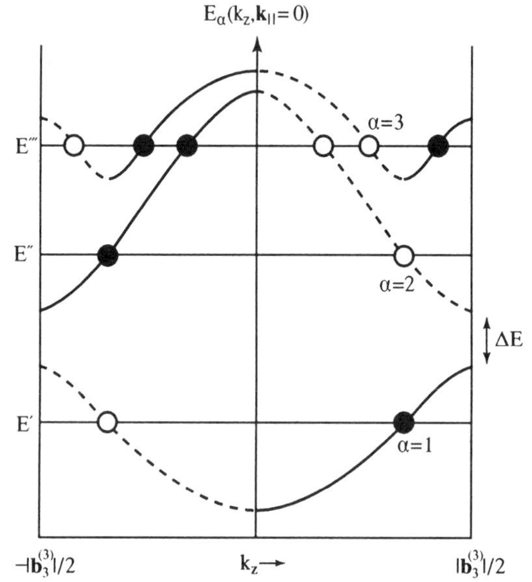

Figure 4.47. The energy bands of the infinite crystal are given as functions of k_z, for a fixed value of $\boldsymbol{k}_\parallel$ (the reduced wavevector in the SBZ of the surface under consideration). In the present case $\boldsymbol{k}_\parallel = 0$

be the same as those at the point $(0, k_y)$ on the positive k_y axis, because x, $y \to x, -y$ is another symmetry operation in this case. Usually one presents the energy bands for points on the so-called symmetry lines of the SBZ, such as the $\bar{\Sigma}$ line (along the y direction) and the $\bar{\Delta}$ line (along the x direction) of Fig. 4.46, and at a few other points if necessary.

We must emphasise that the energy bands of the infinite crystal presented in the above manner (in relation to a particular surface) are not in the least different from those we have described in Sections 4.3 to 4.6. Only the way we present them is different. The reader should be convinced about this before proceeding further.

When the plane under consideration is a plane of mirror symmetry of the crystal, then, if for given $\boldsymbol{k}_{\|}\alpha$ there is an electron state, a Bloch wave (4.121), propagating to the right ($v_z(k_z, \boldsymbol{k}_{\|})$, as defined in Section 4.8, is positive), then there is also an electron state $\psi_{-k_z,\boldsymbol{k}_{\|}\alpha}$ of the same energy propagating to the left ($v_z(-k_z, \boldsymbol{k}_{\|}) < 0$); and vice-versa, if there is a Bloch wave propagating to the left, then there is also one of the same energy and opposite k_z propagating to the right. The above is demonstrated schematically in Fig. 4.47. The solid lines of the energy bands shown in this figure correspond to Bloch waves propagating to the right (towards $z = +\infty$) and the broken lines correspond to Bloch waves propagating to the left (towards $z = -\infty$).

There is more to observe in Fig. 4.47. We note that for the given $\boldsymbol{k}_{\|}$, there is an energy gap, denoted by ΔE, between the first ($\alpha = 1$) and the second ($\alpha = 2$) bands. At other $\boldsymbol{k}_{\|}$ points of the SBZ the corresponding gap may be smaller, larger or simply shifted along the energy axis. Given a surface, there can be regions of energy where one-electron states exist only for $\boldsymbol{k}_{\|}$ within a part of the SBZ. Let us take as an example the (100) surface of copper. The centre ($\boldsymbol{k}_{\|} = 0$) of the corresponding SBZ corresponds to the direction ΓX (Δ line) in Fig. (4.14). The variation of the energy bands with k_z for $\boldsymbol{k}_{\|} = 0$, in a diagram such as that of Fig. 4.47, will be given by the dispersion curves along the ΓX line of Fig. 4.15. We see that for $\boldsymbol{k}_{\|} = 0$ (and by continuity in the region about it) there is a gap from $E \simeq 0$ to $E \simeq 0.2$ Hartrees (between the energy levels X_4' and X_1 respectively), but this gap obviously disappears for $\boldsymbol{k}_{\|}$ away from $\boldsymbol{k}_{\|} = 0$ (note the dispersion curve Σ_1 along the ΓK line of Fig. 4.15). The above example, from a copper surface, is typical of metals in the energy region about E_F. On the other hand, in a semiconductor there is a region of energy about E_F, as we know, where no electron states exist at all, and this means that given a semiconductor surface, there will be a region of energy about E_F where no electron state (4.121) exists whatever the value of $\boldsymbol{k}_{\|}$. An energy region over which there are no electron states (4.121) of the infinite crystal, whatever the value of $\boldsymbol{k}_{\|}$, is called (in surface physics) an absolute energy gap.

So far we have dealt with infinite crystals extending from $z = -\infty$ to

$z = \infty$ and what we have said is, really, a restatement of things we have known, in terms which will facilitate our description of surface effects and phenomena, and this we are now ready to do. Let us cut our infinite crystal in two halves; we keep the half extending from $z = -\infty$ to $z = 0$ and replace the other half ($z > 0$) by a slowly varying potential barrier as shown in Fig. 4.44. We restrict our discussion to unreconstructed metal surfaces and, moreover, we shall assume that the potential field $U(\boldsymbol{r})$ the electron sees is the same as that in the infinite crystal right up to the metal–vacuum interface at $z = 0$, and that $U = U(z)$ for $z > 0$.

Let us then consider an electron state of the *infinite* crystal, a Bloch wave $\overset{>}{\psi}_{k_z,\boldsymbol{k}_\parallel\alpha}(\boldsymbol{r}, t)$ given by (4.121) propagating to the right (this fact is indicated by the superscript $>$ on ψ). Referring to the example of Fig. 4.47, this could be one of the states represented by the solid lines of the dispersion curves. To this corresponds an electron state $\Psi_{k_z,\boldsymbol{k}_\parallel\alpha}(\boldsymbol{r}, t)$ of the semi-infinite crystal of the same (parallel to the surface) wavevector $\boldsymbol{k}_\parallel$ and the same energy $E = E_\alpha(k_z, \boldsymbol{k}_\parallel)$. When $E < E_{\mathrm{F}} + \varphi$ (see Fig. 4.44) we can write $\Psi_{k_z,\boldsymbol{k}_\parallel\alpha}$ as follows

$$\Psi_{k_z,\boldsymbol{k}_\parallel\alpha} = \overset{>}{\psi}_{k_z,\boldsymbol{k}_\parallel\alpha} + \sum_{k_z',\alpha'} a(k_z', \alpha')\overset{<}{\psi}_{k_z',\boldsymbol{k}_\parallel\alpha'}$$

$$+ \sum_{\boldsymbol{g}} \overset{<}{\varphi}_{\boldsymbol{g}}(z) \exp[\mathrm{i}(\boldsymbol{k}_\parallel + \boldsymbol{g}) \cdot \boldsymbol{r}_\parallel] \exp(-\mathrm{i}Et/\hbar), \text{ for } z < 0 \qquad (4.122\text{a})$$

$$= \sum_{\boldsymbol{g}} c_{\boldsymbol{g}} \overset{>}{\chi}_{\boldsymbol{g}}(z) \exp[\mathrm{i}(\boldsymbol{k}_\parallel + \boldsymbol{g}) \cdot \boldsymbol{r}_\parallel] \exp(-\mathrm{i}Et/\hbar), \text{ for } z > 0 \qquad (4.122\text{b})$$

The first term of (4.122a), representing a Bloch wave propagating to the right, incident upon the surface barrier, generates one or more Bloch waves, denoted by $\overset{<}{\psi}_{k_z',\boldsymbol{k}_\parallel\alpha'}$, propagating to the left; the second term of (4.122a) is a linear sum of such waves; it includes all Bloch waves (propagating to the left) with the same energy, $E_{\alpha'}(k_z', \boldsymbol{k}_\parallel) = E$, and the same reduced wavevector $\boldsymbol{k}_\parallel$ as the incident wave (the $a(k_z', \alpha')$ are appropriate coefficients). For example, referring to Fig. 4.47, if the incident wave is that corresponding to the full circle of energy E', the second term in (4.122a) includes only one Bloch wave, that corresponding the open circle of the same energy; if the incident Bloch wave corresponds to one of the full circles of energy E''', the second term in (4.122a) includes the three Bloch waves corresponding to the three open circles of the same energy. Besides the above propagating waves (defined by (4.121)), the incident wave generates a number of evanescent waves on the metal side of the metal–vacuum interface ($z < 0$), represented by the third term of (4.122a), and a number of evanescent waves on the vacuum side ($z > 0$) of the interface, represented by (4.122b). In principle, the sums over $\boldsymbol{g}$ appearing in (4.122a) and

(4.122b) contain infinite-many terms, one for every vector $\boldsymbol{g}$ of the 2D reciprocal lattice (4.116); in practice, however, a small number of terms (a few tens at most) suffice (these are the terms corresponding to the smallest in magnitude $\boldsymbol{g}$ vectors). The functions $\varphi_{\boldsymbol{g}}^{<}(z)$, which depend on the energy E and wavevector $\boldsymbol{k}_{\parallel}$ of the electron, diminish rapidly away from the surface, so that, practically, the last term of (4.122a) vanishes at a depth of a few atomic planes from the surface. Similarly, when $E = E_{\alpha}(k_z, \boldsymbol{k}_{\parallel}) < E_{\mathrm{F}} + \varphi$, all functions in (4.122b) diminish exponentially away from the surface:

$$\chi_{\boldsymbol{g}}^{>}(z) \propto \exp(-[2m(E_{\mathrm{F}} - E + \varphi)/\hbar^2 + (\boldsymbol{k}_{\parallel} + \boldsymbol{g})^2]^{1/2} z), \text{ when } z \gg 0$$

The wavefunctions in (4.122a) are essentially determined, for given E and $\boldsymbol{k}_{\parallel}$, by the potential field inside the metal ($z < 0$), which, in our approximation, is the same with that of the infinite crystal. However, there are a number of hidden coefficients in the third term of (4.122a) which are not determined by the potential field; and of course the coefficients $a(k_z', \alpha')$ in the second term of (4.122a) are not determined by the potential field. Similarly the functions $\chi_{\boldsymbol{g}}^{>}(z)$ are determined by the potential field on the vacuum side of the interface but the $c_{\boldsymbol{g}}$ coefficients remain to be determined. Given the incident (on the surface) Bloch wave (which has the form (4.121)), all the above coefficients are determined uniquely from the requirement that $\Psi_{k_z, \boldsymbol{k}_{\parallel}\alpha}$ and its z-derivative be continuous at every point of the metal vacuum interface (the plane $z = 0$). The reader may wish to note the similarity and the differences between (4.122) describing the scattering of a Bloch wave by the surface barrier and the wavefunction which describes the scattering of a plane wave (in one dimension) by a potential wall (see, e.g., problem 1.12a). In the one-dimensional problem there is only one reflected wave propagating towards $z = -\infty$) and only one evanescent wave, the latter diminishing exponentially into the barrier ($z > 0$); the coefficients of the above two waves are determined by the continuity of the wavefunction and its derivative at the point $z = 0$. In the 3D problem we have instead of the one reflected wave, a small number of waves propagating towards $z = -\infty$ (second term in 4.122a) and, in principle, an infinite number of evanescent waves decaying as $z \to -\infty$; and instead of the one evanescent wave into the barrier, we have infinite-many such waves (4.122b), and, naturally, instead of two coefficients at our disposal we have infinitely many, to be determined uniquely by the continuity of the wavefunction and its z-derivative *at every point* of the metal–vacuum interface (the plane $z = 0$). In practice we have a finite number of terms in (4.122) and a finite number of coefficients to be determined ($2n$ of them, if we keep $n\boldsymbol{g}$ vectors in the $\boldsymbol{g}$-sums of (4.122)).

One can also look at (4.122) as the limiting form of a wavepacket (see

Sections 1.8 and 4.8): we imagine a wavepacket of Bloch waves of the same $\boldsymbol{k}_{\|}\alpha$ as $\psi^{>}_{k_z,\boldsymbol{k}_{\|}\alpha}$ and of approximately the same k_z (and therefore of approximately the same energy) localised, at $t = 0$, in the interior of the metal at some distance from the surface and propagating towards it; at some time $t \simeq t_1$ it reaches the surface barrier and is scattered by it; as a result we obtain a number of wavepackets (one for each of the Bloch waves in the second term of (4.122a)) propagating away from the surface towards the interior of the metal. The probability that the reflected electron will be in one or the other of these states is determined by what happens at the metal–vacuum interface.

There are a few additional comments to be made about (4.122). We should note that $\Psi_{k_z,\boldsymbol{k}_{\|}\alpha}$ satisfies Bloch's theorem parallel to the surface, i.e., $\Psi_{k_z,\boldsymbol{k}_{\|}\alpha}(\boldsymbol{r}_{\|} + \boldsymbol{R}_n^{(2)}, z; t) = \exp(\mathrm{i}\boldsymbol{k}_{\|} \cdot \boldsymbol{R}_n^{(2)})\Psi_{k_z,\boldsymbol{k}_{\|}\alpha}(\boldsymbol{r}_{\|}, z; t)$ as expected. We have already noted that $\Psi_{k_z,\boldsymbol{k}_{\|}\alpha}$ is determined by the corresponding Bloch wave $\psi^{>}_{k_z,\boldsymbol{k}_{\|}\alpha}$ of the infinite crystal. At this stage we must say a few words about the normalisation constant A in the definition of the latter (4.121). In reality we never deal with a semi-infinite crystal but with a slab. Let the slab have area $A = L^2$ parallel to the surface under consideration and a thickness (normal to the surface) equal to L. The corresponding semi-infinite crystal has volume $V = L^3$ $(L \to \infty)$ and the infinite crystal has volume $2V = 2L^3$ $(L \to \infty)$. With the above in mind, the normalisation of $\psi^{>}_{k_z,\boldsymbol{k}_{\|}\alpha}$ is obtained by putting $A = (2V)^{-1/2}$ in (4.121) in accordance with (4.25) and with the understanding that $u_{k_z,\boldsymbol{k}_{\|}\alpha}(\boldsymbol{r})$ is normalised in the usual manner[7]. Accordingly (see Section 4.3) the wavevector $\boldsymbol{k} = (\boldsymbol{k}_{\|}, k_z)$ takes the following values within the reduced $\boldsymbol{k}$-zone (4.120)

$$\boldsymbol{k}_{\|} = (k_x, k_y) = \frac{2\pi}{L}(n_x, n_y),\ k_z = \frac{\pi}{L}n_z \qquad (4.123)$$

where $n_x, n_y, n_z = 0, \pm 1, \pm 2, \ldots$ We note that we have from (4.123) twice as many points in the reduced zone than are obtained from (4.24). On the other hand only half of the points in (4.123) will correspond to waves propagating towards the surface, and only these enter into our description of the one-electron states of the semi-infinite crystal. In that way one obtains the correct number of states per unit volume of the crystal.

Apart from the states described by (4.122), which we shall call bulk states, there may be one-electron states of the semi-infinite crystal which do not derive from one-electron states of the infinite crystal. These so-called surface states can be understood as follows. Take a certain $\boldsymbol{k}_{\|}$ in the SBZ of a given surface and look for an energy gap of this $\boldsymbol{k}_{\|}$, such as the one denoted by ΔE in the diagram of Fig. 4.47. For this $\boldsymbol{k}_{\|}$ there is no state of the electron in the infinite crystal with energy within the gap and, therefore, there can be no electron state of the semi-infinite crystal of the form (4.122) either. But one cannot exclude the possibility of an electron eigenstate of the following form

$$\Psi^S_{\boldsymbol{k}_\parallel\gamma} = \sum_{\boldsymbol{g}} \varphi^<_{\boldsymbol{g}}(z) \exp[\mathrm{i}(\boldsymbol{k}_\parallel + \boldsymbol{g}) \cdot \boldsymbol{r}_\parallel] \exp[-\mathrm{i}E_\gamma(\boldsymbol{k}_\parallel)t/\hbar], \; z < 0 \quad (4.124a)$$

$$= \sum_{\boldsymbol{g}} c_{\boldsymbol{g}} \chi^>_{\boldsymbol{g}}(z) \exp[\mathrm{i}(\boldsymbol{k}_\parallel + \boldsymbol{g}) \cdot \boldsymbol{r}_\parallel] \exp[-\mathrm{i}E_\gamma(\boldsymbol{k}_\parallel)t/\hbar], \; z > 0$$

where $E_\gamma(\boldsymbol{k}_\parallel)$ is some energy within the gap (γ is a surface-band index to be explained below). The $\varphi^<_{\boldsymbol{g}}(z)$ are evanescent waves as in (4.122a), diminishing towards the interior of the metal (for any given $\boldsymbol{k}_\parallel$ and energy E we can always obtain a set of such waves), and $\chi^>_{\boldsymbol{g}}(z)$ are the functions defined earlier which decay exponentially away from the surface, provided the energy $E = E_\gamma(\boldsymbol{k}_\parallel) < E_\mathrm{F} + \varphi$. There is a number of hidden coefficients in the sum of (4.124a) (n of them, if we include n terms in the sum over $\boldsymbol{g}$) and there are also n coefficients $\{c_{\boldsymbol{g}}\}$ in the sum of (4.124b), which are not determined by the potential field. It may just be possible that these coefficients can be determined in such a way that $\Psi^S_{\boldsymbol{k}_\parallel\gamma}$, as defined by (4.124), and its z-derivative are continuous at every point of the metal–vacuum interface. Only then can (4.124) be accepted as a physical one-electron state of the crystal. There are a number of possibilities. It is possible that the above condition will not be satisfied for E within the given energy gap of the given $\boldsymbol{k}_\parallel$. Then there will be no surface states in this gap. On the other hand the above condition may be satisfied at one, two or even three energies within a given gap. If that is the case a band index γ will serve to distinguish between the different levels $E_\gamma(\boldsymbol{k}_\parallel)$, where $\gamma = 1$, 2 or $\gamma = 1$, 2, 3 as the case may be.

From (4.124) it is obvious that a surface state extends, in the manner of a Bloch wave, in the directions parallel to the surface under consideration, but is localised about the metal–vacuum interface in the direction normal to the surface. This is demonstrated schematically in Fig. 4.48 which is based on a model calculation for the (100) surface of tungsten. Each curve represents a probability density $|\Psi|^2$ averaged over a surface unit cell (since $|\Psi|^2$ is periodic parallel to the surface, averaging over a unit cell is equivalent to averaging over the entire plane at the given z). (The calculation was done for values of z halfway between consecutive atomic planes (crosses and circles on the figure). The lines joining these points are there to facilitate the reading of the figure, and to remind us of the continuity of $|\Psi|^2$.) The closed circles represent a bulk state (4.122), the open circles a surface state (4.124), and the crosses a surface resonance (to be explained below). In all cases $|\Psi|^2$ decreases exponentially on the vacuum side of the surface ($z > 0$).

Obviously, an energy gap changes continuously with $\boldsymbol{k}_\parallel$; in metals it extends over a limited part of the SBZ. If one or more surface states exist in

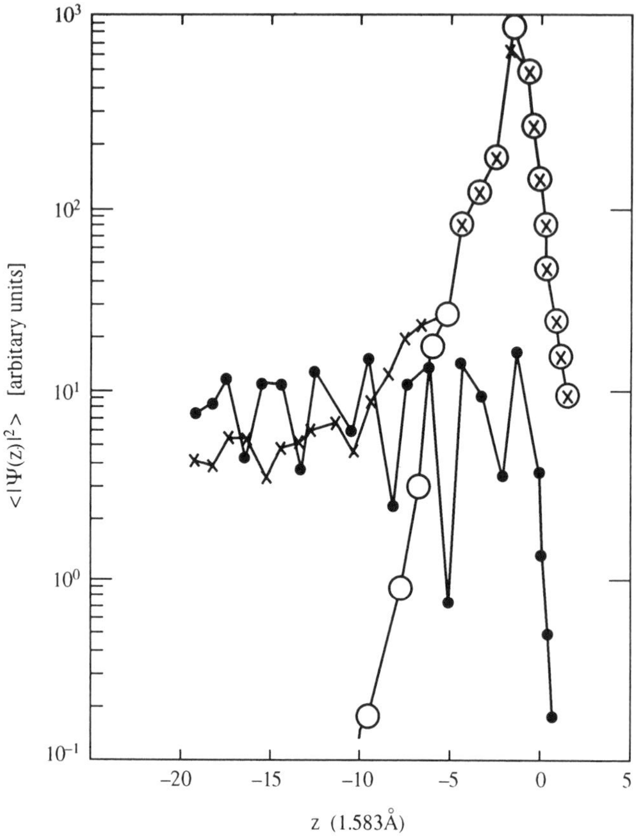

Figure 4.48. A bulk state (closed circles), a surface state (open circles) and a surface resonance (crosses) at a tungsten (100) surface. $\langle|\Psi|^2\rangle$ represents the probability density averaged over a surface unit cell

such a gap, their energies will also change continuously with $\boldsymbol{k}_{\|}$: we obtain bands of surface states, corresponding to surface energy-bands $E_\gamma(\boldsymbol{k}_{\|})$ defined over a region within that part of the SBZ for which the energy gap exists. Usually, a surface state can not coexist with a bulk state of the same energy E and wavevector[28] $\boldsymbol{k}_{\|}$; a hybridisation between the two leads, as a rule, to the disappearance of the surface state. In many instances, however, we obtain an 'extension' of the surface band into a region of $\boldsymbol{k}_{\|}$ and E where bulk states do exist. The bulk state of given E and $\boldsymbol{k}_{\|}$ interacts with a would-be surface state of the same E and $\boldsymbol{k}_{\|}$ forming a so-called surface resonance: the amplitude of the wavefunction of the bulk state peaks at the surface in a manner reminiscent of a surface state, as demonstrated schematically in Fig. 4.48. Although a surface resonance does not constitute an independent one-electron state, in many practical respects it has the effect of a true surface state and for that reason we often present a band of surface resonances together with a band of (true) surface states when the former appears as an extension of the latter in the manner explained above.

The above discussion, in so far as it was based on the assumption of an unreconstructed surface applies to metal surfaces for which this is not a bad approximation; but our general analysis applies, at least qualitatively, to semiconductor surfaces as well, whether they reconstruct or not. In Fig. 4.49 we show, as an example, the surface bands on a reconstructed Si (100) surface along various symmetry lines of the corresponding SBZ. At $T = 0$ the states below the top of the valence band (zero of energy) are occupied and those above it are empty.

The existence of surface states and resonances on metal and semiconductor surfaces has been confirmed experimentally in many cases. It has also been established that they play an important role in many physical phenomena (especially in semiconductors) and chemical phenomena (e.g., in relation to chemical reactions on certain metal surfaces).

We close this section with a brief comment relating to the so-called local density of states defined by

$$\rho(\boldsymbol{r}, E)\,\mathrm{d}E = 2 \sum_{E<E'<E+\mathrm{d}E} |\Psi_{E'}(\boldsymbol{r};\, t)|^2 \tag{4.125}$$

The sum includes *all* states, i.e. bulk states (4.122) and surface states (4.124) with energy between E and $E + \mathrm{d}E$. For the sake of simplicity we have

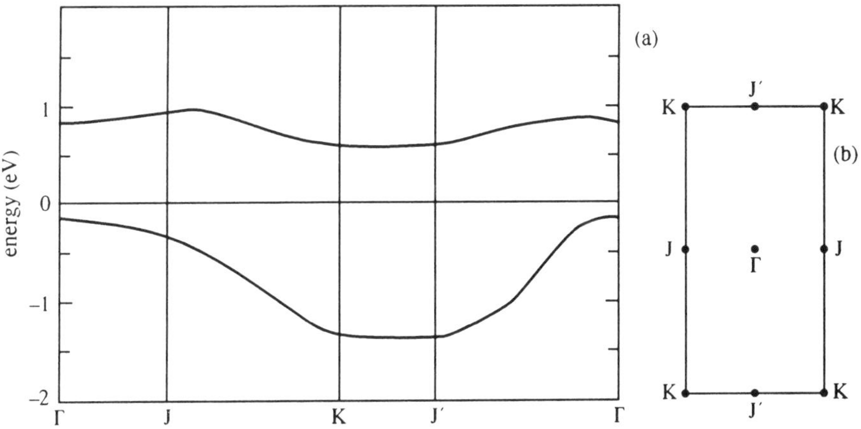

Figure 4.49. (*a*) Bands of surface states/resonances on a reconstructed Si (100) surface along various symmetry lines of the SBZ shown in (*b*) [D. J. Chadi, *Physical Review Letters*, **43** (1979) 43]

suppressed the quantum numbers k_z, $\boldsymbol{k}_\parallel$ and the band-index (α or γ) and denoted the wavefunction in (4.125) only by its energy. The factor of two takes into account the spin-degeneracy (we have assumed throughout our analysis that the potential the electron sees is independent of the spin-orientation of the electron). The average of the above quantity over a layer of atoms is shown schematically in Fig. 4.50 for the surface layer, first, second

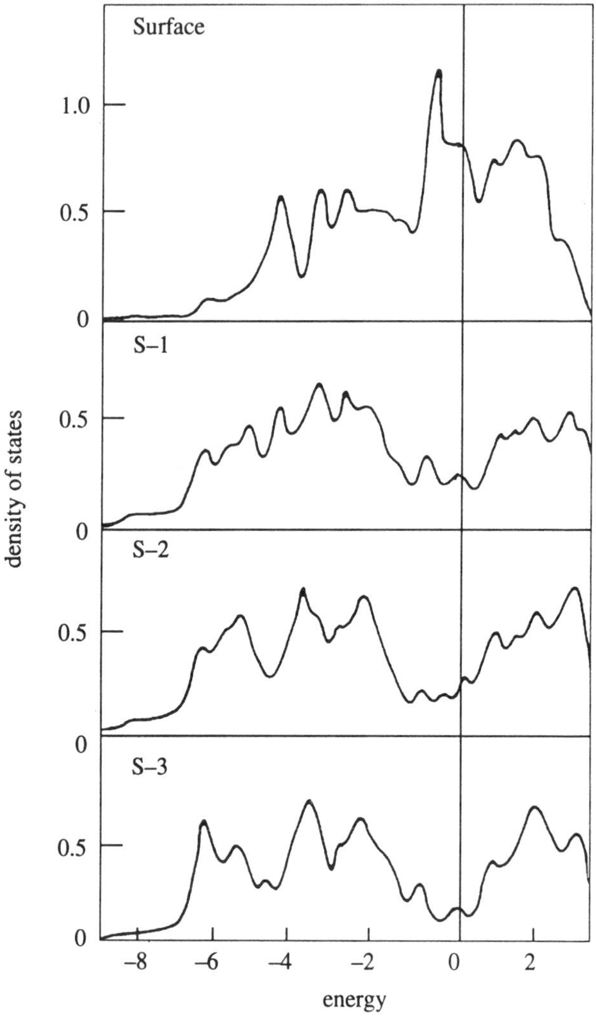

Figure 4.50. Theoretical layer density of states for a tungsten (100) slab. S-1, S-2 and S-3 are (respectively) the first, second and third layer in from the surface. [M. Posternak, H. Krakauer, A. J. Freeman and D. D. Koelling, *Physical Review B*, **21** (1980) 5601]

and third layer in from the surface of a tungsten (100) slab. The local density of states on the third layer (S − 3) is practically identical with that in the bulk of the crystal. We note, however, the large difference between the local density of states on the top (surface layer) and the one immediately beneath it, which suggests that electronic processes (e.g., chemical interactions) on metal surfaces (or semiconductor surfaces) may have properties which are specific to the surface (and not just the metal or semiconductor) under consideration.

4.13.3 ELECTRON EMISSION

It is evident from Fig. 4.44 that if an electron of the metal finds itself, by some means or other, in a state of energy $E > E_F + \varphi$, then it can escape from the metal with certain probability which depends, mainly, on its transport velocity (it is obviously necessary that $v_z > 0$).

At high temperatures ($T \gtrsim 1000$ K) the high-energy tail of the Fermi distribution (4.42), shown in Fig. 4.23, extends beyond $E_F + \varphi$ and a small fraction of the electrons incident on the surface from within the metal do escape from it. The phenomenon is known as thermionic emission of electrons. Using the Sommerfeld model (Section 4.9) to describe the one-electron states of the metal, and assuming that all electrons incident on the surface barrier (Fig. 4.44) with kinetic energy (associated with the motion normal to the surface) $\hbar^2 k_z^2/2m > E_F + \varphi$ escape from it, we obtain the following formula for the density of the emitted current (number of electrons emitted per unit area of the surface, per unit time, times the electronic charge)

$$J(T) = A_R T^2 \exp(-\varphi/k_B T) \tag{4.126}$$

$$A_R = \frac{emk_B^2}{2\pi^2\hbar^3}$$

The above equation is known as Richardson's equation. It describes correctly the variation of the emitted current density with the temperature and its dependence on the work function φ. The experimental value of the constant A_R is of the same order of magnitude as its theoretical value above, but it varies from surface to surface reflecting the peculiarities of each surface, which are not taken into account by the Sommerfeld model.

Electron emission from a metal surface can also occur at low temperatures ($T \to 0$) when a strong electric field ($F \gtrsim 0.3$ V/Å) is applied to the surface. The electric field, adding a term $-eFz$ to the potential field (4.115) changes the surface potential barrier as shown schematically in Fig. 4.51. It now becomes possible for electrons with energy $E \leqslant E_F$ to tunnel out of the metal into vacuum in accordance with what we have said in Section 1.8. The

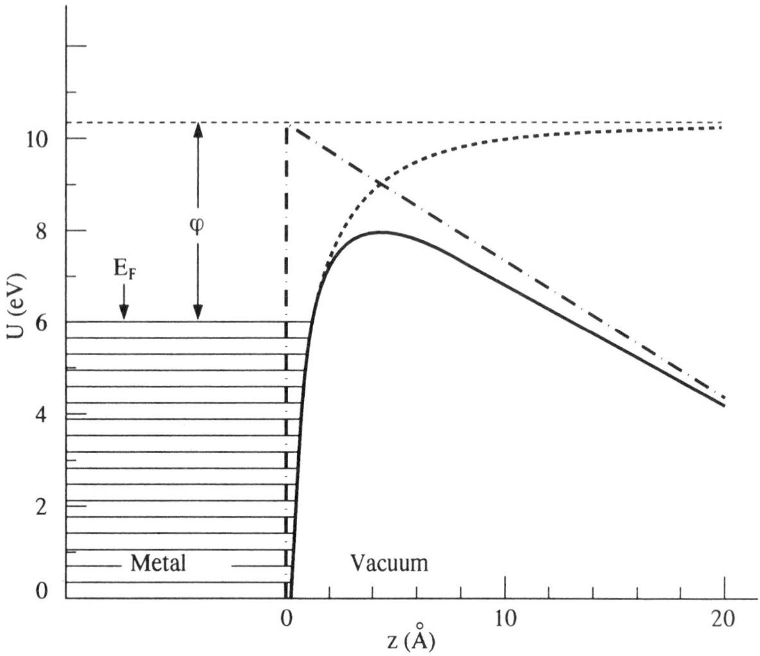

Figure 4.51. Surface potential barrier when an electric field is applied to the surface (solid line). The broken line describes the potential field prior to the application of the electric field, and the chain-like one represents the term $-eFz$. The zero of the energy coincides with the bottom of the conduction band

phenomenon is known as field-emission of electrons. Using the Sommerfeld model to describe the one-electron states of the metal, and formula (1.75) to calculate the transmission coefficient of an electron incident on the surface barrier (the energy E in (1.75) is identified, in the present case, with the energy $\hbar^2 k_z^2/2m$ of the electron motion parallel to the z-axis (see Fig. 4.51)), we obtain the following formula for the emitted current density, at absolute zero temperature,

$$
\begin{aligned}
J(F) &= AF^2 \exp(-B\varphi^{3/2}/F) \\
A &\simeq e^3/(16\pi^2\hbar\varphi) \qquad (4.127) \\
B &\simeq \left(\frac{2m}{\hbar^2}\right)^{1/2} \bigg/ e
\end{aligned}
$$

The above equation, known as the Fowler–Nordheim equation, describes correctly the variation of the emitted current density with the applied field and its dependence on φ. When it comes to the pre-exponential constant A, we find, as in the case of the thermionic emission current density, that this constant varies from surface to surface; and the reason is the same: the pre-exponential constants in both equations (4.126 and 4.127) are sensitive to electronic properties of the emitting surface which are not taken into account by the Sommerfeld model. On the other hand, the exponential terms do not depend critically on these properties; they depend, essentially, only on the shape of the surface barrier away from the metal–vacuum interface ($z \gtrsim 3$ Å) which, as we have seen, has the same form (4.115) for all metal surfaces.

We close this section with a very brief reference to photoemission: an electron can be excited from its initial state (with energy $E < E_F$) to a state of energy $E + h\nu > E_F + \varphi$ by absorbing the energy $h\nu$ of an incident photon. The excited electron can then escape from the metal with a certain probability which depends on its transport velocity and its lifetime (the electron may lose energy in collisions, mainly with other electrons, before reaching the surface as described in the next section).

4.14 Electronic excitations

The description we have presented of the electronic properties of crystals is based on the assumption (see Section 4.3) that each electron moves, independently of all others, in a mean field $U(\boldsymbol{r})$. To this field corresponds a set of one-electron eigenstates, a fraction of which (those with energy below E_F) are occupied and the rest are empty (we assume that $T = 0$). We know that the mean field total energy of the crystal is given not by the sum of the one-electron energies $E_\alpha(\boldsymbol{k})$ of the occupied states but by the more complicated formula (4.95). Nevertheless, it turns out that the transition of an electron from an occupied one-electron state with energy $E_\alpha(\boldsymbol{k})$ ($< E_F$) to an initially empty state with energy $E_{\alpha'}(\boldsymbol{k}')$ ($> E_F$) does correspond, at least approximately, to a real transition of the crystal from its ground state to an excited state with (total) energy greater than the ground-state energy by the amount $E_{\alpha'}(\boldsymbol{k}') - E_\alpha(\boldsymbol{k})$. (We have, in fact, implicitly assumed that this is the case in our discussion of the specific heat, of transport phenomena, etc.; see also Section 2.15.5.) However, one cannot assume that one-electron states with energy above E_F are *true* stationary states of an electron moving independently in the solid. Consider an electron in a metal at zero temperature and say that it occupies (at a given time) the orbital ψ_j (a Bloch wave, $j \equiv \boldsymbol{k}'\alpha'$) with energy $E_j > E_F$. If ψ_j were a true stationary state, its development with time would be given by $\psi_j(\boldsymbol{r})\exp(-iE_jt/\hbar)$. This is not true in practice; residual electrostatic interaction (correlation) between the electrons will lead to a finite lifetime τ_j of this state. (The situation is similar to the scattering of conduction electrons by impurities and lattice vibrations; it is simply the case that correlation (collisions between electrons) becomes the dominant cause of scattering at energies away from E_F). We may describe the time development of ψ_j as follows

$$\psi_j(\boldsymbol{r}, t) = \psi_j(\boldsymbol{r}) \exp[-(i/\hbar)(E_j - i\Gamma_j/2)t] \tag{4.128a}$$

so that

$$|\psi_j(\boldsymbol{r}, t)|^2 = |\psi_j(\boldsymbol{r})|^2 \exp(-t/\tau_j) \tag{4.128b}$$

$$\tau_j = \hbar/\Gamma_j \tag{4.129}$$

Equation (4.128) tells us that as the time goes on the electron has a finite probability of leaving the jth state. This occurs, for example, when the electron (in the jth state) collides with another electron in the crystal losing energy in the process. In such a collision the first electron 'jumps' into a state of lower energy (we have assumed that $E_j > E_F$) and the second electron gains energy moving from an energy level below E_F to an energy level above E_F. It can be shown by very general considerations, based on the conservation of energy and momentum in an electron–electron collision, that in metals and for energies near E_F,

$$\Gamma(E) = \text{const}\,(E - E_F)^2 \tag{4.130}$$

In a typical case (say that of copper), $\Gamma = 0.092\ (E - E_F)^2$ Ryd, when Rydberg is taken as the energy unit. We note that (4.130) is valid also for E below (but near) E_F, which implies that a vacant state (a hole) created by removing an electron from an occupied state with energy $E < E_F$ has a finite lifetime. The above formula is valid only for metals and near E_F. The variation of Γ over an extended region of energy depends on a number of factors including inelastic scattering of the electron by plasmons (see Section 4.16), but it appears to be more or less the same in most common materials. This is demonstrated in Fig. 4.52 which shows the inelastic mean free path of excited electrons, defined by $L(E) = v(E)\tau$, where $\tau(E) = \hbar/\Gamma(E)$ and $v(E)$ is the average over all possible directions of the transport velocity of the electron (to estimate this quantity we can put $v = (2E/m)^{1/2}$ where E is the energy of the electron measured from the bottom of the conduction band). We see that many materials show mean free paths of five angstrom or so in the range of electron energies from about 20 eV to several hundred eV. These short mean free paths affect greatly the observed spectra we shall be discussing below, and one may justifiably wonder whether the one-electron picture employed so far remains applicable in such cases.

A finite lifetime implies an uncertainty (a spread) in the energy of the electron state as shown schematically in Fig. 4.53. The full width at half maximum of the broadened state equals Γ as defined above (4.129). As a result of the broadening of the one-electron energy levels, fine features in density-of-states-curves (such as that of Fig. 4.20) are smoothed out to some

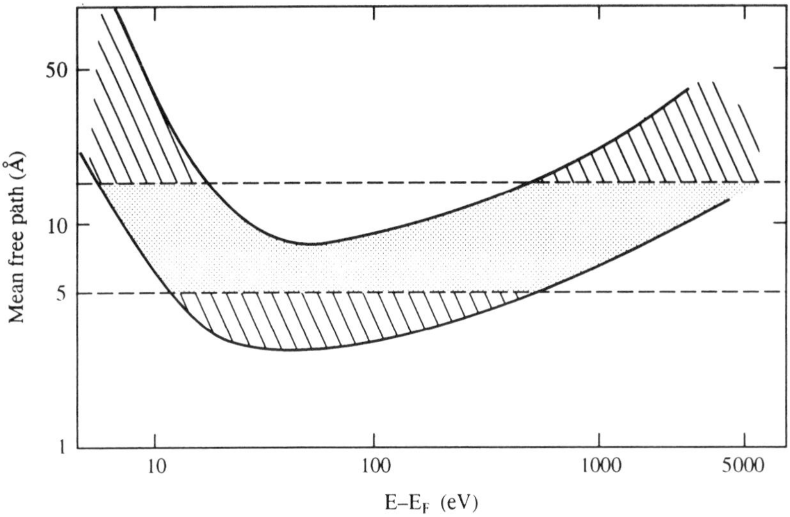

Figure 4.52. Inelastic mean free path for excited electrons in solids as a function of electron energy. The shaded band encompasses most common materials. The upper and lower regions demarcate the extreme cases of low damping (top) and high damping (bottom). (B. Feuerbacher and R. F. Willis, *J. Phys. C: Solid State Phys.*, **9** (1976) 169. Redrawn with permission of IOP Publishing Ltd)

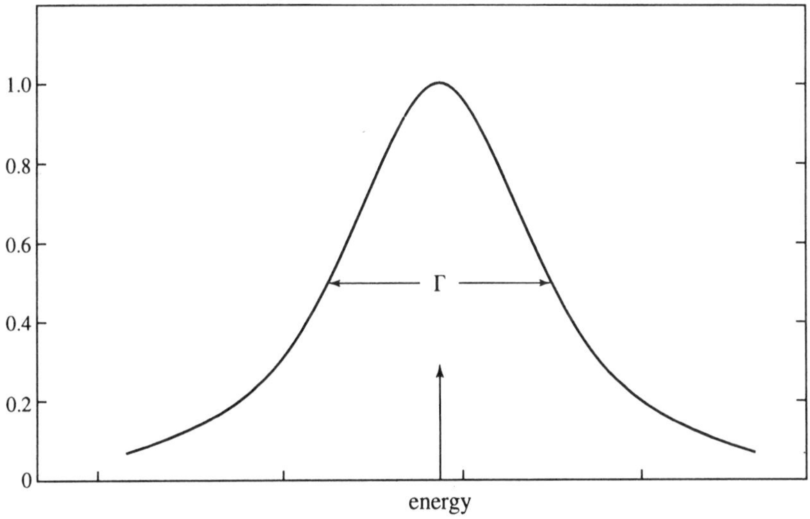

Figure 4.53. A sharp energy level, where the arrow is, is broadened because of residual interactions. The value on the vertical axis corresponding to a given E is the probability that the electron will have energy E

degree, and narrow energy gaps may disappear in energy regions where Γ is considerable. The density of states $\rho(E)$, obtained with $\Gamma = 0$, is effectively replaced by an averaged density of states $\bar{\rho}(E)$, given by

$$\bar{\rho}(E) \simeq \frac{1}{\pi}\int \frac{\Gamma/2}{(E' - E)^2 + \Gamma^2/4}\rho(E')\,\mathrm{d}E' \tag{4.131}$$

where Γ is of course a function of E.

It is clear from the above discussion that nearly stationary electron (hole) states exist only in the immediate vicinity of the Fermi level[29]. However, in spite of the fact that the one-electron states, corresponding to the mean potential field $U(\boldsymbol{r})$, can not be taken as truly independent (stationary) one-electron states, their knowledge is extremely useful even at energies well removed from the Fermi level, provided such knowledge is interpreted in the spirit we have just described. In what follows we shall, for the sake of simplicity, and following common practice, disregard the damping of excited one-electron states (we put $\Gamma = 0$) in the above equations).

In Fig. 4.54 we show, schematically, transitions of an electron from a core state to a state with energy above E_{F}. The final state may belong to the conduction band of the metal (or semiconductor, if we are dealing with a semiconductor) or to a higher energy band. Such a transition is possible as a result of bombarding the solid with fast particles (protons, electrons, etc.). The incident particle colliding with a core electron transfers part of its energy to it. The excitation can also be the result of irradiation by photons of high frequency (X-rays). Whatever the source of the excitation energy, it is clear that if the energy of the excited electron is $E' > E_{\mathrm{F}} + \varphi$, where φ denotes the work function of the crystal, it will be possible for it to escape from the crystal with certain probability depending on its velocity and its lifetime. Obviously, only an electron generated within a distance from the surface less than its inelastic mean free path can escape from the crystal without losing energy. The collection of these electrons is possible experimentally and allows us to determine accurately the initial energy E_0 of the electron. For example, if the ionisation of the core state is the result of irradiation by photons of energy $h\nu$, we have $E_0 = E' - h\nu$, so that by measuring E' we can determine E_0. Measurements of this kind make it possible to determine, among other things, the presence of foreign atoms (impurities) in a crystal, given that E_0 is characteristic of the free atom. We recall that the core states of an electron in a crystal differ only a little from those in the free atom, so that the ionisation energy of a core state will be very nearly the same with the ionisation energy of the corresponding inner shell of the free atom (listed in Table 2.5).

In Fig. 4.55 we show, schematically, transitions of an electron from the conduction band of a metal (a), and from the valence band of a semiconductor (b). We have transitions which conserve $\boldsymbol{k}$ (the initial and

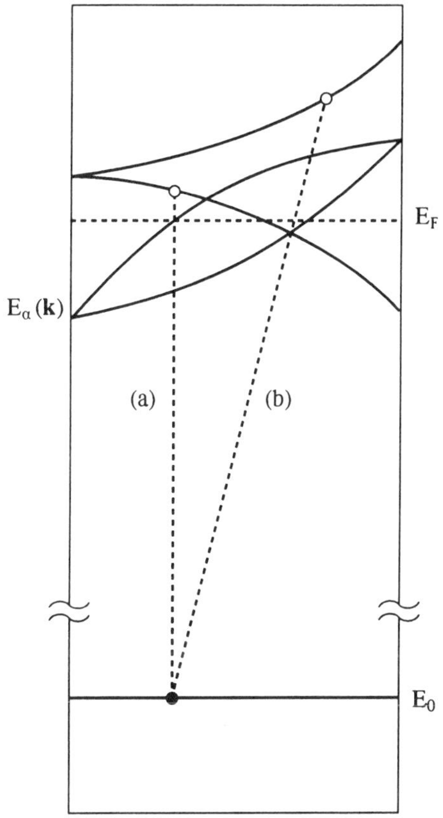

Figure 4.54. Transitions of an electron from a core state. (*a*) vertical transition; (*b*) non-vertical transition

final states of the electron have the same wavevector) and transitions which do not conserve $\boldsymbol{k}$. We refer to them as vertical and non-vertical transitions respectively, because they appear as such in diagrams like those of Figs. 4.54 and 4.55. Transitions due to absorption of photons of a large (relative to the lattice constant) wavelength are always vertical transitions. The translational symmetry of the system 'electron in the crystal + incident photon' demands that the sum of the wavevectors of the electron and the photon (see Section (2.8) for a definition of the latter) be constant. The magnitude $2\pi/\lambda$ of the photon wavevector can be neglected when the wavelength λ is much larger than the lattice constant (as is the case, say, in the optical region of the electromagnetic radiation) and therefore, practically, the wavevector of an electron does not change in a transition due to photon absorption. We note also that the spin of the electron (whether up or down) does not change in such transitions.

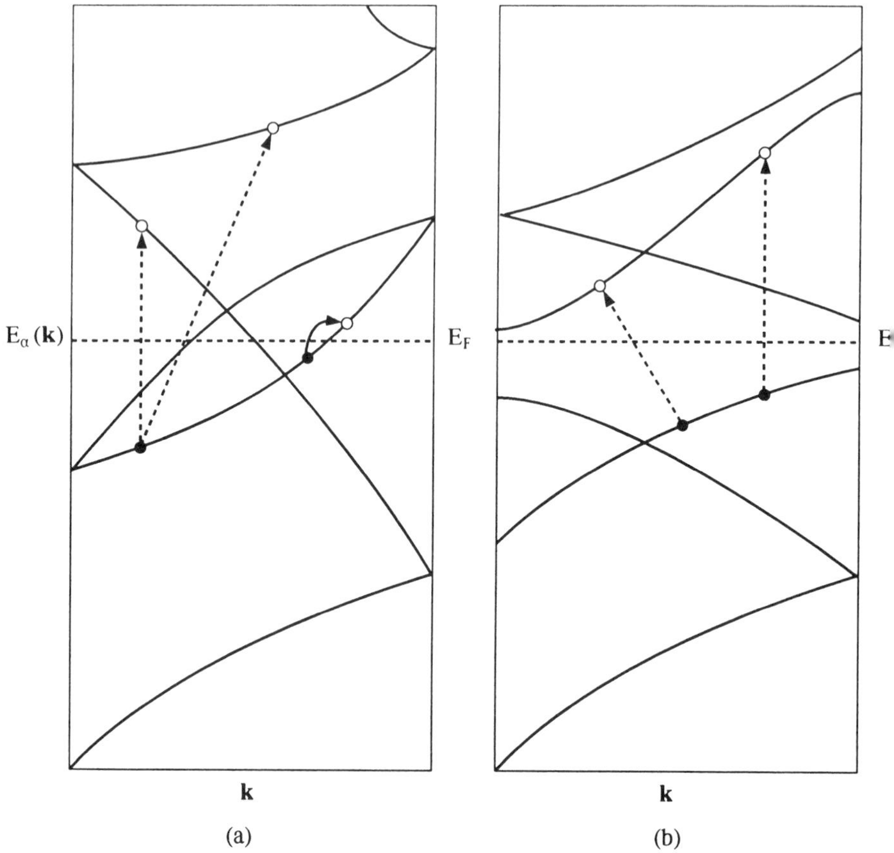

Figure 4.55. Electronic transitions. (*a*) from the conduction band of a metal; (*b*) from the valence band of a semiconductor

It is evident from the above that the absorption of light of frequency ν by a crystal is determined by the possibility of vertical transitions from occupied levels $E_\alpha(\boldsymbol{k})$ to corresponding empty levels $E_{\alpha'}(\boldsymbol{k}) = E_\alpha(\boldsymbol{k}) + h\nu$, where $\boldsymbol{k}$ can be any point in the BZ of the crystal and α and α' any two bands. If many such transitions are possible for given ν, then the absorption of photons of that frequency will be relatively large. (In practice one measures the so-called absorption coefficient, which is defined as the energy absorbed by unit thickness of the material divided by the incident energy.) It is obvious, for example, that photons whose energy $h\nu$ does not exceed the energy gap E_g of a semiconductor or insulator can not be absorbed by that semiconductor or insulator.

If the energy of the electron in its final state is greater than $E_F + \varphi$, the electron has a finite probability (depending on its velocity and its lifetime) to escape from the crystal without loss of energy. Measuring the energy $E_{\alpha'}(\boldsymbol{k})$ and wavevector $\boldsymbol{k}$ of the emitted electron and knowing the energy $h\nu$ of the absorbed photon, we can obtain the initial energy $E_\alpha(\boldsymbol{k})$ of the electron from the relation $E_\alpha(\boldsymbol{k}) = E_{\alpha'}(\boldsymbol{k}) - h\nu$. The above and the fact that in recent years it has become possible to obtain (from synchrotron sources) photons of any frequency, has enabled researchers to check experimentally calculated energy bands such as those of Figures 4.15, 4.18, etc.; and in some instances to determine the broadening of energy levels (due to the finite lifetime of the corresponding states) as well.

In the above discussion we have assumed that both the initial and the final state of the electronic transition are bulk states. There are also transitions, induced by photon absorption, where the initial state of the transition or the final one or both of them are surface states. In such transitions only the $\boldsymbol{k}_\parallel$-component of the wavevector is conserved.

Non-vertical transitions (see Figs. 4.54 and 4.55) occur in consequence of irradiation with small-wavelength photons or, as we have already noted, when an incident particle (electron, ion) collides with the electrons of the solid. Such transitions provide the means, the stopping power, by which a fast particle loses its energy as it penetrates into the solid.

In the above description of electronic transitions we have assumed that the initial and final states of the electron are independent Bloch waves. An electron is excited to a state above E_F leaving an empty state (a hole) below E_F; we say that an electron–hole pair has been created; but following their creation, the electron and the hole move about independently. (We can build Bloch-wavepackets of holes in the same way as we have done for electrons. In our picture of electron–hole excitation, the two wavepackets move apart as there is no interaction between them.) The above picture is certainly valid in metals (the mobility of the conduction electrons is such that it screens out any interaction between electron and hole), but complications may arise in semiconductors and insulators. An electron excited into the bottom of the conduction band from the top of the valence band of the semiconductor may be attracted and bound to the positive hole it leaves behind; the two, then, will orbit around their centre of mass, while it moves through the crystal. The bound electron–hole pair is called an exciton. In a typical semiconductor, like germanium, the average separation between the electron and the hole of the exciton turns out to be quite large (about 50 Å or so) and this is consistent with a reduced (screened) Coulomb interaction between the two, due to the polarisation of the electronic cloud of the crystal. The situation is very similar to the binding of an electron to an impurity centre discussed in Section 4.10.1. We find that the binding energy of the exciton is very small, about 0.01 eV, which means that the electron–hole pair which constitutes the exciton will dissociate except at very low temperatures. Therefore excitons do not play a significant role in semiconductors and our picture of independent one-electron states is valid.

The situation is different in insulators. In the first instance the dielectric constant is smaller, about 5 (to be compared with 16 for germanium), and, also, the dispersion curves $E_\alpha(\boldsymbol{k})$ at the top of the valence band and the bottom of the conduction band are such as to favour the formation of relatively stable excitons. A description of these excitons is complicated by the fact that the average separation between the electron and the hole tends to be the same, more or less, with the interatomic distance, in which case the use of a dielectric constant to describe the interaction between them (as in (4.78)) becomes inappropriate. We shall say no more about this in the present volume.

4.15 The dielectric function

The electronic cloud of a crystal polarises to some degree under the influence of the electric field of an incident electromagnetic wave, in the same way that the electronic cloud of an atom does (see Section 2.7.1); the polarisation $\boldsymbol{P}$ is defined as the induced dipole moment per unit volume of the crystal. Let us assume that an electromagnetic wave propagates through the crystal in the z direction with wavevector q and frequency $\nu = \omega/2\pi$, in which case its electric field component will be given by

$$\boldsymbol{F} = \boldsymbol{F}_O(q)\cos(qz - \omega t) \tag{4.132}$$

with the vector $\boldsymbol{F}_O(q)$ in the xy plane. When the wavelength $\lambda = 2\pi/q$ is much greater than the lattice constant, the polarisation is given by

$$\boldsymbol{P} = (\epsilon(\omega) - 1)\epsilon_O\boldsymbol{F}_O(q)\cos(qz - \omega t) \tag{4.133}$$

where ϵ_O denotes, as usual, the permittivity of free space, and $\epsilon(\omega)$ is the so-called dielectric function of the crystal; it depends only on ω, it does not depend on q, for large wavelengths; and we have further assumed that it is independent of the polarisation direction (the direction of $\boldsymbol{F}_O(q)$) of the electric field. This latter assumption is valid for crystals of cubic symmetry, but it does not hold generally.

The polarisation of the electronic cloud of the crystal can be described, approximately, by analogy to what we have said in Section 2.7.1 in relation to the polarisation of an one-electron atom, as the sum of independent oscillating dipoles. To every allowed vertical transition (see Figs. 4.54 and 4.55) corresponds an 'oscillator' of natural frequency

$$\omega_{\alpha'\alpha}(\boldsymbol{k}) = 2\pi\nu_{\alpha'\alpha}(\boldsymbol{k}) \equiv (E_{\alpha'}(\boldsymbol{k}) - E_\alpha(\boldsymbol{k}))/\hbar \tag{4.134}$$

where $E_{\alpha'}(\boldsymbol{k}) > E_{\mathrm{F}}$ and $E_\alpha(\boldsymbol{k}) < E_{\mathrm{F}}$ are, respectively, the energy levels of the final and initial states of the transition (we assume for simplicity absolute zero temperature). We have a large number of such transitions: the wavevector $\boldsymbol{k}$ in (4.134) can be any one of the allowed values (4.24) in the BZ of the crystal; for any given $\boldsymbol{k}$, the initial state can be any occupied state $\psi_{\boldsymbol{k}\alpha}$ of spin up or down and the final state any empty state $\psi_{\boldsymbol{k}\alpha'}$ of the same wavevector and spin. The contribution of the above oscillators to the dipole moment of a unit-volume (we put $V = 1$ in all formulae where the volume appears) is given by (4.133) with

$$\epsilon_O(\epsilon(\omega) - 1)_{osc} = 2 \sum_{\alpha'\alpha} \sum_{\boldsymbol{k}} \frac{\omega_{\alpha'\alpha}(\boldsymbol{k}) p_{\alpha'\alpha}(\boldsymbol{k}) f(E_\alpha(\boldsymbol{k}))(1 - f(E_{\alpha'}(\boldsymbol{k})))}{\omega^2_{\alpha'\alpha}(\boldsymbol{k}) - \omega^2} \tag{4.135}$$

where $p_{\alpha'\alpha}(\boldsymbol{k})$ are coefficients (positive quantities) which determine the contributions to the dielectric function of the individual oscillators; the Fermi–Dirac distribution functions in the numerator of (4.135) tell us that the above mentioned oscillators correspond to 'transitions' between occupied and empty states. The factor of 2 takes into account spin degeneracy. We observe the similiarity of the above result ((4.133) and (4.135)) with (2.17) for the induced dipole moment of an atom (the coefficients $p_{\alpha'\alpha}(\boldsymbol{k})$ have the same meaning as the f-coefficients of (2.17)).

We note that for given ω and with α, α' fixed, the denominator of the corresponding term in (4.135) vanishes, when $\omega_{\alpha'\alpha}(\boldsymbol{k}) = \omega$, and the term becomes infinite! We *exclude* these points in evaluating the sum over $\boldsymbol{k}$; this can be done in a systematic manner (by taking the principal part of the sum, as the mathematicians would say) so that the expression (4.135) remains meaningful. We note that the condition $\omega_{\alpha'\alpha}(\boldsymbol{k}) = \omega$ defines the allowed vertical transitions, as these have been defined in the previous section, for the given ω; when this condition is satisfied a photon of energy $\hbar\omega$ can be absorbed with a certain probability which is, in fact, proportional to $p_{\alpha'\alpha}(\boldsymbol{k})$.

Formula (4.135) holds for metals and semiconductors. In the case of metals, however, there is an additional contribution to the polarisation which is due to the mobility of the conduction-band electrons (we recall that this mobility derives from the fact that in a metal there are empty electron states with energy next to that of occupied states). We can say that this contribution corresponds to oscillators of vanishing natural frequencies. The contribution to the dielectric function from these oscillators can be approximated (in the optical and ultraviolet region of the spectrum) by the following formula

$$(\epsilon(\omega) - 1)_p = -\frac{\omega_p^2}{\omega^2} \tag{4.136}$$

$$\omega_p^2 \equiv \frac{ne^2}{m\epsilon_O} \tag{4.137}$$

where m is the mass of the electron and n is an effective number of free electrons per unit volume which we can estimate in the manner of Section 4.9 (see text following (4.76)). Therefore, the dielectric function of a metal is given by

$$\epsilon(\omega) = 1 - \frac{\omega_p^2}{\omega^2} + (\epsilon(\omega) - 1)_{\rm osc} \tag{4.138}$$

Knowing the dielectric function of a crystal and the absorption coefficient of light (which is determined by the allowed vertical transitions at any given frequency), we can calculate, using standard formulae of electromagnetism, the fraction of a radiation incident on a slab of the crystal that will be reflected, absorbed or transmitted through it (see, e.g., problem 4.18). The study of these phenomena lies, however, beyond the scope of the present book.

4.16 Plasmons

The model of independent electrons which move in a mean field $U(\boldsymbol{r})$ explains easily, as we have seen, many of the properties of crystalline solids (equilibrium properties, transport, optical, magnetic properties, etc.). It describes complicated phenomena, such as the shape of the Fermi surface of a metal, the variation with temperature of the conductivity of a semiconductor, the spontaneous magnetisation of ferromagnetic metals at zero temperature, electron emission from surfaces etc., with remarkable accuracy in many cases. There are, however, phenomena which it cannot explain. Plasma oscillations in metals, called also plasmons, is such a phenomenon. They owe their existence to residual long-range Coulomb interaction (correlation) between the electrons of the metal.

Plasmons are collective oscillations of the electrons of the conduction band of a metal; they can be described classically as follows.[30] Let us assume that the electronic cloud due to all electrons in the conduction band of a metallic slab (parallel to the xy plane and of certain thickness Δz) is displaced by x parallel to the x-axis, as shown in Fig. 4.56. The displacement of the electronic cloud leaves a layer of positive charge due to the stationary ions (nuclei + core electrons) on the left end of the slab and creates a layer of negative charge on its right end. The density of electric charge per unit area (in the yz plane) is the same on both ends, apart from the sign, and equals

$$\sigma = eNx \tag{4.139}$$

where N is the number of electrons per unit volume in the conduction band of the metal. The situation is similar to that of a charged parallel-plate

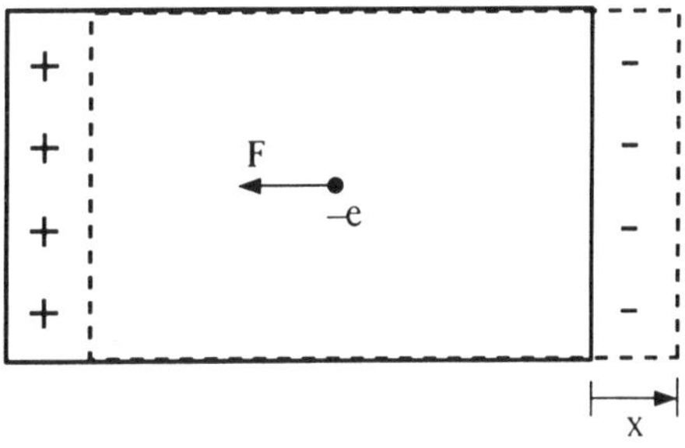

Figure 4.56. Classical picture of plasma oscillation

capacitor. The electric field (along the x direction) inside the metal is given, by analogy to the field between the plates of the capacitor, by the relation

$$F = \frac{\sigma}{\epsilon_O} = \frac{eNx}{\epsilon_O} \tag{4.140}$$

which means that a force

$$-eF = -e^2Nx/\epsilon_O \tag{4.141}$$

acting on every one of the electrons, tends to bring them (the entire electronic cloud in unison) back to equilibrium ($x = 0$). The force (4.141) corresponds to a potential energy field

$$\begin{aligned} U &= kx^2/2 \\ k &\equiv Ne^2/\epsilon_O \end{aligned} \tag{4.142}$$

which is that of a harmonic oscillator (see Section 3.3.2). The natural frequency of this oscillator is obtained from:

$$\omega_P = (k/m)^{1/2} = (Ne^2/m\epsilon_O)^{1/2} \tag{4.143}$$

where m is the mass of an electron; and the energy eigenvalues corresponding to it, the plasmon energy-levels, are given by (see 3.21))

$$E_{P_v} = \hbar\omega_P(v + \tfrac{1}{2}),\ v = 0, 1, 2, \ldots \tag{4.144}$$

It is obvious that to excite a plasmon a minimum of energy equal to $\hbar\omega_P$ is required. For most metals $10 < \hbar\omega_P < 20$ eV in agreement with (4.143). The existence of plasmons is confirmed experimentally in energy loss spectra of high-energy (several thousand eV) electrons passing through thin metal foils (less than 500 Å or so thick). In such an experiment a monoenergetic (to within a fraction of an eV) beam of electrons is incident on the foil, and one measures the energy distribution of the electrons after they have passed through the foil, to obtain the so-called energy loss spectrum: number of scattered electrons versus energy lost. A typical such spectrum is shown in Fig. 4.57 for the metal aluminum; we interpret it as follows. An electron passing through the foil may collide inelastically with plasmons losing energy to them. On exit, it has lost $\Delta E = \hbar\omega_P$, $2\hbar\omega_P$, $3\hbar\omega_P$, and so on, depending on the number of collisions it had.

The plasmons we have described have an infinite wavelength ($\lambda = \infty$, $q \equiv 2\pi/\lambda = 0$). We do have plasmons of finite wavelength, although $|\boldsymbol{q}|$ cannot exceed a certain value (see below). These correspond to periodic (in time and space) variations of the density of the electronic cloud (similar to the variation of mass density in a sound wave) with frequencies $\omega_P(q)$ which do not differ significantly from (4.143).

The coherent motion of a multitude of electrons, which constitutes a plasmon, is possible only as long as the probability of an electron scattering out (through some collision or other) of this collective motion is not possible, which implies that plasmons exist only when their period $T = 2\pi/\omega_P$ is much smaller than the average time between electron collisions; we must have

$$T \ll \tau \tag{4.145}$$

which can always be satisfied at sufficiently low temperatures. In this respect plasma waves are different from sound waves which happen in systems where local equilibrium exists as a result of frequent collisions between the particles (this being a pre-requisite of hydrodynamic motion on a larger scale), which means that for sound waves to exist, their period must be much larger than the average time between particle collisions, i.e. the opposite of (4.145) must hold in their case.

A plasmon excitation will decay rapidly into an electron–hole pair (an electron is removed from a level below to one above E_F leaving an empty

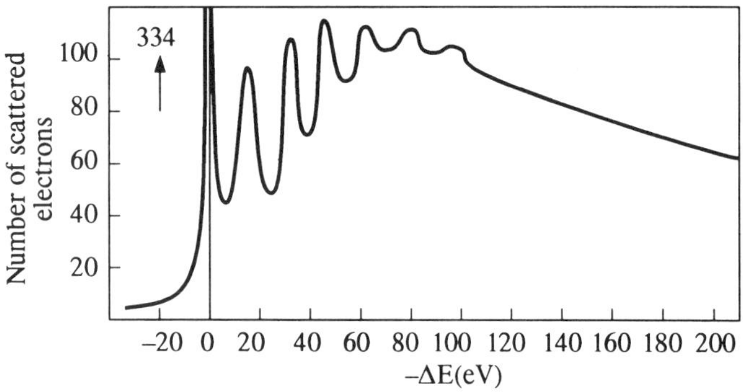

Figure 4.57. Energy loss spectrum of Al. [G. Ruthermann, *Annals of Physics* **2** (1948) 113]

state (hole) below E_F), if that is allowed by energy and wavevector conservation, i.e. if the following equation is satisfied

$$E(\boldsymbol{k} + \boldsymbol{q}) - E(\boldsymbol{k}) = \hbar\omega_P(q) \tag{4.146}$$

where $E(\boldsymbol{k})$ is an one-electron level below E_F and $E(\boldsymbol{k} + q)$ one above E_F. Using the Sommerfeld model (Section 4.9) to estimate the left-hand side of the above equation and putting $\hbar\omega_P(q)$ equal to (4.143), one finds (problem 4.20) that (4.146) will be satisfied if $q > \omega_P/v_F$, where v_F is the velocity of an electron at the Fermi level; which means that plasmon excitations can exist only for $q < \omega_P/v_F$.

We have been discussing plasma oscillations in metals; but such exist also in semiconductors. The corresponding plasma frequency is given by (4.143) where N is now the number of electrons per unit volume in the valence band of the semiconductor. The energy gap (we note that $E_g \ll \hbar\omega_P$) does not seem to affect the collective oscillation of the electrons.

We close this section with a remark concerning the contribution of plasmons to the total energy of the crystal. We know that the ground energy of an oscillator is not zero. In the present case, it is equal to $\hbar\omega_P/2$ according to (4.144). We expect, therefore, that there will be a contribution to the total energy of the crystal (in its ground state) from plasmons; and indeed there is; a small amount, but it is there. It forms a part of the so-called correlation energy mentioned in Section (4.11.3).

4.17 Superconductivity

Another phenomenon which the model of independent electrons cannot explain is that of superconductivity. We expect, from what has been said in Section 4.8, that the resistivity $\rho = 1/\sigma$ of a pure (without impurities) metal will go to zero, as $T \to 0$, continuously. In many metals, however, the resistivity vanishes abruptly at a temperature above zero as seen in Fig. 4.58, which shows the variation with temperature of the resistivity of mercury in the region of the so-called critical temperature (denoted by T_c). This is the temperature (a characteristic property of the metal) at which the transition occurs from the normal phase ($T > T_c$) to the superconducting phase of the metal ($T < T_c$). The experimental data of Fig. 4.58 are due to Onnes who observed the phenomenon for the first time in 1911. Since then, the phenomenon has been observed in many metals (see Table 4.1). The vanishing of the resistivity of a superconductor (by which we shall mean a metal in its superconducting phase) implies that an electric current can be sustained in it for a long (in principle, infinite) time without an electromotive force. There is experimental evidence of currents persisting, in some cases, for two years or so, which is indeed a long time.

The superconductor also has interesting magnetic properties: it will not

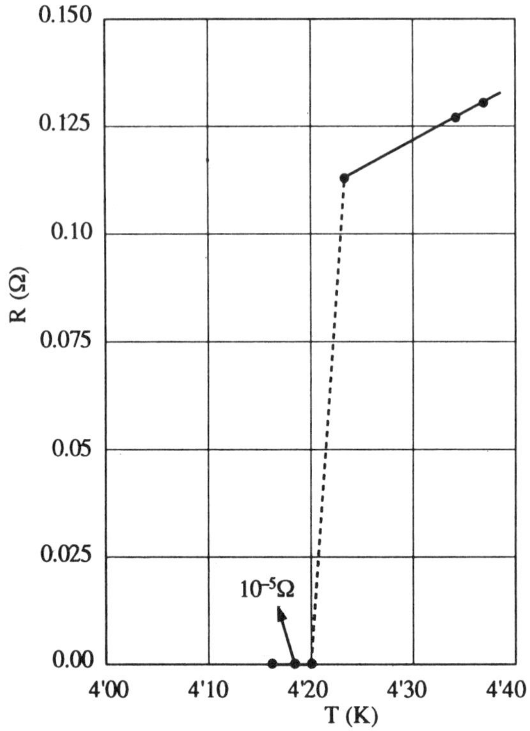

Figure 4.58. Variation of the resistance of mercury with the temperature in the region of the critical temperature (Onnes, 1911)

allow a magnetic field in its interior. Placed in a magnetic field whose magnitude does not exceed a critical value H_c, it modifies it in such a way as to expel it from its interior. The phenomenon, known as the Meissner effect[31], is demonstrated schematically in Fig. 4.59.

An understanding (in terms of a microscopic theory) of the vanishing resistivity, of the Meissner effect and other thermodynamic properties of the superconductor (for example, that the temperature dependence of its specific heat is quite different from that of a normal conductor) turned out to be quite difficult and took a long time to develop. One may appreciate the difficulties of the problem by noting that at $T = 0$ the difference in total energy (per atom) between the superconducting and the normal phase of a metal is orders of magnitude smaller than the total energy per atom of either phase. This means, in fact, that one is looking for a mechanism of superconductivity sustained by a reduction $\delta\varepsilon$ in the total energy ε of the normal phase, when we can only estimate ε within an error $\Delta\varepsilon \gg \delta\varepsilon$.

The BCS theory (proposed by Bardeen, Cooper and Schrieffer in 1957)

Table 4.1. T_c and H_0 of various superconductors[a]

	T_c (K)	H_0 (gauss)
Aluminium	1.19	99
Cadmium	0.56	30
Gallium	1.09	51
Indium	3.4	283
Lead	7.18	803
Mercury-α	4.15	411
Mercury-β	3.95	340
Niobium	9.46	1944
Rhenium	1.70	201
Ruthenium	0.49	66
Tantalum	4.4	830
Technetium	7.8	—
Thallium	1.37	162
Tin	3.7	306
Titanium	0.40	100
Vanadium	5.30	1310
Zinc	0.88	53

[a]The list does not include all known superconductors.

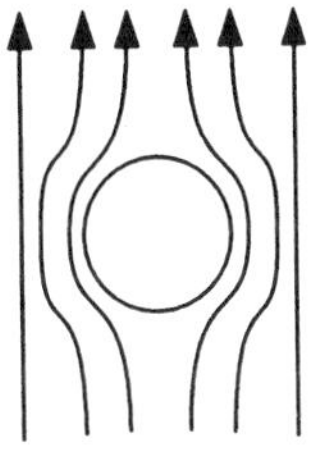

Figure 4.59. The Meissner effect

provided an explanation of superconductivity along the following lines.[32] To begin with, we assume than in the normal phase ($T > T_c$) the conduction-band electrons behave as if they are free; they are described by plane waves of certain wavevector $\boldsymbol{k}$ and energy (measured from the bottom of the conduction band) $E(\boldsymbol{k}) = \hbar^2 k^2/2m$. (Replacement of the plane waves by Bloch waves does not affect the essential results of the theory.) We further assume that all interactions, except that which is responsible for superconductivity, produce the same correction to the total energy in either of the two phases (normal and superconducting) and therefore, for the purpose of understanding superconductivity, we can disregard them. In this way it becomes possible to estimate the energy difference between the two phases without it being necessary to know accurately the total energy of either phase. The interaction responsible for superconductivity, according to the BCS theory, can be described as follows: Two electrons with energy $E(\boldsymbol{k})$ and $E(\boldsymbol{k}')$, respectively, within the region

$$E_F - \hbar\omega_D < E(\boldsymbol{k}),\ E(\boldsymbol{k}') < E_F + \hbar\omega_D \tag{4.147}$$

where ω_D is the Debye frequency of the metal introduced in Section 4.12[33], may be attracted to each other in spite of the residual Coulomb interaction (repulsion) that exists between them. This is possible when the lattice of atomic cores is polarised by a pair of electrons (4.147) in such a way as to produce an effective attraction between the electrons of the pair which is greater than the residual Coulomb repulsion between them. We shall not attempt a proof of the above statement here; suffice it to say that it has turned out to be quite a common occurrence in metals, as evidenced by the fact that so many of them become superconductors at sufficiently low temperatures. Let us assume, then, that there exists a negative (attractive) net interaction between any two electrons (4.147) and let us denote the corresponding potential energy term by U'. In the BCS theory U' is approximated by $U' = U'_O/L^3$, where L^3 is the volume of the crystal and U'_O a constant. This interaction favours the formation of mutually bound pairs of electrons which are called Cooper pairs. Because the interaction U' between the electrons of a Cooper pair is so small, the average distance ξ between them is quite large, of the order of $\xi \simeq 10^{-4}$ cm. Between and beyond the 'electrons of a given pair' lies the electronic cloud due to all other electrons paired and unpaired, which are of course indistinguishable from the electrons of the given pair. It is obvious that we are dealing with the dynamics of a complex many-electron system, whereby the electrons exchange their roles in the collective action all the time; it is as if we are watching a formation dance by thousands of dancers and we perceive order to be there, because the dancing roles are given, though the dancers exchange their roles all the time.

We admit, then, that in thermodynamic equilibrium (at $T = 0$) the electrons that would (when $U' = 0$) be occupying the one-electron states with energy between $E_F - \hbar\omega_D$ and E_F, are now (when $U' < 0$) accommodated with certain probability in pairs $(\boldsymbol{k}, \boldsymbol{k}')$ within the energy region (4.147). We may assume that electrons with energy below $E_F - \hbar\omega_D$ are not affected by U'. Obviously, there are more possible pairs $(\boldsymbol{k}, \boldsymbol{k}')$ within the region (4.147) than there are electrons to occupy them, and the question arises as to which pairs are to be occupied?

It is obvious from what we said earlier, that the interaction U' exists between any two electrons in the region (4.147) and, therefore, operates not only between the electrons of a given pair but also between electrons belonging to different pairs. The pairing of the electrons must be arranged in such a way as to minimise the total energy of the superconducting phase (to lower it as much as possible relative to the energy of the normal phase). According to the BCS theory, this is achieved when the electrons we are considering are in Cooper pairs of the following form

$$(\boldsymbol{k}\uparrow, -\boldsymbol{k}\downarrow) \tag{4.148}$$

which tells us that the electrons of a Cooper pair will have opposite wavevectors (momenta) and opposite spins. We observe that we still have more Cooper pairs (4.148) of electrons with $E(\boldsymbol{k})$ in the region (4.147) than we have electrons to fill them with. The probability $0 < w(\boldsymbol{k}) < 1$ that a Cooper pair (4.148) will be occupied (at $T = 0$) is finally determined by the requirement that the total energy of the system be a minimum. We shall not need the explicit expression for $w(\boldsymbol{k})$ in the context of our brief exposition of the BCS theory.

It turns out that the ground-state energy of the superconductor, determined in the above manner, is separated from the first excited state of the superconductor by a finite energy gap given by

$$\Delta_O = 4\hbar\omega_{\mathrm{D}} \exp\left(-\frac{1}{\rho(E_{\mathrm{F}})|U'_O|}\right) \tag{4.149}$$

where $\rho(E_{\mathrm{F}})$ is the one-electron density of states, per spin, at the Fermi level, of the normal phase. Δ_O equals the energy required to break a Cooper pair (we should point out that because the electrons of different Cooper pairs interact with each other, the said energy cannot be calculated by considering one pair in isolation). We have: $(\rho(E_{\mathrm{F}})|U'_O|)^{-1} \simeq 3.5$, which means that $\Delta_O \ll \hbar\omega_{\mathrm{D}}$. The energy gap between the ground state and the first excited state of the superconductor plays a crucial role in our understanding of superconductivity.

The momentum of a pair (4.148) equals zero and therefore the total momentum of the electrons in the ground state of the superconductor vanishes; and so does the electric current in a superconducting ring (such as that of Fig. 4.28). The persistence of an electric current in a superconductor is explained as follows. To a given current corresponds a metastable state[34] constituted in very much the same way as the ground state we have described, except that now all electrons in the region (4.147) are to be found in Cooper pairs of the form

$$(\boldsymbol{k}\uparrow, -\boldsymbol{k}'\downarrow) \tag{4.150}$$

where $\boldsymbol{k} - \boldsymbol{k}' = \boldsymbol{K}$ is a non-zero vector, the same for every pair. It is obvious that in this case the total momentum of the electrons is an integer multiple of $\hbar\boldsymbol{K}$, which implies a non-vanishing electric current, which persists because the state we have described has a very long life-time (as long as a few years, in some cases, as we have noted in our introduction). The reason is this: The state of the superconductor can change if one or more Cooper pairs break up, but for this a minimum of energy is required equal approximately to the energy gap (4.149), and the vibrating nuclei with which the paired electrons might collide cannot provide that much energy at very low temperatures ($T < T_c$). The state of the superconductor could also change, if the momentum (the value of $\boldsymbol{K}$) of a large number of pairs could change in a collision, but the probability of this happening is negligible.

The BCS theory can also explain the Meissner effect. It turns out that a state of mutually interacting Cooper pairs expels the magnetic field from the interior of the superconductor. We shall not deal with this problem here. The theory describes reasonably well other thermodynamic properties as well.

When $T > 0$, a fraction of the electrons in the region (4.147) are not to be found in pairs, but moving independently like normal electrons (electrons in the normal phase). As the temperature rises this fraction increases until all electrons become normal, when $T > T_c$. It can be shown that

$$3.5k_B T_c \simeq \Delta_O \tag{4.151}$$

It is, of course, the coupled electrons which are responsible for the observed superconducting properties of the system when $0 < T < T_c$, the normal electrons behave as in an ordinary metal. We should also note that the above mentioned energy gap is reduced from its zero-temperature value (4.149) to zero as the temperature rises from $T = 0$ to T_c.

Finally, it is worth noting that when $T < T_c$, the application of an external magnetic field $H \geqslant H_c(T)$ destroys the superconducting phase. The function $H_c(T)$ is shown schematically in Fig. 4.60. The critical value $H_0 \equiv H_c(0)$ for the superconductors listed in Table 4.1 appears in the last column of that table.

We close this section with a brief reference to high temperature superconductors which have been discovered during the last ten years or so. These superconductors are crystals with a chemical composition of the type $YBa_2Cu_3O_7$, $Bi_2Sr_2CaCu_2O_8$, etc., and a correspondingly large unit cell. The critical temperature T_c of these superconductors can be as high as 120 K in some cases. At the present time the mechanism of high-temperature superconductivity is not properly understood.

4.18 Non-crystalline solids

4.18.1 AMORPHOUS FILMS AND GLASSES

Imagine a space lattice (4.4) and assume that only a small fraction of its sites are occupied by atoms. If the occupation of a site is a random event, the

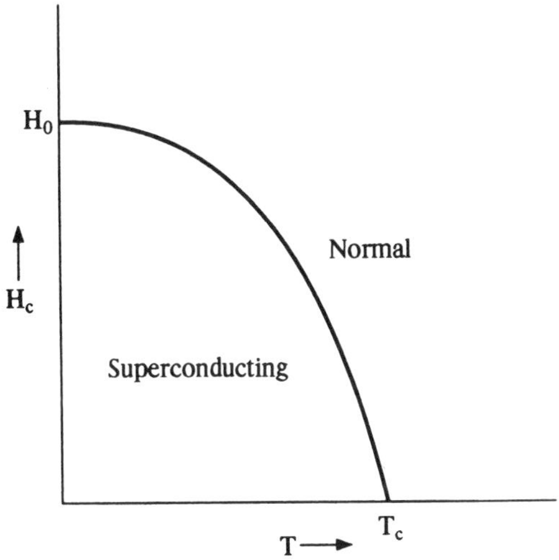

Figure 4.60. Phase diagram of a superconductor

result is a random (disordered) distribution of atoms in space, which models reasonably well the distribution of impurities in a crystalline semiconductor (see Section 4.10.1). Similarly, if a fraction of the sites of a given lattice, chosen randomly, are occupied by atoms A and the remaining sites are occupied by atom B, the result is a disordered distribution in space of atoms A and B, which would be a good model for a binary alloy of two metals. The above are examples of what is called substitutional disorder. In nature we find a number of different types of disordered structures.

When germanium, for example, is deposited by condensation from the vapour state on a substrate the result is often a so-called amorphous film. The long-range order which characterises crystalline Ge is replaced in the film by a random network. However, the atomic nuclei sitting at the knots of the network have, on the whole, four nearest neighbours to whom they are bonded in a way similar to that in crystalline Ge. We can say that much of the short-range structure (or order) of the crystalline Ge is retained in the amorphous state, although variations in the bond-length between atoms and in the angle between bonds do exist; and there are 'defects' in the structure, voids of some kind or other, Ge atoms with three rather than four nearest neighbours and, of course, there may be impurities.

Ge and other materials such as Si, Te, B and InSb which are obtained in amorphous form by deposition, but cannot be obtained in this form by cooling from the melt (see below), have properties which tend to be sensitive to the conditions of deposition and to subsequent annealing treatment. Moreover, such amorphous films may be unstable and special care (e.g., a low substrate temperature) is often needed to avoid crystallisation.

Amorphous materials which can be obtained by cooling from the molten state are known as glasses and they are in general more stable than those, mentioned above, which can be obtained only be deposition. They include Se, As_2Se_3 and similar chalcogenide compounds (these contain one or more of the chalcogenide elements S, Se or Te) and the common borosilicate glasses. The formation of a glass by cooling from the melt is described schematically in Fig. 4.61. In some cases the passage from the molten state to the glass state is smooth with no apparent discontinuity in the properties of the system. In other cases, the system goes through a so-called second-order phase transition (transformation) at the so-called glass-transition temperature T_g. At this temperature the thermal expansion coefficient (the rate of change of volume with temperature) changes to a lower value as shown in Fig. 4.61, and so does the specific heat of the material, which implies a 'hardening' of the material: fewer configurational energy states or degrees of freedom are available to the system below T_g. We note, however, that T_g, unlike the melting temperature T_m, is not so well defined and depends on the rate of cooling, or the rate of heating when the material is subjected to the reverse process. There is, of course, always the possibility of crystallisation at any temperature above or below T_g, as indicated by the vertical arrows in Fig. 4.61. For certain materials, such as As_2Te_3, the possibility of crystallisation is very high so that very fast cooling must be used to prepare the glass. In contrast, the crystallisation process in As_2Se_3 is

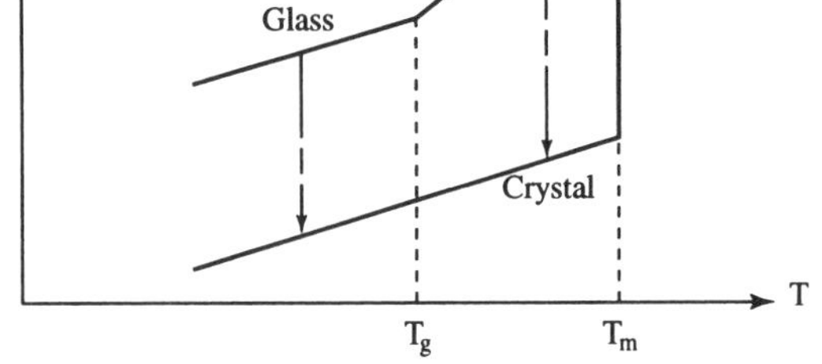

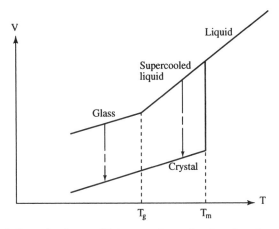

Figure 4.61. Variation of volume with temperature of a glass-forming material. T_m is the melting temperature and T_g the glass-transition temperature. The vertical arrows indicate the possibility of crystallisation from the supercooled liquid or the glass

very slow and therefore glass formation can be achieved by cooling the melt at a relatively low rate.

An example of glass formation which one can easily visualise is that of selenium. In its crystalline state, Se is constituted of parallel helical chains as shown in Fig. 4.62. The bonds between the atoms of a chain are strong so that a chain is not easily broken, but the bonding between different chains is much weaker and one may assume that in the molten state these are randomly oriented. When the liquid is cooled sufficiently fast the chains do not have the time to align themselves before the viscosity becomes too high for this to happen, so the material is trapped in a glassy state. The situation, as far as the stability of the system is concerned can be described, schematically, as in Fig. 4.63. The glassy state is a metastable state, thermodynamically less stable, under ordinary conditions of temperature and pressure, than the crystalline state. However, an activation energy W is required for the transition from the glassy to the crystalline state and as long as this is large enough, the glass is practically stable.

4.18.2 LOCALISATION OF ONE-ELECTRON STATES DUE TO DISORDER

The periodicity of the potential field in a crystalline solid makes it possible, as we have seen in Section 4.3, to describe the eigenstates of an electron in that field in a systematic manner using Bloch's theorem. The eigenstates of an electron in a non-periodic field are much more difficult to describe and our present knowledge of the electronic properties of non-crystalline solids is not anything as good as that of crystalline ones, although considerable

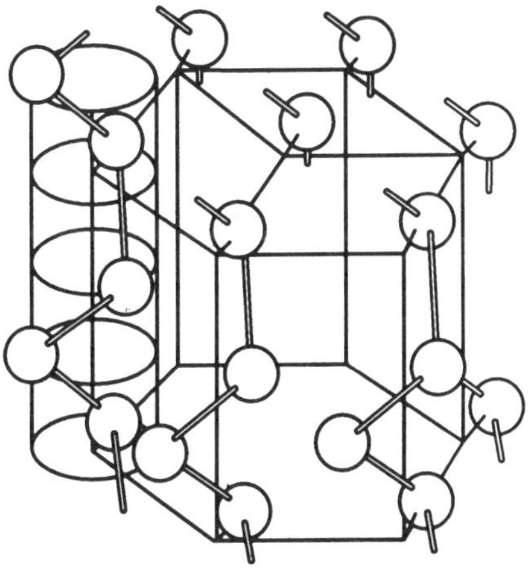

Figure 4.62. Crystalline selenium consists of parallel chains

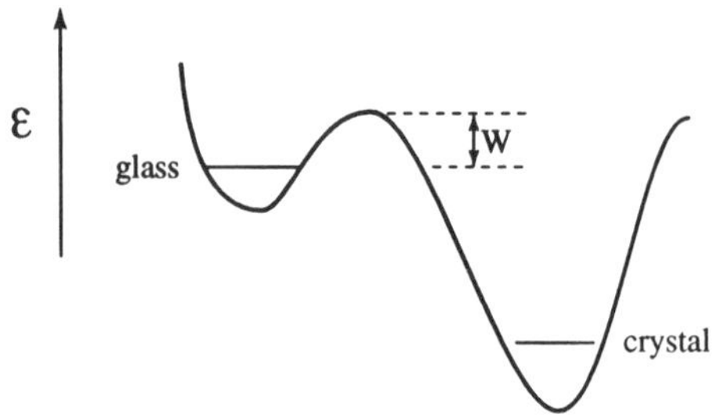

Figure 4.63. The glassy state is practically stable if the activation energy W is sufficiently large

progress has been made in this direction during the last three decades or so. It is easy to see why the problem is a difficult one. For a crystalline solid we can write down a one-electron mean potential field (such as the schematic one of Fig. 4.64(*a*))[35] and proceed to find the one-electron states in it, in the manner described in Section 4.3, but the moment we allow, for example, the potential wells at different sites to have different depths (as in Fig. 4.64(*b*)) the best we can say, if we wish to be realistic, is that the potential well at any given site has a depth which can take a range of values according to some probability distribution. It is obvious that there are infinitely many realisations of a potential field, defined in this way, each of them corresponding to a specific assignment of different potential wells to different sites. These infinitely many realisations constitute (are the members of) a so-called ensemble. The traditional way to proceed, at least in principle, is to

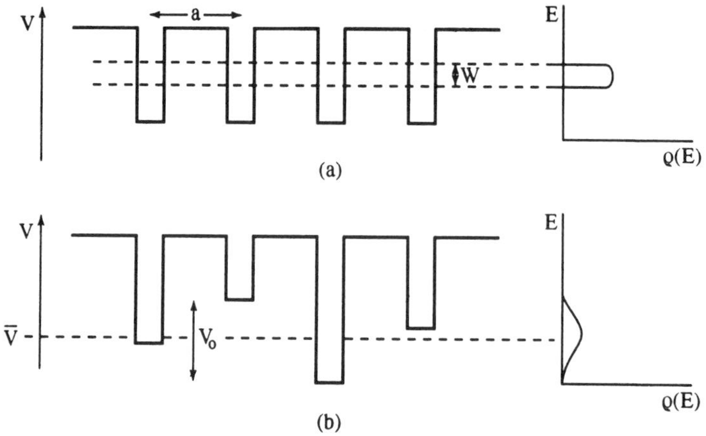

Figure 4.64. Periodic (*a*) and non-periodic (*b*) arrays of potential wells, and the corresponding densities of one-electron states ($\rho(E)$)

calculate the physical quantity we are interested in (the density of one-electron states, the specific heat, the conductivity, etc.) for every member of the ensemble and then take the average of these values; we call it the ensemble average of the quantity A and denote it by $\langle\langle A \rangle\rangle$. We then expect, according to traditional statistical physics, the measured value of A to be the same as $\langle\langle A \rangle\rangle$, provided that the real system under consideration is sufficiently large (contains many atoms).

Our task is therefore defined. It is to calculate the physical quantity we are interested in for every member of the ensemble (this meaning all possible realisations of the system under the given experimental conditions) and then to take the ensemble average of this quantity. Let us say that we are interested in the density of one-electron states of a non-crystalline solid represented by an array of potential wells of randomly varying depth. The problem of finding the one-electron states for a specific realisation of this potential field is extremely difficult (except of course for that realisation which corresponds to a periodic array of identical wells). To solve the same problem for many different realisations in order to obtain an ensemble average appears to be an impossible task. Nevertheless, considerable progress has been made, and a lot is presently known, at least at a semi-quantitative level, about the electronic properties of non-crystalline solids.

The Anderson model[36] of a disordered (non-crystalline) solid can be described as follows. We begin with a periodic potential field (as shown schematically in Fig. 4.64(a)). We assume a simple cubic lattice

$$\begin{aligned} \boldsymbol{R}_n &= n_1\boldsymbol{t}_1 + n_2\boldsymbol{t}_2 + n_3\boldsymbol{t}_3 \\ \boldsymbol{t}_1 &= a\hat{x},\ \boldsymbol{t}_2 = a\hat{y},\ \boldsymbol{t}_3 = a\hat{z} \\ n_1,\ n_2,\ n_3 &= 0,\ \pm 1,\ \pm 2,\ \ldots \end{aligned} \tag{4.152}$$

which means that the corresponding BZ, is a cube (in $\boldsymbol{k}$-space) of volume $(2\pi)^3/a^3$. We consider in particular an s-band of one-electron states in the given field: the corresponding one-electron levels are given, acording to (4.28), by

$$E(\boldsymbol{k}) = E_0 + \Delta E(\boldsymbol{k}) \tag{4.153}$$

where E_0 stands for the first two terms of (4.28). We assume that $E(\boldsymbol{k})$ has been calculated and that the result is a band of width W. The corresponding density of states, per spin, is shown schematically in Fig. 4.64(a). We recall (see (4.27)) that the eigenfunctions (their orbital parts) which describe the one-electron states of the band are Bloch waves

$$\psi_{\boldsymbol{k}}(\boldsymbol{r}, t) = \frac{A}{\sqrt{N}} \left\{ \sum_{\boldsymbol{R}_n} \exp(\mathrm{i}\boldsymbol{k}\cdot\boldsymbol{R}_n)\varphi(\boldsymbol{r} - \boldsymbol{R}_n) \right\} \exp(-\mathrm{i}E(\boldsymbol{k})t/\hbar) \tag{4.154}$$

where $\varphi(\boldsymbol{r} - \boldsymbol{R}_n)$ denotes an atomic s-state centred on the $\boldsymbol{R}$th site; and we note once more that Bloch waves extend over the entire crystal.

We then introduce disorder into the system by adding a random potential ΔV to each potential well, as shown in Fig. 4.64(*b*). We assume that ΔV takes all values at random between $-V_o/2$ and $V_o/2$, so that the depth of the potential well takes at random all values between $\bar{V} - V_o/2$ and $\bar{V} + V_o/2$ where $\bar{V}$ denotes the mean value of this quantity. A calculation of the one-electron states in the above random potential field is not possible, but Anderson was able to prove, by an indirect method, the following properties (we emphasise that these refer always to an ensemble average of the system under consideration):

(i) The band of states spreads over a greater energy region as compared to the ordered solid. This is shown schematically by the density of states curve in Fig. 4.64(*b*).

(ii) When V_o/W exceeds a certain critical value $(V_o/W)_{\text{crit}}$ all electron states in the band are localised, i.e. they are described by eigenfunctions of the following form[37]

$$\psi_E = \left(\sum_n A_n \varphi(\boldsymbol{r} - \boldsymbol{R}_n)\right) \exp(-\bar{r}/\lambda) \exp(-\mathrm{i}Et/\hbar) \qquad (4.155)$$

where $\bar{r}$ is the distance of $\boldsymbol{r}$ from some site of the lattice (this site being different for different ψ_E); $\varphi(\boldsymbol{r} - \boldsymbol{R}_n)$ are atomic orbitals as in (4.154) and A_n are complex coefficients which vary more or less randomly. The coefficient λ, known as the localisation length, determines the extension of the wavefunction which, according to the above equation, is localised within a volume of diameter equal approximately to 2λ. The localisation length is in general a function of the energy E. Finally, it turns out that $(V_o/W)_{\text{crit}} \simeq 2$.

The above applies to an s-band; but a similar criterion seems to apply to p-bands and other bands, except that $(V_o/W)_{\text{crit}}$ may have a different value.

The concept of a localised state is not new in itself; we have already met such states (e.g., electron states localised about impurity-atoms in crystalline semiconductors). What is new and remarkable is the fact that one can have a band of states, described by a continuous density of states $\rho(E)$ (as in Fig. 4.64(*b*)), in which all states are localised, although there can be considerable overlap between the corresponding eigenfunctions.

The question arises as to what happens when $(V_o/W) < (V_o/W)_{\text{crit}}$. It turns out that when $(V_o/W) \simeq (V_o/W)_{\text{crit}}$, the eigenstates corresponding to energies in the middle of the band become delocalised; they are described by (4.155) but the localisation length λ now tends to infinity ($\lambda \to \infty$). When $(V_o/W) < (V_o/W)_{\text{crit}}$ we obtain a distribution of localised ($\lambda < \infty$) and extended ($\lambda = \infty$) states as shown schematically in Fig. 4.65. The states in the band tails (shaded areas) are localised, while those in the middle of the

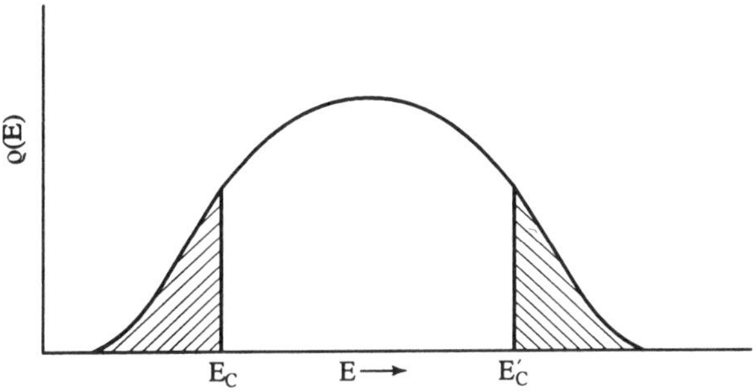

Figure 4.65. Schematic description of the density of states in the Anderson model when $(V_o/W) < (V_o/W)_{crit}$. The states in the band tails are localised, those in the middle of the band are extended

band are extended states. It appears that weak disorder leads to localisation of the eigenfunctions where the density of states is sufficiently small, as in the tails of a band. In the middle of the band where the density of states is relatively large, a weak random potential cannot localise the electron states to finite regions of space. The amplitudes of the corresponding wavefunctions, fluctuating more or less randomly, will be different from zero practically everywhere.

An important thing to note is that there cannot be localised and extended states with the same energy. If that were the case, a small change in the random potential would be sufficient to couple the extended with localised states, transforming the latter into extended ones. We find that the energy region of extended states is separated by sharp edges (denoted by E_C and E'_C in Fig. 4.65) from the regions of localised states. Because extended states contribute to the conduction of electricity (we refer to *dc*-conduction at $T = 0$) while localised states do not (see Section 4.8; though a formal description in terms of wavepackets is not possible in the present case, one can show that extended states contribute to electron transport in more or less the same way that Bloch waves do so in crystals), E_C and E'_C are called mobility edges.

In what follows we shall describe briefly some phenomena relating to electron transport which are due to the localisation of electrons because of disorder.

4.18.3 THE ANDERSON TRANSITION

Doping of crystalline semiconductors, such as germanium and silicon, accompanied by appropriate compensation can produce in the energy gap of the semiconductor a half-filled band of donor states[38], possibly overlapping

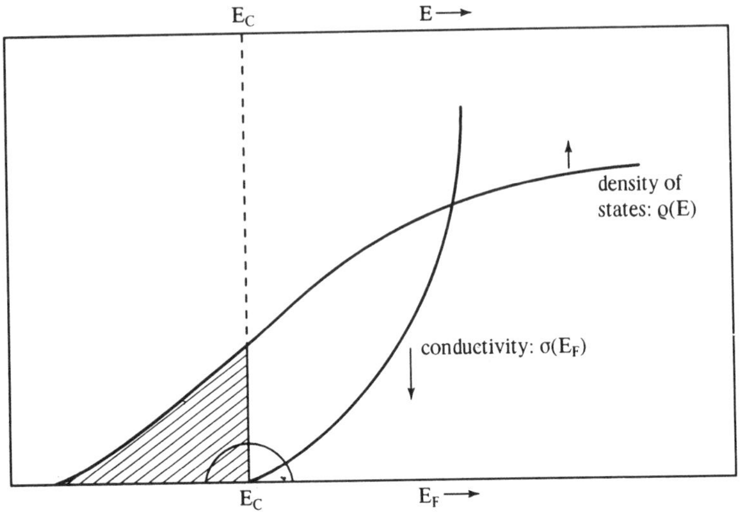

Figure 4.66. The electron states below the mobility edge E_C (shaded region) are localised, those above E_C are extended. If the Fermi level E_F lies below E_C, the system is insulating at $T = 0$; if $E_F > E_C$, it is metallic

with a band of lower-energy acceptor states. The donor-impurity band may also overlap to some degree with the conduction band of the semiconductor. In any case a band of states may arise in the crystal with the characteristics of the band shown schematically in Fig. 4.66. The states in the low-energy tail (shaded region) up to the mobility edge E_C are localised and those above E_C are extended. Now it may happen that the Fermi energy E_F lies below E_C, in which case the conductivity σ vanishes at $T = 0$; this is so because localised states do not contribute to electron transport (see Section 4.8).

Let us now assume (this is experimentally possible) that by some means, e.g., by further doping and/or compensation or by applying a stress to the system, we shift the Fermi energy (the temperature is always zero) towards the mobility edge. As long as $E_F < E_C$ the conductivity will be zero; the moment E_F crosses E_C into the region of extended states, the conductivity will begin to rise as shown schematically in Fig. 4.66. The above transition from insulating (due to disorder) to metallic behaviour is called the Anderson transition and has attracted a lot of attention in recent years. A remarkable feature of this transition is the fact that within the so-called critical region about E_C (indicated by a semi-circle in Fig. 4.66) the variation of σ with E_F appears to be independent of the specific characteristics (type of disorder, etc.) of the system under consideration; it turns out that

$$\sigma \propto (E_F - E_C)^{\nu}, \; E_F > E_C \tag{4.156a}$$

where ν, the so-called critical exponent, has the same (or nearly the same) value, $\nu = 1$, for many different materials. And it appears that this 'universal' behaviour extends below E_C: the localisation length λ defined in the previous section is given near E_C (within the critical region) by the following formula

$$\lambda \propto (E_C - E)^{-\nu},\ E < E_C \tag{4.156b}$$

where ν has the same value as in (4.156a).

4.18.4 HOPPING TRANSPORT

Another interesting aspect of electron transport in Fermi-glasses (the name applies to those materials which do not conduct at $T = 0$, although $\rho(E_F) \neq 0$, because the states at E_F are localised due to disorder) relates to the finite dc-conductivity of these materials at non-zero but still very low temperatures. Let us then assume that in a given material the Fermi level lies below the mobility edge of a band, such as the one shown in Fig. 4.66, in which case all states in its vicinity are localised, and the conductivity vanishes at $T = 0$. When the temperature is raised above zero, electron transport becomes possible via phonon-assisted hops of electrons between localised states, whose wavefunctions overlap to some degree and their energies do not differ by much more than $k_B T$. The rate of transport is determined by processes where an electron hops from an occupied localised state with energy below E_F to an empty localised state with energy above E_F; therefore the number of electrons per unit volume which can contribute to conduction equals approximately $2\rho(E_F)k_B T$, where $\rho(E_F)$ denotes as usual the density of states per spin at the Fermi level. Now let the two localised states be centred at a mean distance $\bar{R}$ from each other and let the mean difference between their energy levels be $\bar{\Delta}$. The energy required for the hopping of the electron from one to the other is provided by excited lattice oscillators and is, therefore, determined by the availability of phonons of energy $\hbar\omega \simeq \bar{\Delta}$, which is proportional to $\exp(-\bar{\Delta}/k_B T)$ when $\bar{\Delta} > k_B T$, according to (4.105). On the other hand, electron transfer is possible to the degree that the electron can tunnel from one state to the other, and the tunnelling probability, determined by the overlap of the corresponding wavefunctions, decays exponentially with the mean distance $\bar{R}$ between the centres of the localised states. Therefore, the probability p of a hop, per unit time, between two states which are energetically apart by $\bar{\Delta}$ and spatially apart by $\bar{R}$ is given by

$$p = \nu_{ph} \exp(-2\alpha\bar{R} - \bar{\Delta}/k_B T) \tag{4.157}$$

where α is proportional to the inverse of the localisation length of the states, and ν_{ph} is a frequency factor depending on the phonon spectrum (it has a value between 10^{12} and $10^{13}\ \text{s}^{-1}$).

Next we observe, following Mott and Davis[39], that application of an electric field F increases $\bar{\Delta}$ in the direction of the field by an amount $eF\bar{R}$ and lowers it by the same amount in the opposite direction, producing a difference in the hopping probabilities in the two directions given by

$$\nu_{\text{ph}} \exp(-2\alpha\bar{R} - \bar{\Delta}/k_{\text{B}}T)[\mathrm{e}^{eF\bar{R}/k_{\text{B}}T} - \mathrm{e}^{-eF\bar{R}/k_{\text{B}}T}] = \nu_{\text{ph}}(2eF\bar{R}/k_{\text{B}}T)\exp(-2\alpha\bar{R} - \bar{\Delta}/k_{\text{B}}T) \quad (4.158)$$

where we have used the fact that $\exp(\pm eF\bar{R}/k_{\text{B}}T) = 1 \pm eF\bar{R}/k_{\text{B}}T$, when $eF\bar{R} \ll k_{\text{B}}T$. The current density is given by the number of electrons per unit volume contributing to the transport process (which equals $2\rho(E_{\text{F}})k_{\text{B}}T$, as we have seen), times the charge of an electron, times the hopping distance $\bar{R}$, times (4.158). We obtain

$$j = \sigma F = 4e^2\bar{R}^2\nu_{\text{ph}}\rho(E_{\text{F}})\exp(-2\alpha\bar{R} - \bar{\Delta}/k_{\text{B}}T)F \quad (4.159)$$

We must now estimate $\bar{\Delta}$ and $\bar{R}$. We take these to be the mean values for transitions from a localisation centre at the centre of a sphere of radius R to other localisation centres within the sphere, and determine $\bar{\Delta}$ and $\bar{R}$ as functions of R. Assuming a homogeneous distribution in space of the localisation centres one easily finds

$$\bar{R} = 3R/4 \quad (4.160)$$

$$\bar{\Delta} = 3/(4\pi R^3\rho(E_{\text{F}})) \quad (4.161)$$

Equation (4.160) requires no comment. Equation (4.161) tells us that when the range of hopping increases it is more likely that a final state will be found nearer energetically to the initial state, which is what we should expect. Substituting (4.160) and (4.161) in the exponential term of (4.159) we obtain

$$\text{Exp. term of (4.159)} = \exp\left(-\frac{3\alpha R}{2} - \left(\frac{3}{4\pi\rho(E_{\text{F}})k_{\text{B}}T}\right)\frac{1}{R^3}\right) \quad (4.162)$$

We see that the first term in the exponent of (4.162) decreases (becomes more negative) and the second term increases (becomes less negative) as R increases. It is reasonable to assume that the rate of hopping processes will be determined by that value of R which maximises (4.162). One can show that (4.162) obtains its maximum when

$$R = \left(\frac{3}{2\pi\alpha\rho(E_{\text{F}})k_{\text{B}}T}\right)^{1/4} \quad (4.163)$$

Substituting the above in (4.162) and the resulting expression in (4.159) one finally obtains the following formula for the conductivity

$$\sigma = 4e^2\rho(E_F)\bar{R}^2\nu_{ph}\exp(-B/T^{1/4}) \tag{4.164}$$

where $B = 2[3\alpha^3/(2\pi\rho(E_F)k_B)]^{1/4}$ and $\bar{R}$ is given by (4.160) and (4.163). When comparing with experimental data, many workers write the above formula in the following form

$$\sigma(T) = \sigma_O \exp(-(T_O/T)^{1/4}) \tag{4.165}$$

and treat σ_O and T_O as adjustable parameters. In this form, Mott's formula has been verified by many experiments relating to impurity conduction in crystalline semiconductors and, also, to conduction in amorphous semiconductors (Ge, Si and others). As an example, we show in Fig. 4.67 the conductivity of amorphous silicon as a function of the temperature. It can be seen that Mott's law is obeyed at sufficiently low temperatures.

At higher temperatures the main contribution to conduction comes from electrons which have been thermally excited from E_F to the mobility edge E_C; the conductivity is then proportional to the number of such electrons per unit volume, which is given by

$$n \propto \exp[-(E_C - E_F)/k_BT] \tag{4.166}$$

Equation (4.166) is obtained in the same way as (4.79) which gives the number of electrons thermally excited into the conduction band of a crystalline semiconductor from localised electron states (on impurity centres) with energy below the conduction band edge. At intermediate temperatures both mechanisms (hopping and thermal activation) will contribute to conduction to some degree. The behaviour of the experimental curves of Fig. 4.67 at the higher temperatures are explained in this way.

4.18.5 ENERGY BANDS IN AMORPHOUS MATERIALS

We would like to close our discussion of non-crystalline materials with a brief qualitative description of the energy bands one meets, or expects to find, in amorphous semiconductors and glasses. We know that energy gaps exist in amorphous semiconductors because these are transparent to incident electromagnetic radiation (in the infrared or visible region). We note, for example, that the optical absorption coefficient for amorphous germanium is not very different from that of crystalline germanium; this is demonstrated in Fig. 4.68. It is, of course, possible that because of disorder, of one kind or the other, electron states are pushed into the gap from the valence and conduction bands (as these would be in the corresponding crystal), as shown schematically in Fig. 4.69(a). It may, also, happen that defect-generated

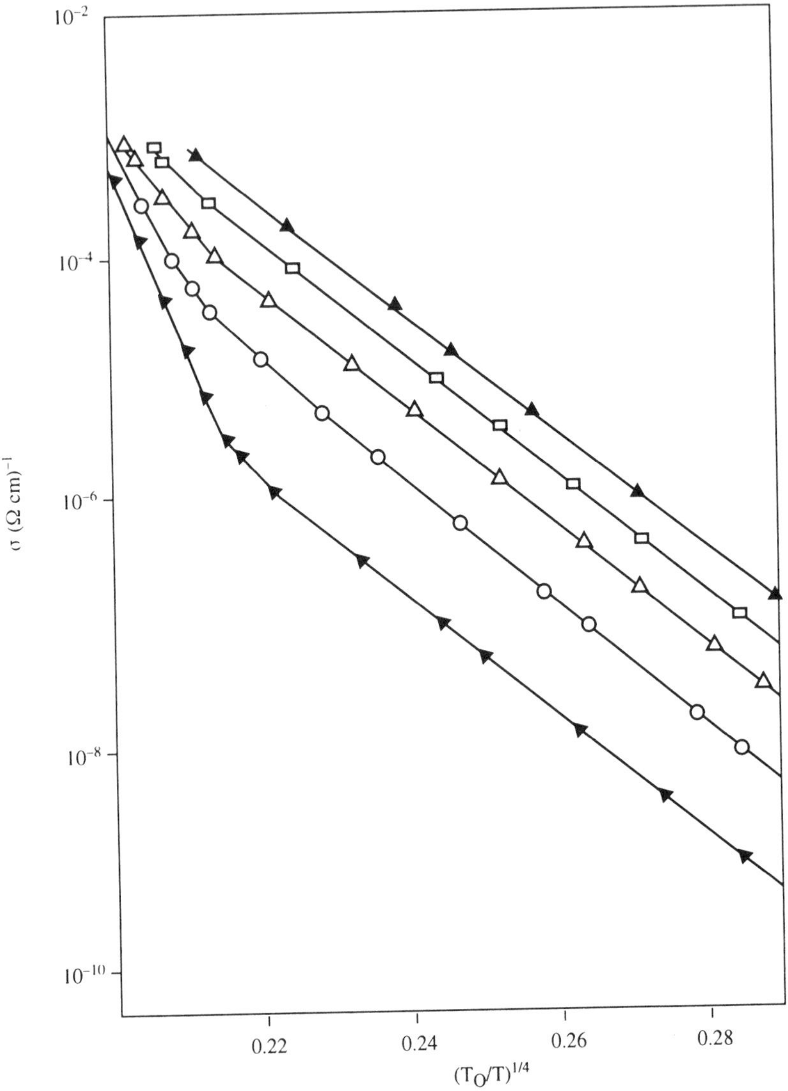

Figure 4.67. Hopping transport in amorphous silicon. The different curves correspond to different deposition temperatures. (W. Beyer, Ph.D. thesis, Universität Marburg, 1974. Reproduced by permission of W. Beyer)

bands exist in the middle of the gap as shown schematically in Fig. 4.69(b). (In the case of germanium or silicon, for example, the defects could be vacancies or divacancies in the network of the amorphous material.) If that is the case, and if $\rho(E_F) \neq 0$, optical excitation from E_F to the conduction

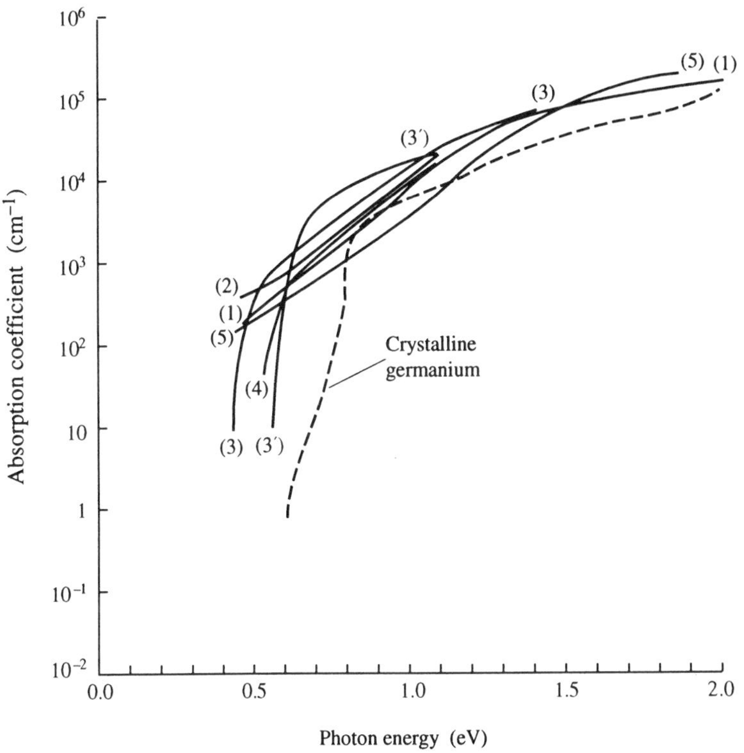

Figure 4.68. Room-temperature optical absorption edges in evaporated films of germanium. The curves denoted by different numbers represent data by different authors. (N. F. Mott and E. A. Davies, Electronic Processes in Non-Crystalline Materials (Oxford: OUP) 1979. Reproduced by permission of Oxford University Press)

band is possible, as well as the hopping conduction we discussed in the previous section.

A detailed discussion of matters arising from the density-of-states curves shown in Fig. 4.69 lies beyond the scope of the present book. However, we should perhaps mention that, while the defect-generated bands referred to above appear to be stable and to have an important effect on the properties of the material, attempts to modify the properties of an amorphous semiconductor like Ge or Si by doping it with dopants like P, Sb, etc., have failed, in contrast to what, as we know, happens in crystalline semiconductors. The reason appears to be the relatively flexible coordination in the amorphous material, which allows, for example, a P atom to have five Ge neighbour atoms and by covalently bonding to them, keep its five electrons firmly about it, rather than donating one to the conduction band. In other

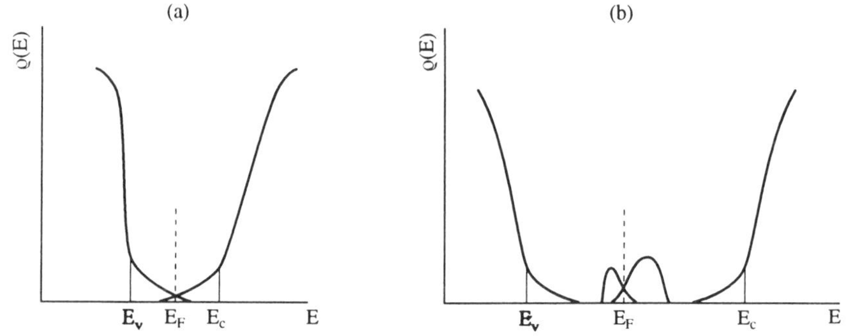

Figure 4.69. Possible density-of-states distributions in amorphous semiconductors. E_v and E_c denote respectively the valence and conduction band edges of the corresponding crystals and indicate where, approximately, the corresponding mobility edges are in the amorphous material

cases (in many glasses) doping fails because the above mentioned defect-generated bands at the centre of the gap act as acceptors or donors to implanted dopants, and in this way pinch E_F within these bands, as shown in Fig. 4.69(b).

Problems

***4.1** Consider an electron in the one-dimensional periodic potential field (Kronig–Penney):

$$U(x) = 0, 0 < x < a$$
$$= U_O, a < x < b$$
$$U(x + b) = U(x), -\infty < x < +\infty$$

Show that in the energy region $E < U_O$ the energy-bands consist of those energies which satisfy the following equation

$$\frac{\beta^2 - \alpha^2}{2\alpha\beta} \sinh \beta(b - a) \sin \alpha a + \cosh \beta(b - a) \cos \alpha a = \cos kb \tag{1}$$

where $\alpha^2 \equiv 2mE/\hbar^2$, $\beta^2 \equiv 2m(U_O - E)\hbar^2$, and k denotes the reduced wavevector: $-\pi/b < k < \pi/b$.
Hint: We seek solutions of the Bloch type: $\psi_k = u_k(x) \exp(ikx)$, where $u_k(x + b) = u_k(x)$, and demand that the wavefunction and its derivative be everywhere continuous.

***4.2** Show that when $(b - a) \to 0$ and $U_O \to \infty$ in such a way that $(b - a)U_O \equiv A$ is constant, equation (1) of problem 4.1 reduces to the following equation

$$\frac{mA}{\hbar^2} \frac{\sin \alpha a}{\alpha} + \cos \alpha a = \cos ka$$

Using the above equation, show how one can obtain graphically the energy bands of the electron in the given potential field.

*__4.3__ A crystal has a simple cubic lattice (see Fig. 4.7) with one atom per lattice site. Its primitive vectors are:

$$\boldsymbol{t}_1 = a\hat{x},\ \boldsymbol{t}_2 = a\hat{y},\ \boldsymbol{t}_3 = a\hat{z} \tag{1}$$

(a) Show that the reduced $\boldsymbol{k}$-zone (BZ) of the above crystal is

$$-\pi/a < k_x,\ k_y,\ k_z < \pi/a \tag{2}$$

(b) Consider an s-band of the crystal. Assuming that only the orbitals of nearest neighbours overlap, show that the corresponding one-electron states are given by (4.27) with

$$|A|^2 \simeq 1 - 2I(\cos k_x a + \cos k_y a + \cos k_z a)$$

$$I = \int \varphi_s(\boldsymbol{r} - \boldsymbol{R}_n)\varphi_s(\boldsymbol{r} - \boldsymbol{R}_m)\,\mathrm{d}V \tag{3}$$

where $\boldsymbol{R}_m - \boldsymbol{R}_n = \pm\boldsymbol{t}_1, \pm\boldsymbol{t}_2, \pm\boldsymbol{t}_3$; and we have assumed that $6I \ll 1$.

(c) We can calculate $E_s(\boldsymbol{k})$, defined by (4.28), from the following formula (see Appendix B.10)

$$E_s(\boldsymbol{k}) = \frac{|A|^2}{N}\int\left(\sum_{\boldsymbol{R}_n}\exp(-\mathrm{i}\boldsymbol{k}\cdot\boldsymbol{R}_n)\varphi_s(\boldsymbol{r} - \boldsymbol{R}_n)\right)\hat{H}\left(\sum_{\boldsymbol{R}_m}\exp(\mathrm{i}\boldsymbol{k}\cdot\boldsymbol{R}_m)\varphi_s(\boldsymbol{r} - \boldsymbol{R}_m)\right)\mathrm{d}V$$

$$\hat{H} = -\frac{\hbar^2}{2m}\nabla^2 + U(\boldsymbol{r}) \tag{4}$$

where $U(\boldsymbol{r})$ is the mean potential field the electron sees in the crystal. Introducing

$$\int \varphi_s(\boldsymbol{r} - \boldsymbol{R}_n)\hat{H}\varphi_s(\boldsymbol{r} - \boldsymbol{R}_n)\,\mathrm{d}V = E_s^{(0)} + \Delta E_s \tag{5}$$

where $E_s^{(0)}$ and ΔE_s have the same meaning as in (4.28), and assuming that, when $\boldsymbol{R}_n \neq \boldsymbol{R}_m$;

$$\begin{aligned}\int \varphi_s(\boldsymbol{r} - \boldsymbol{R}_n)\hat{H}\varphi_s(\boldsymbol{r} - \boldsymbol{R}_m) &= v', \text{ if } \boldsymbol{R}_n - \boldsymbol{R}_m = \pm\boldsymbol{t}_1, \pm\boldsymbol{t}_2, \pm\boldsymbol{t}_3\\ &= 0, \text{ otherwise}\end{aligned} \tag{6}$$

show that

$$E_s(\boldsymbol{k}) = E_s^{(0)} + \Delta E_s + 2v(\cos k_x a + \cos k_y a + \cos k_z a)$$

where

$$2v = 2v' - (2I)E_s^{(0)}$$

and, therefore, show that the energy spread of the above band equals $12v$.

4.4 Verify that the line UX in the BZ of the fcc lattice (shown in Fig. 4.14) is equivalent to a continuation of the ΓK line beyond the Brillouin zone.
Hint: Look carefully at the diagram of Fig. 4.6.

4.5 (a) Verify equation (4.74).

(b) The conduction band of a metal is approximated by a free-electron model (Section 4.9). The density of the metal is $9 \times 10^3\ \mathrm{kg\,m^{-3}}$, and every atom (of mass $M = 1.07 \times 10^{-22}$ g) contributes one free-electron to the conduction band. Calculate the Fermi energy E_F, the density of one-electron states at E_F and the transport velocity v_F of an electron with this energy.

4.6 The measured resistivity of copper at $T \simeq 300$ K is $\rho \simeq 1.7 \times 10^{-8}\ \Omega$ m. Estimate the relaxation time τ and the mean free path L of an electron with energy $E \simeq E_F$, for the same temperature. Use a free-electron approximation and assume one free-electron per atom. The density of copper is $8930\ \mathrm{kg\,m^{-3}}$.

4.7 Using the density of one-electron states of Fig. 4.20 and (4.45), estimate the electronic specific heat of tungsten when $T \to 0$, and evaluate the constant c_1 of (4.46a) for a metal of unit volume.

4.8 The density of one-electron states (per spin) near the bottom (E_c) of the conduction band of a semiconductor is given by

$$\rho_c(E) = \frac{4\pi}{h^3}(2m_c^3)^{1/2}(E - E_c)^{1/2} \tag{1}$$

and the corresponding quantity near the top (E_v) of the valence band is given by

$$\rho_v(E) = \frac{4\pi}{h^3}(2m_v^3)^{1/2}(E_v - E)^{1/2} \tag{2}$$

(a) Show that the chemical potential is given by

$$\mu \simeq E_v + E_g/2 + \frac{3}{4}k_B T \ln \frac{m_v}{m_c}$$

where $E_g = E_c - E_v$ is the energy gap.

(b) Show that the number (n) of electrons per unit volume in the conduction band and the number (p) of holes per unit volume in the valence band are given by

$$n = p = 2\left(\frac{2\pi k_B T}{h^2}\right)^{3/2}(m_v m_c)^{3/4} \exp[-E_g/(2k_B T)]$$

4.9 The constant-energy surfaces of germanium near the bottom of the conduction band can be approximated by eight spherical surfaces (in reality they are ellipsoids) the centres of which lie where the axes normal to the planes (111), ($\bar{1}$11), (1$\bar{1}$1) and

$(1\,1\,\bar{1})$ meet the BZ (point L and its equivalent points in the BZ of Fig. 4.14). Each of the above spherical surfaces is described by

$$E = E_c + \frac{\hbar^2}{2m^*}(\boldsymbol{k} - \boldsymbol{k}_L)^2$$

$m^* = 0.1375$ m, where m denotes the mass of the electron, and $\boldsymbol{k}_L$ is the centre of the sphere. (The reader should note the consistency of the above picture with the energy bands shown in Fig. 4.21.) Show that the density of one-electron states near the bottom of the conduction band is given by (1) of problem (4.8) with $m_c = 0.55$ m.

4.10 (a) Derive equation (4.60) starting from the equation

$$\mathrm{d}E/\mathrm{d}t = -e\boldsymbol{F} \cdot \boldsymbol{v}$$

which tells us that the increase in the energy of the electron is equal to the work done by the external force.

(b) Verify equations (4.68) and (4.69).

4.11 (a) Explain how the energy gap of an intrinsic semiconductor can be obtained from measurements of the conductivity of the semiconductor.

(b) Using the results of problem 4.8, estimate the resistivity of intrinsic germanium when $T \simeq 300$ K. For Ge we have: $m_c = 0.55$ m, $m_v = 0.37$ m; and when $T \simeq 300$ K, $\mu_n \simeq 3900$ cm^2/Vs and $\mu_p \simeq 1900$ cm^2/Vs.

(c) Estimate roughly the resistivity of n-type germanium, assuming a concentration of donor impurities equal to 10^{17}/cm^3.

4.12 (a) Assuming that the conduction band of a semiconductor is described by

$$E(\boldsymbol{k}) = E_c + \frac{\hbar^2 k^2}{2m_c}$$

show that the average velocity of the conduction-band electrons of an intrinsic semiconductor at a temperature $T \gtrsim 300$ K is proportional to $T^{1/2}$, i.e.

$$\overline{v^2} \propto T$$

(b) The inverse of the relaxation time of the conduction-band electrons is proportional to the scattering cross-section of a vibrating nucleus, times the number of nuclei the electron meets per unit time which is proportional to the average velocity of the electron. Using the above and equations (4.76) and (4.77), show that for an intrinsic semiconductor at $T \gtrsim 300$ K we obtain $\mu_n \propto T^{-3/2}$.

4.13 In the presence of an external magnetic field, the energy of an electron in a paramagnetic metal is reduced by $\mu_B B$ (μ_B is Bohr's magneton and B the magnitude of the field) when the magnetic moment associated with the spin of the electron is parallel to the direction (z) of the field, and it is increased by $\mu_B B$ when its magnetic moment is antiparallel to the field. Show that a magnetisation (magnetic moment per unit volume of the metal) is, therefore, induced given, when $T \to 0$, by

$$\boldsymbol{M} = \chi \boldsymbol{B}$$
$$\chi = 2\mu_{\mathrm{B}}^2 \rho(E_{\mathrm{F}})$$

where $\rho(E_{\mathrm{F}})$ denotes the density of one-electron states, per spin, at the Fermi level. Calculate χ for the free-electron metal of problem 4.5.

4.14 Using the results of Fig. 4.38 estimate the magnetisation (magnetic moment per atom) of nickel at the absolute zero temperature.

4.15 Explain how we should plot the experimental data in order to obtain more accurately the work function φ of a metal surface:

(a) from measurements of thermionic emission using equation (4.126).

(b) from field-emission measurements using equation (4.127).

4.16 Let $x_i = ia$, $i = 0, \pm 1, \pm 2, \ldots$ be the equilibrium positions of a linear periodic chain of atoms. The classical equation of motion, for the displacement X_i of the ith atom from its equilibrium position, is

$$M\frac{\mathrm{d}^2 X_i}{\mathrm{d}t^2} = -K(X_i - X_{i+1}) - K(X_i - X_{i-1})$$

where M is the mass of the atom and K is an elastic constant; and we have assumed that an atom couples with its nearest neighbours only. Show that in this case, putting

$$X_i(t) = \mathrm{Re}\,\{X_0^{(k)} \exp(\mathrm{i}kx_i + i\omega(k)t)\}$$

we obtain (instead of (I) of footnote 27) the following equation

$$\omega^2(k) M X_0^{(k)} = 2[1 - \cos ka] K X_0^{(k)}$$

and that, therefore,

$$\omega(k) = 2\omega_0 \left| \sin\left(\frac{k\alpha}{2}\right) \right|$$

where $\omega_0 = (K/M)^{1/2}$; and k takes the values

$$k = \frac{2\pi n}{Na},\ n = 0, \pm 1, \pm 2 \ldots \text{ in the BZ: } -\frac{\pi}{a} < k < \frac{\pi}{a},$$

where N is the number of atoms in the chain.

4.17 Verify equation (4.103).

4.18 The reflection coefficient R of electromagnetic radiation of frequency $\nu = \omega/2\pi$ incident normally on a metal surface is given by

$$R = \frac{(n-1)^2 + \kappa^2}{(n+1)^2 + \kappa^2}$$

where $n + i\kappa \equiv \sqrt{\epsilon(\omega)}$. For some metals (Cs, Rb, K, Na, Li) the last term of (4.138) is very small, in which case

$$\epsilon(\omega) \simeq 1 - \frac{\omega_P^2}{\omega^2}$$

The measured frequency ν_0, below which the incident electromagnetic wave is entirely reflected ($R = 1$) from the above metals, is obtained from the following table (the frequency ν_0 is related to the wavelength λ_0 noted in the table by $\lambda_0 = c/\nu_0$, where c denotes the speed of light).

Metal	Cs	Rb	K	Na	Li
λ_0(Å)	4400	3600	3150	2100	2050

Assuming that the free-electron model applies to them, use the above results to determine an effective number of free electrons for each of the above metals.

4.19 From the energy bands shown in Figs. 4.21 and 4.22 estimate the minimum frequency of an incident electromagnetic radiation capable of exciting an electron from the valence band to the conduction band of Ge, Si, GaAs, ZnS and KCl respectively.

4.20 A plasmon excitation decays into an electron–hole pair when (4.146) is satisfied. Using the Sommerfeld model to estimate $E(\boldsymbol{k} + \boldsymbol{q}) - E(\boldsymbol{k})$ and assuming that $\hbar\omega_P(q)$ is practically independent of q (i.e. it is given by (4.143)), show that plasmon excitations can exist only when $q < \omega_P/v_F$, where v_F is the velocity of an electron at the Fermi level. Note that $q \ll k_F$.

4.21 Verify the equations (4.160), (4.161) and (4.163).

Footnotes of Chapter 4

1. Shorthand notation: bcc.
2. Shorthand notation: fcc.
3. The sites of the fcc lattice may be taken at the centre of the cube and the midpoints of the edges of the cube as in Fig. 4.8, or at the corners of the cube and the centres of the faces of the cube. The two descriptions are equivalent.
4. A systematic description of the vibrating lattice is given in Section 4.12.
5. In what follows we assume, except when the opposite is explicitly stated, that the nuclei are stationed at their equilibrium positions.
6. We do not state explicitly the spin part of the eigenfunction. Given that the potential field (4.16) does not depend on the spin, we have in fact two eigenstates, with the same orbital part (4.17) and the same energy $E_\alpha(\boldsymbol{k})$,

corresponding to the two orientations (up and down) of the spin. These are described in the usual way (see Section 2.12) as follows

$$\Psi_{\boldsymbol{k}\alpha+} = \psi_{\boldsymbol{k}\alpha}(\boldsymbol{r}, t)\begin{pmatrix}1\\0\end{pmatrix} \quad \text{and} \quad \Psi_{\boldsymbol{k}\alpha-} = \psi_{\boldsymbol{k}\alpha}(\boldsymbol{r}, t)\begin{pmatrix}0\\1\end{pmatrix}$$

7. We put $A = 1\sqrt{V}$ in (4.17). In this way, the normalisation (4.19) of $\psi_{\boldsymbol{k}\alpha}$ reduces to

$$\frac{1}{V_o}\int_{V_o} |u_{\boldsymbol{k}\alpha}(\boldsymbol{r})|^2\,\mathrm{d}V = 1$$

where the integral extends over the volume of a single primitive cell.
8. The point where the two curves $\alpha = 1$ and $\alpha = 2$ cross is without significance (one point out of many thousands).
9. This holds true for every non-magnetic crystal (see also Sections 4.3 and 4.11).
10. The radius of the 1s atomic shell (0.018 Å according to Table 2.6) is much smaller than the interatomic distance (2.56 Å) in crystalline copper.
11. We denote the three 2p bands as follows $\alpha = $ 2p:1, 2p:2, 2p:3. And similarly we denote the 3p bands. The way one constructs the appropriate φ_α in each case will not concern us here (see Appendix B.10).
12. In a heavy metal like tungsten corrections to these energy bands due to spin-orbit coupling and relativistic effects in general are to be expected; these do not change the over-all picture we are presenting here and we shall disregard them.
13. The choice of the zero of the energy is arbitrary. Putting it at the top of the valence bands is convenient.
14. We refer to the conduction bands of a metal collectively as the conduction band.
15. Our calculation of n and p has been based on the assumption that the density of states in the regions of the valence band maximum and the conduction band minimum is constant ($\bar{\rho}(E_\mathrm{v})$ and $\bar{\rho}(E_\mathrm{c})$ respectively). For a more realistic calculation of these quantities see problem 4.8. We note that in any case the exponential dependence of n and p on the temperature remains.
16. In a macroscopic wire, such as the one of Fig. 4.28, the curvature is negligible in comparison with the mean free path of the electron (see Section 4.8.1) and we can, therefore, replace the axis of the wire by a straight line (the x-axis).
17. In the remaining of this section we assume, for the sake of simplicity, that there is only one conduction band (per spin), in which case $\boldsymbol{k}$ by itself defines an electron state and we can drop the band index α, writing, for example, $E(\boldsymbol{k})$ instead of $E_\alpha(\boldsymbol{k})$. The generalisation to many conduction bands is trivial. The contribution of each band to the current can be calculated, to a first approximation, independently. The final formula (4.70) is valid generally.
18. We put the volume of the metal equal to unity. The final formula for the conductivity (which does not depend on the volume) is not affected by this assumption.
19. We use the classical term orbit for convenience. We can think of its radius as corresponding to the mean value of r of a corresponding one-electron state in the above potential field.
20. A different type of impurity-conduction, known as hopping conduction, is also possible when $T > 0$, (see Section 4.18.4).
21. The critical temperature (T_c) is called the Curie temperature. For Fe: $T_\mathrm{c} =$ 1043 K, for Ni: $T_\mathrm{c} = 631$ K. When $T < T_\mathrm{c}$ these metals are characterised by a spontaneous magnetisation (they are, as we say, in their ferromagnetic phase). When $T > T_\mathrm{c}$, there is no spontaneous magnetisation (the metals are in the

so-called paramagnetic phase). A paramagnetic metal (non-magnetic metals are paramagnetic at any temperature) acquires a finite magnetisation (a finite magnetic moment per unit volume) under the influence of an external magnetic field. The coupling of the spin magnetic moment of an electron with the external field leads to a small decrease or increase in its energy depending on whether its magnetic moment is parallel or antiparallel to the direction (z) of the external field. This means, given that E_F is constant, that there will be more electrons with a magnetic moment parallel to the external field than antiparallel and, that way, one obtains an induced magnetisation proportional to the applied field (see problem 4.13).

22. In what follows we assume, for the sake of simplicity, that the crystal is made of same atoms, and that there is only one atom per primitive cell of the crystal, in which case $\boldsymbol{R}_j$ are the sites of a given lattice (4.4).
23. The procedure is analogous to the approximation of the potential field an electron sees in an atom by a spherically symmetric one.
24. The spontaneous magnetisation of nickel decreases with temperature and vanishes for $T \geqslant 631$ K. A discussion of the variation of the spontaneous magnetisation with temperature lies beyond the scope of the present book.
25. V. L. Moruzzi, J. F. Janak and A. R. Williams, *Calculated Electronic Properties of Metals*, loc. cit.
26. If there are more than one atom in the ith primitive cell, the equilibrium position of the nth atom is given by $\boldsymbol{R}_{ni} = \boldsymbol{R}_i + \boldsymbol{\tau}_n$; if there is only one atom in the cell, its equilibrium position can be identified with $\boldsymbol{R}_i$ at the centre of the cell.
27. The case of one atomic nucleus, of mass M, per primitive cell can serve as an example in demonstrating the above statement. In this case one can drop the subindex n in (4.96). The classical equations of motion are given by (I) of footnote 11 in chapter 3. Putting in those equations $\omega = \omega(\boldsymbol{k})$, $X_{oi} = \exp(\mathrm{i}\boldsymbol{k}\cdot\boldsymbol{R}_i)X_o^{(\boldsymbol{k})}$, $Y_{oi} = \exp(\mathrm{i}\boldsymbol{k}\cdot\boldsymbol{R}_i)Y_o^{(\boldsymbol{k})}$, $Z_{oi} = \exp(\mathrm{i}\boldsymbol{k}\cdot\boldsymbol{R}_i)Z_o^{(\boldsymbol{k})}$ (we drop for the moment the index α), and multiplying the resulting equations (left and right) by $\exp(-\mathrm{i}\boldsymbol{k}\cdot\boldsymbol{R}_i)$ we obtain

$$\omega^2(\boldsymbol{k})MX_o^{(\boldsymbol{k})} = \sum_j k_{ij;xx}\exp[\mathrm{i}\boldsymbol{k}\cdot(\boldsymbol{R}_j-\boldsymbol{R}_i)]X_o^{(\boldsymbol{k})} + \sum_j k_{ij;xy}\exp[\mathrm{i}\boldsymbol{k}\cdot(\boldsymbol{R}_j-\boldsymbol{R}_i)]Y_o^{(\boldsymbol{k})}$$
$$+ \sum_j k_{ij;xz}\exp[\mathrm{i}\boldsymbol{k}\cdot(\boldsymbol{R}_j-\boldsymbol{R}_i)]Z_o^{(\boldsymbol{k})}$$
$$\omega^2(\boldsymbol{k})MY_o^{(\boldsymbol{k})} = \sum_j k_{ij;yx}\exp[\mathrm{i}\boldsymbol{k}\cdot(\boldsymbol{R}_j-\boldsymbol{R}_i)]X_o^{(\boldsymbol{k})} + \sum_j k_{ij;yy}\exp[\mathrm{i}\boldsymbol{k}\cdot(\boldsymbol{R}_j-\boldsymbol{R}_i)]Y_o^{(\boldsymbol{k})}$$
$$+ \sum_j k_{ij;yz}\exp[\mathrm{i}\boldsymbol{k}\cdot(\boldsymbol{R}_j-\boldsymbol{R}_i)]Z_o^{(\boldsymbol{k})} \qquad \text{(I)}$$
$$\omega^2(\boldsymbol{k})MZ_o^{(\boldsymbol{k})} = \sum_j k_{ij;zx}\exp[\mathrm{i}\boldsymbol{k}\cdot(\boldsymbol{R}_j-\boldsymbol{R}_i)]X_o^{(\boldsymbol{k})} + \sum_j k_{ij;zy}\exp[\mathrm{i}\boldsymbol{k}\cdot(\boldsymbol{R}_j-\boldsymbol{R}_i)]Y_o^{(\boldsymbol{k})}$$
$$+ \sum_j k_{ij;zz}\exp[\mathrm{i}\boldsymbol{k}\cdot(\boldsymbol{R}_j-\boldsymbol{R}_i)]Z_o^{(\boldsymbol{k})}$$

We note that the elastic constants $k_{ij;xx}$, $k_{ij;xy}$, etc., depend on the separation $(\boldsymbol{R}_j - \boldsymbol{R}_i)$ between the nuclei but not on the position of either of them and therefore the sums over j in equations (I) are independent of i. Consequently, the 3N equations (see (III) of footnote 11 of Chapter 3; $\Lambda = N$ in the present case) which determine the motion of the N nuclei of the crystal are reduced to the above three equations. Equations (I) constitute an algebraic system of three linear homogeneous equations with three unknowns $X_o^{(\boldsymbol{k})}$, $Y_o^{(\boldsymbol{k})}$ and $Z_o^{(\boldsymbol{k})}$. It has three linearly independent solutions, $(X_o^{(\boldsymbol{k}\alpha)}, Y_o^{(\boldsymbol{k}\alpha)}, Z_o^{(\boldsymbol{k}\alpha)})$, $\alpha = 1, 2, 3$ corresponding to the three values of $\omega^2(\boldsymbol{k}) = \omega_\alpha^2(\boldsymbol{k})$, $\alpha = 1, 2, 3$ which make the determinant of the coefficients of (I) zero. $\omega_\alpha(\boldsymbol{k})$ varies continuously with $\boldsymbol{k}$ because the sums over j in (I) and, therefore, the determinant of the coefficients of (I) vary continuously with $\boldsymbol{k}$.

The elastic constants $k_{ij;xx}$, etc., get smaller as $|\boldsymbol{R}_i - \boldsymbol{R}_j|$ gets larger. In many instances it will be sufficient to include the interaction of first neighbours only in the evaluation of the sums over j in (I); in other cases it may be necessary to include the interaction between second and third neighbours, and so on, to obtain accurate results. A calculation from first principles of the elastic constants is in general difficult, and often these are determined empirically, by fitting $\omega_\alpha(\boldsymbol{k})$ to observed data. The simplest application of (I) relates to a linear crystal: a chain of atoms with elastic coupling between first neighbours only (see problem 4.16).

28. This is strictly true for general points of the SBZ. On symmetry lines of the SBZ, they can coexist under certain conditions which do not allow a hybridisation between them.
29. In the case of semiconductors this will be true for electron states near the bottom of the conduction band, and for holes near the top of the valence band.
30. A quantum-mechanical treatment of plasmons exists and provides, in a final analysis, the justification of the classical treatment presented here.
31. The phenomenon was first observed by W. Meissner and R. Ochsenfeld in 1933.
32. Our exposition of the BCS theory is by necessity incomplete. It is but a sketch of the theory.
33. We recall that $\hbar\omega_{\rm D} = k_{\rm B}\Theta_{\rm D}$ is typical of the quanta of lattice vibrational energy, and that the Debye temperature $\Theta_{\rm D}$ has a value for most metals between 200 and 400 K.
34. The energy of a metastable state lies above the ground energy and therefore it is not a stable state in the strict sense that thermodynamics attaches to this term; however, it has a very long lifetime, because a transition to a state of lower energy is extremely difficult.
35. For convenience, Fig. 4.64 shows an 1D-array of rectangular wells; the reader can imagine a 3D-array of spherical wells. We note also that a potential well, in general, is described by more than one parameter, and the depth is noted here only as an example.
36. P. W. Anderson (1958).
37. It must not be thought that the energy suffices to define a state. There are certainly more than one state with the same energy (as in the crystal). These will be described by different wavefunctions, but all will have the same form (4.155).
38. When the concentration of the impurities is small one obtains isolated impurity states (see Section 4.10.1); at higher concentrations, these can broaden into a narrow band.
39. This discussion is based upon the treatment given by N. F. Mott and E. A. Davis 1979 *Electronic Processes in Non-crystalline Materials*, Oxford: OUP, 1979.

Appendix A
VECTORS AND COMPLEX NUMBERS

A.1 Vector quantities

A vector quantity is defined by three numbers which determine its magnitude and its direction. For example, the position of particle A in space (see Fig. A1) is defined by its position vector

$$\boldsymbol{R} = \overrightarrow{OA} \tag{A1}$$

its magnitude $R = |\boldsymbol{R}|$ is identified with the length OA and its direction is defined by the angles Θ and Φ ($0 \leqslant \Theta \leqslant \pi$, $0 \leqslant \Phi < 2\pi$). We can describe the same vector by noting, instead of R, Θ, Φ, its three components: $X = OX$, $Y = OY$, $Z = OZ$. We write

$$\boldsymbol{R} = (X, Y, Z) \tag{A2}$$

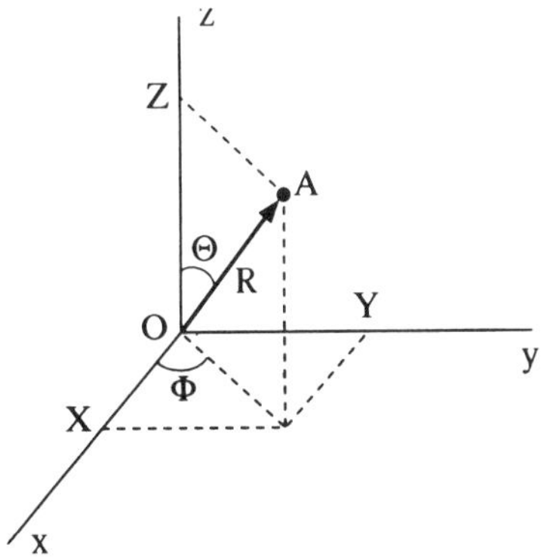

Figure A1. Position vector $\boldsymbol{R}$ of a particle A

and we note that $-\infty < X, Y, Z < +\infty$. It is also evident from Fig. A1 that

$$X = R\sin\Theta\cos\Phi$$
$$Y = R\sin\Theta\sin\Phi \tag{A3}$$
$$Z = R\cos\Theta$$

So if we know the values of (R, Θ, Φ) we can find those of (X, Y, Z) and vice-versa. We note also that

$$R = \sqrt{X^2 + Y^2 + Z^2} \tag{A4}$$

Other vector quantities are the velocity $\boldsymbol{V}$ of a particle, the force $\boldsymbol{F}$ acting on a particle etc. Any vector can be described in exactly the same way as the position vector, by noting its magnitude and direction, or its three components.

In Fig. A2 we present the position vector of a particle at two moments of time t and $t + \Delta t$ (the orbit of the particle is represented by the curve ABΓ). The position vector $\boldsymbol{R}(t + \Delta t)$ is obtained from $\boldsymbol{R}(t)$ by adding to the latter the displacement vector $\Delta\boldsymbol{R}$ as shown in the figure. We have

$$\boldsymbol{R}(t + \Delta t) = \boldsymbol{R}(t) + \Delta\boldsymbol{R} \tag{A5}$$

$$\Delta\boldsymbol{R} = (\Delta X, \Delta Y, \Delta Z) \tag{A6}$$

$$\Delta X \equiv X(t + \Delta t) - X(t)$$
$$\Delta Y \equiv Y(t + \Delta t) - Y(t) \tag{A7}$$
$$\Delta Z \equiv Z(t + \Delta t) - Z(t)$$

The ratios

$$\frac{\Delta X}{\Delta t}, \frac{\Delta Y}{\Delta t}, \frac{\Delta Z}{\Delta t} \tag{A8}$$

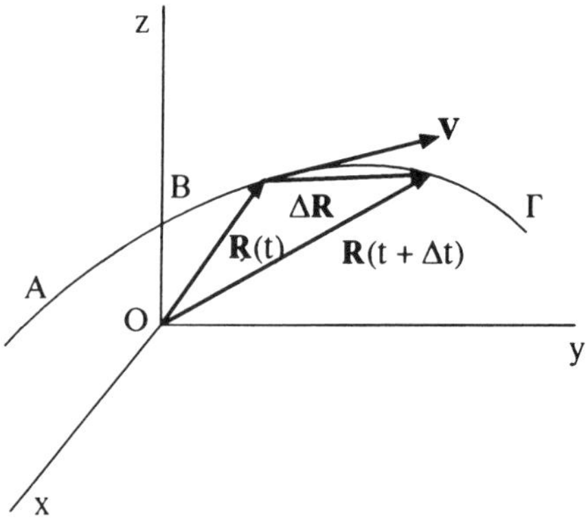

Figure A2. Position vector of a particle at two moments of time

do not depend on Δt when Δt is sufficiently small ($\Delta t \to 0$), and define the velocity of the particle at time t. We have:

$$\boldsymbol{V}(t) = (V_x(t), V_y(t), V_z(t)) \equiv \lim_{\Delta t \to 0} \left(\frac{\Delta X}{\Delta t}, \frac{\Delta Y}{\Delta t}, \frac{\Delta Z}{\Delta t} \right) \tag{A9}$$

One can easily see from Fig. A2 that $\boldsymbol{V}(t)$ has the direction of the tangent to the orbit of the particle at the position $\boldsymbol{R}(t)$. In mathematical language $\boldsymbol{V}(t)$ is the derivative (with respect to the time) of $\boldsymbol{R}(t)$ and is denoted as follows

$$\boldsymbol{V}(t) \equiv \frac{d\boldsymbol{R}}{dt} = \left(\frac{dX}{dt}, \frac{dY}{dt}, \frac{dZ}{dt} \right) \tag{A10}$$

A vector quantity may depend on more than one variable. For example, the force acting on a particle is often a function of its position, i.e.,

$$\boldsymbol{F} = \boldsymbol{F}(X, Y, Z)$$

In Fig. A3, we show the force $\boldsymbol{F}(X, Y, Z)$ acting on a point-particle of charge $-e$ (an electron) due to its Coulomb attraction to a ponit-particle of charge $Q = +Ze$ (a nucleus) at the origin of coordinates. In this case

$$\boldsymbol{F} = -\frac{Ze^2}{4\pi\epsilon_O R^3}\boldsymbol{R} \tag{A11}$$

i.e. the force has magnitude

$$F = \frac{Ze^2}{4\pi\epsilon_O R^2} \tag{A12}$$

and its direction is the opposite of $\boldsymbol{R}$.

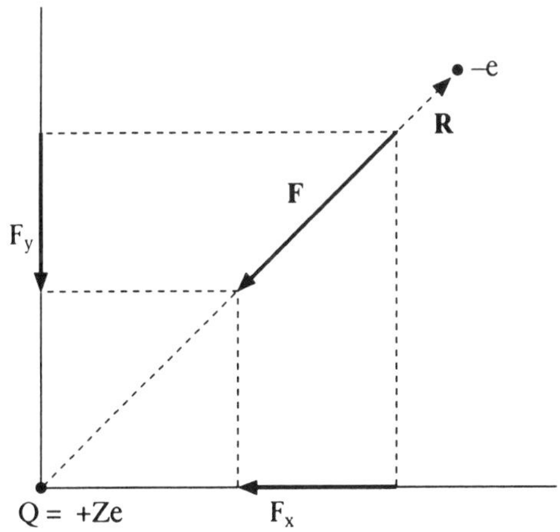

Figure A3. The Coulomb force (A11) when the electron ($-e$) is in the xy-plane. In this case $F_z = 0$

The way we add (subtract) two vectors $\boldsymbol{A}(\tau)$ and $\boldsymbol{B}(\tau)$, where τ denotes one or more variables, is apparent in our example of Fig. A2 and equations (A5) to (A7). If

$$\boldsymbol{A}(\tau) = (A_x(\tau),\ A_y(\tau),\ A_z(\tau)) \tag{A13}$$

and

$$\boldsymbol{B}(\tau) = (B_x(\tau),\ B_y(\tau),\ B_z(\tau)) \tag{A14}$$

then

$$\boldsymbol{A}(\tau) \pm \boldsymbol{B}(\tau) \equiv (A_x(\tau) \pm B_x(\tau),\ A_y(\tau) \pm B_y(\tau),\ A_z(\tau) \pm B_z(\tau)) \tag{A15}$$

Noting that multiplying a vector with a number means mutliplying its every component with this number, we can easily see that any vector $\boldsymbol{A}$ can be written as follows

$$\boldsymbol{A} = A_x\hat{x} + A_y\hat{y} + A_z\hat{z} \tag{A16}$$

where $\hat{x}$, $\hat{y}$, $\hat{z}$ are unit vectors (i.e. each has magnitude equal to unity) parallel to the x, y and z axes respectively.

Finally we define the scalar and vector products of two vectors. The scalar product $\boldsymbol{A} \cdot \boldsymbol{B}$ of two vectors is defined by

$$\boldsymbol{A}(\tau) \cdot \boldsymbol{B}(\tau) = A(\tau)B(\tau)\cos\omega \tag{A17}$$

where $A(\tau)$ and $B(\tau)$ are the magnitudes of the two vectors

$$\begin{aligned} A(\tau) &= \sqrt{A_x^2(\tau) + A_y^2(\tau) + A_z^2(\tau)} \\ B(\tau) &= \sqrt{B_y^2(\tau) + B_y^2(\tau) + B_z^2(\tau)} \end{aligned} \tag{A18}$$

and ω is the angle between them (see Fig. A4). One can easily see that

$$\boldsymbol{A} \cdot (\boldsymbol{B} + \boldsymbol{C}) = \boldsymbol{A} \cdot \boldsymbol{B} + \boldsymbol{A} \cdot \boldsymbol{C} \tag{A19}$$

and by using the above property and (A16), one can easily show that

$$\boldsymbol{A}(\tau) \cdot \boldsymbol{B}(\tau) = A_x(\tau)B_x(\tau) + A_y(\tau)B_y(\tau) + A_z(\tau)B_z(\tau) \tag{A20}$$

The vector product $\boldsymbol{A} \times \boldsymbol{B}$ of two vectors is, by definition, a vector normal to $\boldsymbol{A}$ and $\boldsymbol{B}$ (normal to the plane defined by $\boldsymbol{A}$ and $\boldsymbol{B}$) and points in the direction a

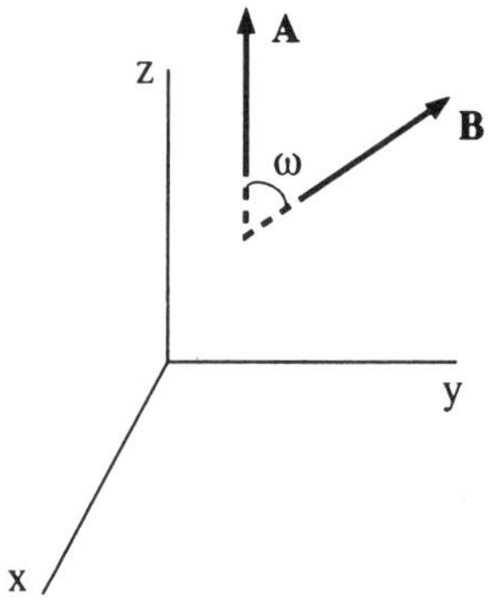

Figure A4. Two vectors $\boldsymbol{A}$ and $\boldsymbol{B}$ with angle ω between them

right-hand screw advances when turned from $\boldsymbol{A}$ toward $\boldsymbol{B}$, and its magnitude is given by

$$|\boldsymbol{A}(\tau) \times \boldsymbol{B}(\tau)| = A(\tau)B(\tau)\sin\omega$$

where ω is the angle between the two vectors (see Fig. A4). We note that according to the above definition $\boldsymbol{B} \times \boldsymbol{A}$ has the same magnitude but opposite direction to that of $\boldsymbol{A} \times \boldsymbol{B}$. One can write the vector product $\boldsymbol{A} \times \boldsymbol{B}$ as follows

$$\boldsymbol{A}(\tau) \times \boldsymbol{B}(\tau) = (A_y(\tau)B_z(\tau) - A_z(\tau)B_y(\tau))\hat{x} + (A_z(\tau)B_x(\tau) - A_x(\tau)B_z(\tau))\hat{y} + (A_x(\tau)B_y(\tau) - A_y(\tau)B_x(\tau))\hat{z} \tag{A21}$$

which we can more easily remember by writing it (formally) as a determinant:

$$\boldsymbol{A}(\tau) \times \boldsymbol{B}(\tau) = \begin{vmatrix} \hat{x} & \hat{y} & \hat{z} \\ A_x(\tau) & A_y(\tau) & A_z(\tau) \\ B_x(\tau) & B_y(\tau) & B_z(\tau) \end{vmatrix} \tag{A22}$$

Problem

If $\boldsymbol{A} = (1, 3, 2)$ and $\boldsymbol{B} = (4, 2, 1)$, find $\boldsymbol{A} \cdot \boldsymbol{B}$ and $\boldsymbol{A} \times \boldsymbol{B}$.

Solution

$$\boldsymbol{A} \cdot \boldsymbol{B} = 1 \times 4 + 3 \times 2 + 2 \times 1 = 12$$

$$\boldsymbol{A} \times \boldsymbol{B} = \begin{vmatrix} \hat{x} & \hat{y} & \hat{z} \\ 1 & 3 & 2 \\ 4 & 2 & 1 \end{vmatrix} = (3 \times 1 - 2 \times 2)\hat{x} + (2 \times 4 - 1 \times 1)\hat{y} + (1 \times 2 - 3 \times 4)\hat{z}$$

$$= -\hat{x} + 7\hat{y} - 10\hat{z}$$

A.2 Complex numbers

The numbers we use in the description of physical quantities are called real numbers; they include all numbers from $-\infty$ to $+\infty$, and they can be thought of as points on a straight line which we call the axis of real numbers. We choose a point as the origin of the axis and correspond to it the number zero (0) and to another point we correspond the number unity (1); in this way we obtain a one-to-one correspondence between the points of the axis and the real numbers: to every point on the axis corresponds a real number and to every real number a point on the axis (see Fig. A5).

We can now introduce the imaginary numbers as follows. We introduce the imaginary unity:

$$\mathrm{i} \equiv \sqrt{-1} \tag{A23}$$

from which follows that

$$\mathrm{i}^2 = -1 \tag{A24}$$

We note, and this is very important, that there is no real number which satifies the equation: $x^2 = -1$. The imaginary numbers are defined by

$$y\mathrm{i}, \ -\infty < y < +\infty \tag{A25}$$

They can be thought of as points on a straight line (the so-called axis of imaginary

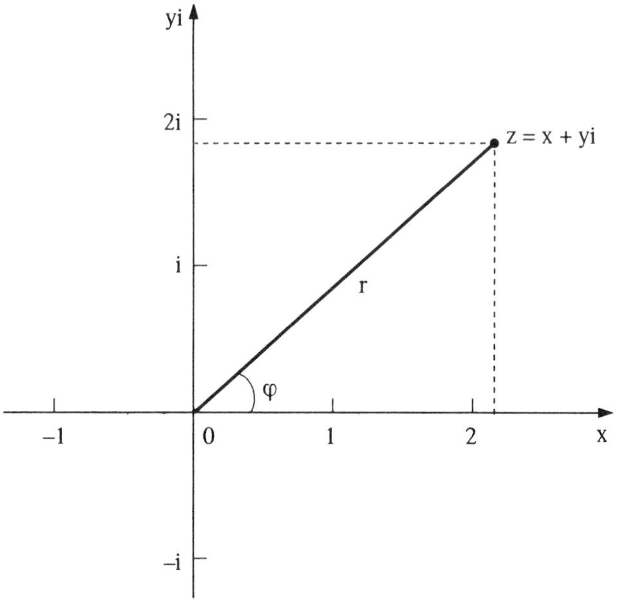

Figure A5. The complex plane. The x-axis is the axis of real numbers: $-\infty < x < +\infty$. The y-axis is the axis of imaginary numbers: yi, $-\infty < y < +\infty$

numbers) normal to the real axis which it crosses at the origin (see Fig. A5). To a certain point of this axis we correspond the imaginary unity i. With the zero at the origin and i at the said point, we obtain a one-to-one correspondence between the imaginary numbers and the points of the imaginary axis.

The complex numbers are defined by:

$$z = x + y\mathrm{i} \tag{A26}$$

where x, y are real numbers ($-\infty < x,\ y < \infty$). We call x the real part of (the complex number) z, and y the imaginary part of z, and write

$$\begin{aligned} x &= \mathrm{Re}\,\{z\} \\ y &= \mathrm{Im}\,\{z\} \end{aligned} \tag{A27}$$

We add and multiply (and correspondingly subtract and divide) complex numbers using the familiar rules of addition and multiplication of real numbers, and the equations (A23) and (A24). Therefore, if

$$\begin{aligned} z_1 &= x_1 + y_1\mathrm{i} \\ z_2 &= x_2 + y_2\mathrm{i} \end{aligned} \tag{A28}$$

then

$$z_1 + z_2 = (x_1 + x_2) + (y_1 + y_2)\mathrm{i} \tag{A29}$$

$$z_1 z_2 = (x_1 + y_1\mathrm{i})(x_2 + y_2\mathrm{i}) = (x_1 x_2 - y_1 y_2) + (x_1 y_2 + y_1 x_2)\mathrm{i} \tag{A30}$$

To every complex number $z = x + y\mathrm{i}$ corresponds a complex conjugate number defined by

$$z^* \equiv x - y\mathrm{i} \tag{A31}$$

The absolute value $|z|$ of a complex number is a positive (real) number defined by

$$|z| \equiv \sqrt{zz^*} = \sqrt{x^2 + y^2} \tag{A32}$$

It is obvious from Fig. A5, that there exists a one-to-one correspondence between the complex numbers (A26) and the points of the so-called complex plane xy (x: axis of the real numbers, y: axis of the imaginary numbers).

We note that a point (x, y) is defined equally well by its polar coordinates r, φ (see Fig. A5), which are obviously related to its Cartesian coordinates x, y as follows

$$\begin{aligned} x &= r\cos\varphi \\ y &= r\sin\varphi \end{aligned} \tag{A33}$$

Substituting the above in (A26) we obtain

$$z = x + \mathrm{i}y = r(\cos\varphi + \mathrm{i}\sin\varphi) \tag{A34}$$

where

$$r = \sqrt{x^2 + y^2} = |z| \tag{A35}$$

φ is called the argument of z. Using the mathematical identity[1]

$$\mathrm{e}^{\mathrm{i}\varphi} = \cos\varphi + \mathrm{i}\sin\varphi \tag{A36}$$

we can write (A34) in the following form

$$z = r\,\mathrm{e}^{\mathrm{i}\varphi} \tag{A37}$$

which is particularly useful when multiplying complex numbers. We have

$$z_1 z_2 = (r_1\,\mathrm{e}^{\mathrm{i}\varphi_1})(r_2\,\mathrm{e}^{\mathrm{i}\varphi_2}) = r_1 r_2\,\mathrm{e}^{\mathrm{i}(\varphi_1+\varphi_2)}. \tag{A38}$$

Footnotes of Appendix A

1. The exponential function e^z of the complex variable z is defined by the sum (power series):

$$\mathrm{e}^z = 1 + z + z^2/2! + z^3/3! + \cdots$$

where $n! \equiv (1)(2)\ldots(n)$. Putting $z = \mathrm{i}\varphi$ in the above series, and noting that $\mathrm{i}^0 = 1$, $\mathrm{i}^1 = \mathrm{i}$, $\mathrm{i}^2 = -1$, $\mathrm{i}^3 = -\mathrm{i}$, $\mathrm{i}^4 = 1$, $\mathrm{i}^5 = \mathrm{i}$, etc., we obtain

$$\begin{aligned} \mathrm{e}^{\mathrm{i}\varphi} &= 1 + \mathrm{i}\varphi - \varphi^2/2! - \mathrm{i}\varphi^3/3! + \varphi^4/4! + \mathrm{i}\varphi^5/5! - \varphi^6/6! - \mathrm{i}\varphi^7/7! + \ldots \\ &= (1 - \varphi^2/2! + \varphi^4/4! - \varphi^6/6! + \cdots) + \mathrm{i}(\varphi - \varphi^3/3! + \varphi^5/5! - \varphi^7/7! + \cdots) \end{aligned}$$

We know that

$$\begin{aligned} \cos\varphi &= 1 - \varphi^2/2! + \varphi^4/4! - \varphi^6/6! + \cdots \\ \sin\varphi &= \varphi - \varphi^3/3! + \varphi^5/5! - \varphi^7/7! + \cdots \end{aligned}$$

therefore we obtain

$$\mathrm{e}^{\mathrm{i}\varphi} = \cos\varphi + \mathrm{i}\sin\varphi.$$

Appendix B
FORMAL QUANTUM MECHANICS

B.1 Schrödinger's equation

The time development of the wavefunction $\psi(\boldsymbol{r}, t)$ which describes the state of a particle of mass m in a potential field $U(\boldsymbol{r})$ is obtained from Schrödinger's equation[1]

$$-\frac{\hbar}{i}\frac{\partial \psi}{\partial t} = -\frac{\hbar^2}{2m}\nabla^2\psi(\boldsymbol{r}, t) + U(\boldsymbol{r})\psi(r, t) \tag{B1}$$

which may be written as follows

$$-\frac{\hbar}{i}\frac{\partial \psi}{\partial t} = \hat{\boldsymbol{H}}\psi \tag{B1$'$}$$

The mathematical operator

$$\hat{\boldsymbol{H}} \equiv \frac{-\hbar^2}{2m}\nabla^2 + U(\boldsymbol{r}) \tag{B2}$$

is the operator of the total energy of the particle; it is also called the Hamiltonian operator or, simply, the Hamiltonian of the particle. $U(\boldsymbol{r})$ denotes, as usual, the potential energy of the particle (we may also think of it as the operator for this quantity). And

$$\hat{T} \equiv \frac{-\hbar^2}{2m}\nabla^2 \tag{B3}$$

is the operator for the kinetic energy of the particle. When the position $\boldsymbol{r}$ of the particle is given by the Cartesian coordinates x, y, z the operator ∇^2 is written as follows

$$\nabla^2 \equiv \frac{\partial^2}{\partial x^2} + \frac{\partial^2}{\partial y^2} + \frac{\partial^2}{\partial z^2} \tag{B4}$$

and when $\boldsymbol{r}$ is given by the spherical coordinates r, θ, φ (see Fig. 2.1) the same operator takes the form

$$\nabla^2 \equiv \frac{\partial^2}{\partial r^2} + \frac{2}{r}\frac{\partial}{\partial r} + \frac{1}{r^2}\left(\frac{\partial^2}{\partial \theta^2} + \frac{\cos\theta}{\sin\theta}\frac{\partial}{\partial\theta} + \frac{1}{\sin^2\theta}\frac{\partial^2}{\partial\varphi^2}\right) \tag{B5}$$

From the general theory of differential equations we can be certain that if we know the wavefunction at $t = 0$, we can determine uniquely the wavefunction for $t > 0$ by solving (B1). We write, symbolically,

$$\psi(\boldsymbol{r}, t = 0) \underset{\substack{\text{Schrödinger's}\\ \text{equation}}}{\Longrightarrow} \psi(\boldsymbol{r}, t), \quad t > 0 \tag{B6}$$

The solution of Schrödinger's equation is intimately connected with the eigenfunctions of the Hamiltonian (B2), which are defined as follows. These are, in general, single-valued complex functions $\{u_\alpha(\boldsymbol{r})\}$, finite and continuous everywhere, and with continuous derivatives with respect to x, y and z; they satisfy the equation

$$\hat{H}u_\alpha(\boldsymbol{r}) = E_\alpha u_\alpha(\boldsymbol{r}) \tag{B7}$$

where $\{E_\alpha\}$ are constants (the eigenvalues of the energy), and appropriate boundary conditions. We distinguish between bound eigenfunctions and non-bound eigenfunctions. A bound eigenfunction must decrease to zero as $|\boldsymbol{r}| \to \infty$, so that[2]

$$\int |u_\alpha(\boldsymbol{r})|^2 \, \mathrm{d}V = 1 \tag{B8}$$

A non-bound eigenfunction, if it is not a plane wave representing a free particle (see Sections 1.7.1 and 2.3.1), must have the form appropriate to a scattering state (see, e.g., Sections 1.8 and 2.3.2).

It turns out that solutions of (B7) which satisfy the boundary condition (B8) exist only for discrete values $\{E_\alpha\}$ of the energy[3]. The corresponding eigenfunctions $\{u_\alpha(\boldsymbol{r})\}$ determine the bound eigenstates of the particle in the given potential field. To the energy eigenvalue E_α corresponds the energy-eigenstate

$$u_\alpha(\boldsymbol{r}, t) = u_\alpha(\boldsymbol{r}) \exp(-\mathrm{i}E_\alpha t/\hbar) \tag{B9}$$

We note that α represents a set of quantum numbers which determine completely the eigenstate of the particle (see, e.g., the description of the eigenstates of a particle in a box, Section 1.3.2). Accepting (B9) as the wavefunction which describes a state of the particle for $t \geqslant 0$, implies that (B9) satisfies the equation of Schrödinger (B1). One can easily prove this as follows. We observe that the Hamiltonian, the operator $\hat{H}$, does not depend on the time and, therefore, we write

$$\hat{H}u_\alpha(\boldsymbol{r}) \exp(-\mathrm{i}E_\alpha t/\hbar) = \exp(-\mathrm{i}E_\alpha t/\hbar)\hat{H}u_\alpha(\boldsymbol{r}) = E_\alpha u_\alpha(\boldsymbol{r}) \exp(-\mathrm{i}E_\alpha t/\hbar)$$

On the other hand we have

$$\frac{-\hbar}{\mathrm{i}}\frac{\partial}{\partial t} u_\alpha(\boldsymbol{r}) \exp(-\mathrm{i}E_\alpha t/\hbar) = E_\alpha u_\alpha(\boldsymbol{r}) \exp(-\mathrm{i}E_\alpha t/\hbar) \tag{B10}$$

which shows that

$$\frac{-\hbar}{\mathrm{i}}\frac{\partial}{\partial t} u_\alpha(\boldsymbol{r}, t) = \hat{H}u_\alpha(\boldsymbol{r}, t)$$

i.e. $u_\alpha(\boldsymbol{r}, t)$ satisfies Schrödinger's equation, as expected.

We note that the above holds true for a scattering eigenstate as well. It, also, has the form (B9) and satisfies the Schrödinger equation (B1).

Depending on the potential field, the energy spectrum may be entirely discrete (as in the field (1.38)), or entirely continuous (as in the case of the free particle); but usually it consists of a discrete part corresponding to bound states, and a continuous part corresponding to scattering states (as in the case of the field (1.39), which we shall consider in more detail in the following section). The discrete part of the spectrum lies below a certain energy (which we can put equal to zero) and above it extends the continuous part of the spectrum.

It can be shown that the eigenfunctions $u_\alpha(\boldsymbol{r})$ of the Hamiltonian constitute an orthogonal and complete set of functions. Orthogonality means that two different eigenfunctions, $u_\alpha(\boldsymbol{r})$ and $u_{\alpha'}(\boldsymbol{r})$, satisfy the relation:

$$\int u^*_{\alpha'}(\boldsymbol{r})u_\alpha(\boldsymbol{r})\,\mathrm{d}V = 0 \tag{B11}$$

where $u^*_{\alpha'}(\boldsymbol{r})$ is the complex conjugate of $u_{\alpha'}(\boldsymbol{r})$. Completeness means that any wavefunction which describes a possible state of the particle, at a given moment, can be written as a linear sum of the eigenfunctions $u_\alpha(\boldsymbol{r})$ of the Hamiltonian. In particular, the state of the particle at time $t = 0$, which is a given quantity in each case, can be written as follows[4]

$$\psi(\boldsymbol{r}, t = 0) = \sum_\alpha c_\alpha u_\alpha(\boldsymbol{r}) \tag{B12}$$

One easily proves (using B8 and B11) that

$$c_\alpha = \int u^*_\alpha(\boldsymbol{r})\psi(\boldsymbol{r}, t = 0)\,\mathrm{d}V \tag{B13}$$

The reader will easily convince himself that the wavefunction $\psi(\boldsymbol{r}, t)$ which describes the state of the above particle as time goes on ($t > 0$) is given by

$$\begin{aligned} \psi(\boldsymbol{r}, t) &= \sum_\alpha c_\alpha u_\alpha(\boldsymbol{r}, t) \\ u_\alpha(\boldsymbol{r}, t) &= u_\alpha(\boldsymbol{r})\exp(-\mathrm{i}E_\alpha t/\hbar) \end{aligned} \tag{B14}$$

The wavefunction (B14) satisfies Schrödinger's equation because every term in the sum of (B14) satisfies this equation, and it satisfies also the initial condition

$$\psi(\boldsymbol{r}, t = 0) = \text{the given function}$$

Assuming that $\psi(\boldsymbol{r}, t = 0)$ has been normalised, one can easily verify, using (B8) and (B11), that

$$\int \psi^*(\boldsymbol{r}, t)\psi(\boldsymbol{r}, t)\,\mathrm{d}V = \sum_\alpha |c_\alpha|^2 = 1$$

and this allows us to interprete $|c_\alpha|^2$ as the probability of finding the particle, at any time, in the eigenstate u_α with energy E_α.

Finally, for the sake of completeness, we write down Schrödinger's equation for a system of N particles (e.g., N electrons in an atom described by the potential field (2.81)):

$$\begin{aligned} -\frac{\hbar}{i}\frac{\partial\psi}{\partial t}(\boldsymbol{r}_1, \ldots, \boldsymbol{r}_N; t) = &-\sum_{i=1}^{N}\frac{\hbar^2}{2m_i}\nabla_i^2\psi(\boldsymbol{r}_1, \ldots, \boldsymbol{r}_N; t) \\ &+ U(\boldsymbol{r}_1, \ldots, \boldsymbol{r}_N)\psi(\boldsymbol{r}_1, \ldots, \boldsymbol{r}_N; t) \end{aligned} \tag{B15}$$

where m_i is the mass of the ith particle and $\boldsymbol{r}_i \equiv (x_i, y_i, z_i)$ its position. The first term on the right of the equation stands for the sum of the kinetic energies of the N particles,

$$\nabla_i^2 \equiv \partial^2/\partial x_i^2 + \partial^2/\partial y_i^2 + \partial^2/\partial z_i^2$$

and $U(\boldsymbol{r}_1, \ldots, \boldsymbol{r}_N)$ represents the potential energy of the system.

The energy-eigenstates of the system are described, by analogy to (B7) and (B9), as follows

$$\psi_\alpha(\boldsymbol{r}_1, \ldots, \boldsymbol{r}_N; t) = \psi_\alpha(\boldsymbol{r}_1, \ldots, \boldsymbol{r}_N) \exp(-\mathrm{i}E_\alpha t/\hbar) \qquad \text{(B16)}$$

where α stands for a set of quantum numbers. The E_α are the eigenvalues of the N-particle Hamiltonian (this is the expression in the curly brackets of (B17)), and $\psi_\alpha(\boldsymbol{r}_1, \ldots, \boldsymbol{r}_N)$ are the corresponding eigenfunctions; they satisfy the equation

$$\left\{-\sum_{i=1}^{N} \frac{\hbar^2}{2m_i}\nabla_i^2 + U(\boldsymbol{r}_1, \ldots, \boldsymbol{r}_N)\right\}\psi_\alpha(\boldsymbol{r}_1, \ldots, \boldsymbol{r}_N) = E_\alpha\psi_\alpha(\boldsymbol{r}_1, \ldots, \boldsymbol{r}_N) \qquad \text{(B17)}$$

and appropriate boundary conditions.

The eigenfunctions $\psi_\alpha(\boldsymbol{r}_1, \ldots, \boldsymbol{r}_N)$ are, also, an orthogonal and complete set of functions (see section B.4).

One may add that the solution of the N-body problem summarised by equations (B15) to (B17) is always very difficult and, therefore, only approximate solutions of this problem are, in general, possible.

B.2 Applications of the time-independent Schrödinger equation

B.2.1 THE RECTANGULAR POTENTIAL WELL

This is defined by (1.39) and shown schematically in Fig. 1.10. The so-called time-independent Schrödinger equation (B7) for a particle of mass m in the potential field (1.39) takes the following form

$$\begin{aligned} -\frac{\hbar^2}{2m}\frac{\mathrm{d}^2u}{\mathrm{d}x^2} &= Eu(x), \quad x < -a \\ -\frac{\hbar^2}{2m}\frac{\mathrm{d}^2u}{\mathrm{d}x^2} - U_O u(x) &= Eu(x), \quad -a < x < +a \\ -\frac{\hbar^2}{2m}\frac{\mathrm{d}^2u}{\mathrm{d}x^2} &= Eu(x), \quad x > a \end{aligned} \qquad \text{(B18)}$$

We have bound states of the particle with discrete energy levels in the region $-U_O < E < 0$ (there cannot be states with $E < -U_O$ because that would mean a negative mean kinetic energy of the particle and that is not possible) and we have scattering states corresponding to a continuous spectrum of energy levels in the region $E > 0$. We consider them separately.

(i) Bound states ($-U_O < E < 0$). We rewrite equations (B18) as follows

$$\frac{d^2u}{dx^2} = \beta^2 u(x), \quad x < -a$$
$$\frac{d^2u}{dx^2} = -\alpha^2 u(x), \quad -a < x < a \tag{B19}$$
$$\frac{d^2u}{dx^2} = \beta^2 u(x), \quad x > a$$

where α and β are positive quantities defined by

$$\alpha \equiv [2m(E + U_O)]^{1/2}/\hbar$$
$$\beta \equiv [2m|E|]^{1/2}/\hbar \tag{B20}$$

The general solution of (B19) is

$$u(x) = A\,e^{\beta x} + B\,e^{-\beta x}, \quad x < -a$$
$$= C\,e^{i\alpha x} + D\,e^{-i\alpha x}, \quad -a < x < a \tag{B21}$$
$$= F\,e^{\beta x} + G\,e^{-\beta x}, \quad x > a$$

The wavefunction cannot increase exponentially into the classically forbidden region and therefore we must put $B = F = 0$. The continuity of $u(x)$ and its derivative at the points $x = -a$ and $x = +a$ demands that

$$(1)\ \ A\,e^{-\beta a} = C\,e^{-i\alpha a} + D\,e^{i\alpha a}$$
$$(2)\ \ \beta A\,e^{-\beta a} = i\alpha C\,e^{-i\alpha a} - i\alpha D\,e^{i\alpha a} \tag{B22}$$
$$(3)\ \ G\,e^{-\beta a} = C\,e^{i\alpha a} + D\,e^{-i\alpha a}$$
$$(4)\ \ -\beta G\,e^{-\beta a} = i\alpha C\,e^{i\alpha a} - i\alpha D\,e^{-\alpha a}$$

Eliminating A from (1) and (2) and G from (3) and (4), we obtain $D = \pm C$. With $D = C$ we obtain $G = A$, in which case (B21) becomes

$$u(x) = A\,e^{\beta x}, \quad x < -a$$
$$= C' \cos \alpha x, \quad -a < x < a \tag{B23}$$
$$= A\,e^{-\beta x}, \quad x > a$$

where A and C' are constants to be determined. We note that (B23) has the form (1.43a); it has even parity. Because $u(x)$ and du/dx must be continuous at $x = \pm a$, so must be their ratio $(du/dx)/u(x)$; which implies that eigenstates of even parity can exist only when the following condition is satisfied

$$\cos \alpha a = (\alpha/\beta) \sin \alpha a \tag{B24}$$

Similarly, with $D = -C$ we obtain $G = -A$, in which case (B21) becomes

$$u(x) = A\,e^{\beta x}, \quad x < -a$$
$$= C' \sin \alpha x, \quad -a < x < a \tag{B25}$$
$$= -A\,e^{-\beta x}, \quad x > a$$

where A and C' are constants to be determined. (B25) has the form (1.43b); it has odd parity. From the continuity of $(du/dx)/u(x)$ at $x = \pm a$, it follows that eigenstates of odd parity can exist only when the following condition is satisfied

$$\sin \alpha a = -(\alpha/\beta) \cos \alpha a \tag{B26}$$

Equations (B24) and (B26) can be written in the simpler form

$$\cot k = \frac{k}{(b^2 - k^2)^{1/2}} \tag{B24)$'$}$$

$$-\tan k = \frac{k}{(b^2 - k^2)^{1/2}} \tag{B26)$'$}$$

where

$$k \equiv a\alpha = a[2m(E + U_O)]^{1/2}/\hbar \tag{B27a}$$

$$b \equiv a[2mU_O]^{1/2}/\hbar \tag{B27b}$$

We note that k lies in the region $0 < k < b$ when $-U_O < E < 0$. The values of k which satisfy (B24)$'$ or (B26)$'$ can be found graphically. This is demonstrated in Fig. B1 for a well whose depth is $U_O = 49h^2/(128ma^2)$. The corresponding energy levels obtained through (B27a) are given by (1.40) and shown in Fig. 1.10. The corresponding eigenfunctions are obtained from (B23) or (B25) depending on their parity (even and odd respectively). For a given energy eigenvalue E_n, one determines the values α_n and β_n of the corresponding parameters from (B20). Then, the constants A and C' are determined as follows. Having secured the continuity of $(du/dx)/u(x)$, we can eliminate C' by using the continuity of $u(x)$ at $x = \pm a$. We obtain $C' = A \exp(-\beta a)/\cos \alpha a$ for the even eigenfunctions and $C' = -A \exp(-\beta \alpha)/\sin \alpha a$ for the odd ones. Finally, we determine $|A|$ by normalisation; we demand that

$$\int_{-\infty}^{+\infty} |u(x)|^2 \, dx = 1 \tag{B28}$$

It is worth noting that according to Fig. B1 there will always be at least one solution of (B24)$'$, however small the value of b (i.e. however small the width and depth of the well), in accordance with the general theorem cited in Section 1.4.2, namely that a particle in an one-dimensional potential well has at least one bound state. We note also that, again in accordance with general properties of bound states in one-dimensional potential fields, none of the discrete energy levels of the particle in the rectangular potential well is degenerate.

(ii) Scattering states ($E > 0$). We rewrite equations (B18) as follows

$$\begin{aligned}
\frac{d^2u}{dx^2} &= -\beta^2 u(x), \quad x < -a \\
\frac{d^2u}{dx^2} &= -\alpha^2 u(x), \quad -a < x < a \\
\frac{d^2u}{dx^2} &= -\beta^2 u(x), \quad x > a
\end{aligned} \tag{B29}$$

where α and β are defined by (B20). We have already explained, in Section 1.8, what the asymptotic form should be of a scattering state corresponding to a particle incident on a scatterer from the left. In the present case the scatterer is a rectangular well, instead of the rectangular barrier of Section 1.8, but otherwise the situation is exactly the same. The solution of (B29) which describes a particle (of mass m) with

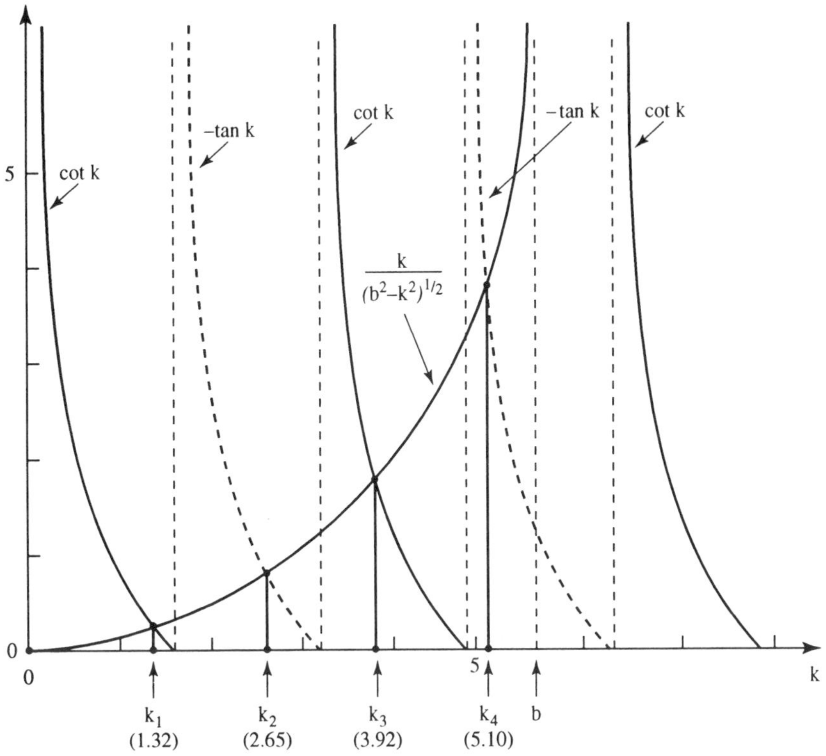

Figure B1. Graphical solution of the transcendental equations (B24)′ and (B26)′. [Redrawn from R. B. Leighton, *Principles of Modern Physics*, (New York: McGraw-Hill) 1959. Reproduced with permission

energy E incident on the rectangular well of Fig. 1.10 from the left and scattered by it, has the following form (away from the scatterer, it is the same, apart from the time factor, as that of (1.71))

$$\begin{aligned} u^+(x) &= \mathrm{e}^{\mathrm{i}\beta x} + B\,\mathrm{e}^{-\mathrm{i}\beta x} \quad x < -a \\ &= D\,\mathrm{e}^{\mathrm{i}\alpha x} + F\,\mathrm{e}^{-\mathrm{i}\alpha x} \quad -a < x < a \\ &= C\,\mathrm{e}^{\mathrm{i}\beta x} \quad x > a \end{aligned} \tag{B30}$$

where B, D, F and C are coefficients to be determined by the requirement that $u^+(x)$ and its derivative be continuous at $x = -a$ and $x = a$. We obtain the following system of inhomogeneous algebraic equations

$$\begin{aligned} -\mathrm{e}^{\mathrm{i}\beta a}B + \mathrm{e}^{-\mathrm{i}\alpha a}D + \mathrm{e}^{\mathrm{i}\alpha a}F &= \mathrm{e}^{-\mathrm{i}\beta a} \\ \mathrm{i}\beta\,\mathrm{e}^{\mathrm{i}\beta a}B + \mathrm{i}\alpha\,\mathrm{e}^{-\mathrm{i}\alpha a}D - \mathrm{i}\alpha\,\mathrm{e}^{\mathrm{i}\alpha a}F &= \mathrm{i}\beta\,\mathrm{e}^{-\mathrm{i}\beta a} \\ \mathrm{e}^{\mathrm{i}\alpha a}D + \mathrm{e}^{-\mathrm{i}\alpha a}F - \mathrm{e}^{\mathrm{i}\beta a}C &= 0 \\ \mathrm{i}\alpha\,\mathrm{e}^{\mathrm{i}\alpha a}D - \mathrm{i}\alpha\,\mathrm{e}^{-\mathrm{i}\alpha a}F - \mathrm{i}\beta\,\mathrm{e}^{\mathrm{i}\beta a}C &= 0 \end{aligned} \tag{B31}$$

the solution of which determines uniquely, for a given energy $E > 0$, the values of B, D, F and C. The physical meaning of $|B|^2$ (reflection coefficient) and of $|C|^2$ (transmission coefficient) is made obvious by our discussion of Section 1.8.

For the same energy one obtains, besides (B30) a second, linearly independent, solution of (B29), which describes the scattering state of the particle when it is incident on the potential well from the right. It has the following form

$$\begin{aligned} u^-(x) &= C\,\mathrm{e}^{-\mathrm{i}\beta x} \quad x < -a \\ &= D\,\mathrm{e}^{-\mathrm{i}\alpha x} + F\,\mathrm{e}^{\mathrm{i}\alpha x} \quad -a < x < a \\ &= \mathrm{e}^{-\mathrm{i}\beta x} + B\,\mathrm{e}^{\mathrm{i}\beta x} \quad x > a \end{aligned} \tag{B32}$$

where B, D, F and C are to be determined, as in (B30), by the continuity of $u^-(x)$ and its derivative at $x = -a$ and $x = a$.

We know that a second-order differential equation, such as (B29), has only two linearly independent solutions. In the present case, for given energy E, these are (B30) and (B32), there can be no other.

The above completes our formal treatment of the energy-eigenstates of a particle in the potential field (1.39).

B.2.2 PARTICLE IN A ONE-DIMENSIONAL BOX

The potential field is given by (1.38). The energy eigenfunctions satisfy Schrödinger's time-independent equation (B7) for a free particle, inside the box, i.e.

$$-\frac{\hbar^2}{2m}\frac{\mathrm{d}^2u}{\mathrm{d}x^2} = Eu(x), \quad 0 < x < a \tag{B33}$$

and the boundary condition ($u = 0$ outside the box and the function must go to zero continuously):

$$u(0) = u(a) = 0 \tag{B34}$$

The general solution of (B33) is

$$u(x) = A_+\,\mathrm{e}^{\mathrm{i}kx} + A_-\,\mathrm{e}^{-\mathrm{i}kx} \tag{B35}$$

where $k = (2mE)^{1/2}/\hbar$. We see that (B35) is the same with u_E given by (1.77) apart from the familiar time factor. To find the energy eigenvalues and corresponding eigenfunctions we proceed as in Section 1.9. We need not add anything here.

B.3 Momentum-eigenstates of a particle

The operator of the momentum of a particle is defined by

$$\hat{\boldsymbol{p}} \equiv \frac{\hbar}{i}\boldsymbol{\nabla} \tag{B36a}$$

$$(\hat{p}_x, \hat{p}_y, \hat{p}_z) = \left(\frac{\hbar}{i}\frac{\partial}{\partial x}, \frac{\hbar}{i}\frac{\partial}{\partial y}, \frac{\hbar}{i}\frac{\partial}{\partial z}\right) \tag{B36b}$$

We have

$$\frac{\hbar}{i}\boldsymbol{\nabla}\,\mathrm{e}^{\mathrm{i}\boldsymbol{k}\cdot\boldsymbol{r}} = \hbar\boldsymbol{k}\,\mathrm{e}^{\mathrm{i}\boldsymbol{k}\cdot\boldsymbol{r}} \tag{B37}$$

which tells us that the eigenvalues of the momentum are given by

$$\boldsymbol{p} = (p_x, p_y, p_z) = (\hbar k_x, \hbar k_y, \hbar k_z) \quad -\infty < k_x, k_y, k_z < \infty \tag{B38}$$

and the corresponding eigenfunctions (they are often called the momentum-eigenstates) by

$$u_{\boldsymbol{k}}(\boldsymbol{r}) = A\,\mathrm{e}^{\mathrm{i}\boldsymbol{k}\cdot\boldsymbol{r}} \tag{B39}$$

where A is a normalisation constant.

It is often convenient to replace the continuous spectrum of eigenvalues (B38) with a discrete one as follows. We replace $\boldsymbol{k}$ in (B38) by

$$\boldsymbol{k} = \frac{2\pi}{L}(n_x, n_y, n_z) \quad n_x, n_y, n_z = 0, \pm 1, \pm 2, \pm 3, \ldots \tag{B40}$$

and write

$$u_{\boldsymbol{k}}(\boldsymbol{r}) = \frac{1}{\sqrt{V}}\mathrm{e}^{\mathrm{i}\boldsymbol{k}\cdot\boldsymbol{r}} \tag{B41}$$

where $V \equiv L^3$. By letting L be very large ($L \to \infty$) we obtain, in any given problem, the same results as with the continuous spectrum (B38). The properties of the functions defined by (B40) and (B41) are well known from the theory of Fourier series. They are an orthogonal set of functions, i.e.

$$\begin{aligned}\int_V u_{\boldsymbol{k}}^*(\boldsymbol{r}) u_{\boldsymbol{k}'}(\boldsymbol{r})\,\mathrm{d}V &= 1, \quad \text{if } \boldsymbol{k} = \boldsymbol{k}' \\ &= 0, \quad \text{if } \boldsymbol{k} \neq \boldsymbol{k}'\end{aligned} \tag{B42}$$

and they are a complete set, i.e. any function $\psi(\boldsymbol{r}, t)$ can be written as a linear sum of (B41):

$$\psi(\boldsymbol{r}; t) = \sum_{\boldsymbol{k}} c(\boldsymbol{k}; t) u_{\boldsymbol{k}}(\boldsymbol{r}) \tag{B43}$$

provided $\psi(\boldsymbol{r}; t)$ is squarely integrable, i.e., it can be (and we assume that it has been) normalised as follows

$$\int |\psi(\boldsymbol{r}; t)|^2\,\mathrm{d}V = 1 \tag{B44}$$

We note that by taking L sufficiently large we can always make the contribution to the integral (B44) from the region outside the volume $V = L^3$ vanishingly small. We obtain a formula for $c(\boldsymbol{k}; t)$ as follows: we multiply (B43) with $u_{\boldsymbol{k}}^*(\boldsymbol{r})$, integrate over the volume V and use (B42); the result is:

$$c(\boldsymbol{k}; t) = \int_V u_{\boldsymbol{k}}^*(\boldsymbol{r}) \psi(\boldsymbol{r}; t)\,\mathrm{d}V \tag{B45}$$

The wavefunction which describes the state of a particle, in any given potential field, satisfies (B44) and, therefore, can be expanded as in (B43). Moreover, one can easily show using equations (B42) to (B44) that

$$\sum_{\boldsymbol{k}} |c(\boldsymbol{k}; t)|^2 = 1 \tag{B46}$$

We can then say that $|c(\boldsymbol{k}; t)|^2$ is the probability that the particle will be found, at time t, in the momentum eigenstate $u_{\boldsymbol{k}}(\boldsymbol{r})$ (with momentum $\hbar\boldsymbol{k}$).

Finally, we note that the momentum-eigenfunctions ((B39) or (B41)) are also the energy-eigenfunctions of a free particle. The Hamiltonian of a free particle is

$$\hat{H}_0 = -\frac{\hbar^2}{2m}\nabla^2 \tag{B47}$$

and we obtain

$$\hat{H}_0 u_{\boldsymbol{k}}(\boldsymbol{r}) = E_{\boldsymbol{k}} u_{\boldsymbol{k}}(\boldsymbol{r}) \tag{B48}$$

where $E_{\boldsymbol{k}} = \hbar^2 k^2/2m$. There are, obviously, infinite-many momentum-eigenstates (corresponding to all possible directions of the momentum) for given energy of the particle.

B.4 Operators

We have already introduced the operators of the total energy (the Hamiltonian $\hat{H}$) of a particle, those of the kinetic and potential energies ($\hat{T}$ and $U(\boldsymbol{r})$ respectively), the momentum operator $\hat{\boldsymbol{p}} = (\hat{p}_x, \hat{p}_y, \hat{p}_z)$ and, though we have not stated this explicitly, the position operator $\hat{\boldsymbol{r}} = \boldsymbol{r} = (x, y, z)$.

The operator of any (classical) physical quantity $\Omega(\boldsymbol{r}, \boldsymbol{p})$ is obtained by replacing $\boldsymbol{r}$ and $\boldsymbol{p}$ in $\Omega(\boldsymbol{r}, \boldsymbol{p})$ by the operators of these quantities, i.e.

$$\hat{\Omega} = \Omega(\hat{\boldsymbol{r}}, \hat{\boldsymbol{p}}) = \Omega\left(\boldsymbol{r}, \frac{\hbar}{i}\boldsymbol{\nabla}\right) \tag{B49}$$

For example, the operator of the potential energy is simply $U(\boldsymbol{r})$; and that of the kinetic energy, whose classical expression is $T = (1/2m)(p_x^2 + p_y^2 + p_z^2)$, is

$$\hat{T} = \left(\frac{1}{2m}\right)(\hat{p}_x\hat{p}_x + \hat{p}_y\hat{p}_y + \hat{p}_z\hat{p}_z) = -\frac{\hbar^2}{2m}\left(\frac{\partial^2}{\partial x^2} + \frac{\partial^2}{\partial y^2} + \frac{\partial^2}{\partial z^2}\right) \equiv -\frac{\hbar^2}{2m}\nabla^2$$

For a system of many (N) particles the relevant physical quantities are functions of the coordinates and momenta of all the particles. If $\Omega(\boldsymbol{r}_1, \boldsymbol{p}_1, \boldsymbol{r}_2, \boldsymbol{p}_2, \ldots, \boldsymbol{r}_N, \boldsymbol{p}_N)$ is such a quantity, the corresponding operator is given by

$$\hat{\Omega} = \Omega(\boldsymbol{r}_1, \hat{\boldsymbol{p}}_1, \boldsymbol{r}_2, \hat{\boldsymbol{p}}_2, \ldots, \boldsymbol{r}_N, \hat{\boldsymbol{p}}_N) \tag{B50}$$

where

$$\hat{\boldsymbol{p}}_i = \frac{\hbar}{i}\boldsymbol{\nabla}_i = \frac{\hbar}{i}\left(\frac{\partial}{\partial x_i}, \frac{\partial}{\partial y_i}, \frac{\partial}{\partial z_i}\right)$$

We note that $\hat{A}\hat{B}$, where $\hat{A}$ and $\hat{B}$ are known operators, denotes an operator whose effect on a function ψ consists of two successive actions: first of the operator $\hat{B}$ on ψ and then of the operator $\hat{A}$ on the function $\hat{B}\psi$. In general $\hat{A}\hat{B}\psi$ is different from $\hat{B}\hat{A}\psi$. For example,

$$x\hat{p}_x\psi = x(\hbar/\mathrm{i})\frac{\partial\psi}{\partial x}, \quad \hat{p}_x x\psi = x(\hbar/\mathrm{i})\frac{\partial\psi}{\partial x} + (\hbar/\mathrm{i})\psi$$

which shows that

$$(\hat{p}_x x - x\hat{p}_x)\psi = (\hbar/\mathrm{i})\psi \tag{B51}$$

for any function ψ; which allows us to write

$$\hat{p}_x x - x\hat{p}_x = \hbar/\mathrm{i} \tag{B51$'$}$$

On the other hand, one can easily show that $\hat{p}_y x - x\hat{p}_y = 0$, $\hat{p}_x\hat{p}_y - \hat{p}_y\hat{p}_x = 0$, etc., and obviously, $xy - yx = 0$, etc..

We define the commutator of two operators $\hat{A}$ and $\hat{B}$ as follows

$$[\hat{A}, \hat{B}] \equiv \hat{A}\hat{B} - \hat{B}\hat{A}$$

Using the above definition we can write

$$\begin{gathered} [\hat{p}_x, x] = (\hbar/\mathrm{i}), [\hat{p}_y, y] = (\hbar/\mathrm{i}), [\hat{p}_z, z] = (\hbar/\mathrm{i}) \\ [\hat{p}_x, y] = 0, [\hat{p}_x, z] = 0, [\hat{p}_x, \hat{p}_y] = 0, \text{ etc.} \end{gathered} \tag{B52}$$

In section B.3 we have introduced the eigenvalues (B38)–(B40) and corresponding eigenfunctions (B41) of the momentum operator. Now, to the operator of any physical quantity corresponds, in similar fashion, a set of eigenvalues (this could be discrete, continuous, or have a discrete and a continuous part) and a corresponding set of eigenfunctions which should be single-valued, finite and continuous everywhere and with continuous derivatives. We have already discussed the eigenvalues and corresponding eigenfunctions of the Hamiltonian operator of a few sytems. But one needs to know the eigenvalues and eigenfunctions of other operators as well.

Let $\{w_\alpha\}$ be the set of eigenvalues and $\{\varphi_\alpha(q)\}$ the corresponding eigenfunctions of an operator $\hat{\Omega}(q, \partial/\partial q)$. We have

$$\hat{\Omega}\varphi_\alpha(q) = w_\alpha\varphi_\alpha(q) \tag{B53}$$

where q represents the coordinates of a given physical system. If we are dealing with just one particle, $q \equiv (x, y, z)$ and $\mathrm{d}q \equiv \mathrm{d}x\,\mathrm{d}y\,\mathrm{d}z$; if we are dealing with N particles $q \equiv (x_1, y_1, z_1, \ldots, x_N, y_N, z_N)$ and $\mathrm{d}q = \mathrm{d}x_1\,\mathrm{d}y_1\,\mathrm{d}z_1 \ldots \mathrm{d}x_N\,\mathrm{d}y_N\,\mathrm{d}z_N$; similarly, $\partial/\partial q$ stands for $(\partial/\partial x, \partial/\partial y, \partial/\partial z)$ in the case of one particle, and for $(\partial/\partial x_1, \partial/\partial y_1, \partial/\partial z_1, \ldots, \partial/\partial x_N, \partial/\partial y_N, \partial/\partial z_N)$ in the case of N particles; α denotes, as usual, a set of quantum numbers which specify the different eigenfunctions (we shall assume for the sake of clarity that all quantum numbers take discrete values). We have (integrations extend over all space)

$$\begin{aligned} \int \varphi_\alpha^*(q)\varphi_{\alpha'}(q)\,\mathrm{d}q &= 1, \text{ if } \alpha = \alpha' \\ &= 0, \text{ if } \alpha \neq \alpha' \end{aligned} \tag{B54}$$

i.e. $\{\varphi_\alpha(q)\}$ is an orthogonal set of functions. It is also a complete set of functions, i.e. any squarely integrable (normalisable) function of q can be written as a linear sum of the $\{\varphi_\alpha\}$. In particular the wavefunction $\psi(q, t)$ which describes the state of the system under consideration, at time t, can be written as

$$\begin{aligned} \psi(q, t) &= \sum_\alpha c_\alpha(t)\varphi_\alpha(q) \\ c_\alpha(t) &= \int \varphi_\alpha^*(q)\psi(q, t)\,\mathrm{d}q \end{aligned} \tag{B55}$$

To obtain the second of the above equations we multiply the first with $\varphi_\alpha^*(q)$, integrate with respect to q, and use (B54).

Given that $\psi(q, t)$ is normalised, i.e.

$$\int |\psi(q, t)|^2\,\mathrm{d}q = 1$$

we can easily obtain, using (B54), that

$$\sum_{\alpha} |c_\alpha(t)|^2 = 1 \tag{B56}$$

We can then say that $|c_\alpha(t)|^2$ is the probability that the system will be, at time t, in the eigenstate $\varphi_\alpha(q)$ of Ω.

We conclude this section with the statement of a most important theorem: If two operators $\hat{A}$ and $\hat{B}$ commute, i.e. if $[\hat{A}, \hat{B}] = 0$, then the two operators have a common set of eigenfunctions $\{\varphi_\alpha\}$:

$$\begin{aligned} \hat{A}\varphi_\alpha(q) &= A_\alpha \varphi_\alpha(q) \\ \hat{B}\varphi_\alpha(q) &= B_\alpha \varphi_\alpha(q) \end{aligned} \tag{B57}$$

where $\{A_\alpha\}$ and $\{B_\alpha\}$ are respectively the eigenvalues of $\hat{A}$ and $\hat{B}$.

If we have three or more operators, $\hat{A}$, $\hat{B}$, $\hat{C}$, . . . and each of them commutes with the rest, then these operators have a common set of eigenfunctions:

$$\begin{aligned} \hat{A}\varphi_\alpha(q) &= A_\alpha \varphi_\alpha(q) \\ \hat{B}\varphi_\alpha(q) &= B_\alpha \varphi_\alpha(q) \\ \hat{C}\varphi_\alpha(q) &= C_\alpha \varphi_\alpha(q) \\ &\text{etc.} \end{aligned} \tag{B58}$$

We have already met with a manifestation of the above theorem. The momentum operators $-i\hbar\partial/\partial x$, $-i\hbar\partial/\partial y$, $-i\hbar\partial/\partial z$ of a particle commute with each other and with the kinetic energy operator $(-\hbar^2/2m)\nabla^2$ (which is also the Hamiltonian operator $\hat{H}_0$ of a free particle). We have seen that the plane waves defined by (B41) constitute a common set of eigenfunctions of $\hat{p}_x$, $\hat{p}_y$, $\hat{p}_z$ and $\hat{H}_0$.

It can also be shown that if two or more operators $\hat{A}$, $\hat{B}$, . . ., $\hat{C}$ have a common set of eigenfunctions then they commute with each other.

B.5 Mean values of physical quantities

Let Ω be a physical quantity and $\hat{\Omega}$ its operator given by (B50). The mean value of Ω at time t is defined by

$$\langle \Omega \rangle_t = \int \psi^*(q, t)\hat{\Omega}\psi(q, t)\,\mathrm{d}q \tag{B59}$$

where $\psi(q, t)$ is the normalised wavefunction of the system at time t. Substituting (B55) into (B59) and using (B53) we obtain

$$\langle \Omega \rangle_t = \sum_{\alpha'}\sum_{\alpha} c^*_{\alpha'}(t)c_\alpha(t)w_\alpha \int \varphi^*_{\alpha'}(q)\varphi_\alpha(q)\,\mathrm{d}q$$

which, with the aid of (B54), can be written as

$$\langle \Omega \rangle_t = \sum_{\alpha} |c_\alpha(t)|^2 w_\alpha \tag{B60}$$

which, having in mind (B56), we can interpret as follows: the outcome of a measurement of Ω, at time t, will be one or the other of the eigenvalues $\{w_\alpha\}$ of $\hat{\Omega}$. The probability that this will be a certain eigenvalue w_α is given by the probability

$|c_\alpha(t)|^2$ of the system being in the eigenstate $\varphi_\alpha(q)$ of $\hat{\Omega}$ at the given time[5]. It is worth noting that $\langle\Omega\rangle_t$ and $|(c_\alpha(t)|^2$ do not change if the wavefunction $\psi(q, t)$ is multiplied by $\exp(\mathrm{i}\theta)$, where θ is a real number, in accordance with what has been stated in this respect in Sections 1.3.1 and 2.14.

The uncertainty $(\Delta\Omega)_t$ about the mean value $\langle\Omega\rangle_t$ is defined by

$$(\Delta\Omega)_t^2 = \int \psi^*(q, t)(\hat{\Omega} - \langle\Omega\rangle_t)^2\psi(q, t)\,\mathrm{d}q \tag{B61}$$

which, by using (B55), (B53) and (B54), can be written as

$$(\Delta\Omega)_t = \left(\sum_\alpha |c_\alpha(t)|^2(w_\alpha - \langle\Omega\rangle_t)^2\right)^{1/2} \tag{B62}$$

It is evident from the above that the uncertainty $(\Delta\Omega)_t$ will be zero, only if the state of the system is an eigenstate of the measured physical quantity; if $\psi(q, t)$ is an eigenfunction of Ω with eigenvalue w_α, then $\langle\Omega\rangle_t = w_\alpha$ and $(\Delta\Omega)_t = 0$. From the above follows, also, that if the physical quantities A, B, ... C are to be known exactly (with zero uncertainty) at the same time, the state of the system must be an eigenstate of every one of the corresponding operators $\hat{A}$, $\hat{B}$, ..., $\hat{C}$: which implies, in general, that these operators have a common set of eigenfunctions, which in turn implies, according to the theorem of the previous section, that $\hat{A}$, $\hat{B}$, ..., $\hat{C}$ commute with each other.

As a corollary of the above statement we have that, if the operators corresponding to two physical quantities do not commute, then the two quantities cannot be known exactly at the same time. The position coordinate x and the corresponding momentum p_x, whose operators do not commute (see (B51)′), cannot be known exactly at the same time. We see at this stage that this famous example is but one manifestation of a more general theorem.

Of particular importance are, as we know, the stationary states (energy-eigenstates) of the system under consideration. They have the form (see Section B.1):

$$u_\lambda(q, t) = u_\lambda(q)\exp(-\mathrm{i}E_\lambda t/\hbar)$$
$$\hat{H}u_\lambda(q) = E_\lambda u_\lambda(q) \tag{B63}$$

where $\hat{H}$ is the Hamiltonian of the system and λ denotes, as usual, a set of quantum numbers. Putting $\psi = u_\lambda(q, t)$ in (B59), and noting that the operators of physical quantities (position, momentum, angular momentum (see next section), kinetic energy, potential energy (except in the case of a time-dependent field), etc.) do not depend on the time, we obtain

$$\langle\Omega\rangle = \int u_\lambda^*(q, t)\hat{\Omega}u_\lambda(q, t)\,\mathrm{d}q = \int u_\lambda^*(q)\hat{\Omega}u_\lambda(q)\,\mathrm{d}q \tag{B64}$$

which proves that the mean values of the above observables are, for a stationary state, independent of time.

Now let $\hat{A}$, $\hat{B}$, ..., $\hat{C}$ be operators of physical quantities which commute with each other and with the Hamiltonian $\hat{H}$ of the system under consideration. Then we can find a set $\{u_\lambda(q)\}$ which constitutes a common set of eigenfunctions of $\hat{A}$, $\hat{B}$, ..., $\hat{C}$ and $\hat{H}$, which means that an eigenfunction of the energy of the system can also be an eigenfunction of $\hat{A}$, $\hat{B}$, ..., $\hat{C}$; and then in the corresponding stationary state, A, B, ..., C will be known exactly besides the energy. We note, however, that it need not be so in the case of degenerate energy levels. For example, if two of the common eigenfunctions correspond to different eigenvalues of $\hat{A}$ but to the same eigenvalue of $\hat{H}$, a linear sum of the two will be an eigenfunction of $\hat{H}$ but not of $\hat{A}$.

When an operator $\hat{\Omega}$ does not commute with $\hat{H}$, its mean value over a stationary state $u_\lambda(q, t)$ of the system can be written as

$$\langle \Omega \rangle_\lambda = \sum_\alpha |c_{\lambda;\alpha}|^2 w_\alpha$$
$$c_{\lambda;\alpha} = \int \varphi_\alpha^*(q) u_\lambda(q)\,\mathrm{d}q \tag{B65}$$

which is obtained by substituting for $u_\lambda(q)$ in (B64) its expansion in terms of the eigenfunctions $\{\varphi_\alpha(q)\}$ of $\hat{\Omega}$, and using (B53) and (B54). The uncertainty associated with $\langle \Omega \rangle_\lambda$ is given by

$$(\Delta\Omega)_\lambda = \left(\sum_\alpha |c_{\lambda;\alpha}|^2 (w_\alpha - \langle \Omega \rangle_\lambda)^2\right)^{1/2} \tag{B66}$$

and it, like $\langle \Omega \rangle_\lambda$, is independent of time.

Finally, let us show that the expectation value $\langle \Omega \rangle_t$ of an operator $\hat{\Omega}$ which commutes with the Hamiltonian (the Hamiltonian itself is obviously such an operator) is *always* independent of time; it is conserved. In this case the operators $\hat{\Omega}$ and $\hat{H}$ have a common set of eigenfunctions $\varphi_\alpha(q)$. Now, by an obvious extension of (B14), the most general state of the system at time t is given by (B55) with $c_\alpha(t) = c_\alpha \exp(-\mathrm{i}E_\alpha t/\hbar)$, which substituted in (B60) gives

$$\langle \Omega \rangle_t = \sum_\alpha |c_\alpha|^2 w_\alpha$$

which is indeed independent of time.

B.6 Angular momentum

B.6.1 ORBITAL ANGULAR MOMENTUM

The orbital angular momentum of a particle, in classical mechanics, is the vector defined in Section 2.2. It can be written as follows

$$\boldsymbol{l} = \boldsymbol{r} \times \boldsymbol{p} \tag{B67}$$

Its components along the x, y, z directions are

$$l_x = yp_z - zp_y, \quad l_y = zp_x - xp_z, \quad l_z = xp_y - yp_x \tag{B67a}$$

We obtain the quantum-mechanical operators of l_x, l_y, and l_z by substituting in the above expressions the operators for $\boldsymbol{r}$ and $\boldsymbol{p}$ which are: $\hat{\boldsymbol{r}} = \boldsymbol{r}$ and $\hat{\boldsymbol{p}} = (\hbar/\mathrm{i})\boldsymbol{\nabla}$ respectively. We obtain

$$\hat{l}_x = \frac{\hbar}{\mathrm{i}}\left(y\frac{\partial}{\partial z} - z\frac{\partial}{\partial y}\right), \quad \hat{l}_y = \frac{\hbar}{\mathrm{i}}\left(z\frac{\partial}{\partial x} - x\frac{\partial}{\partial z}\right), \quad \hat{l}_z = \frac{\hbar}{\mathrm{i}}\left(x\frac{\partial}{\partial y} - y\frac{\partial}{\partial x}\right) \tag{B68}$$

From the operators (B68) we can form the operator

$$\hat{l}^2 = \hat{l}_x^2 + \hat{l}_y^2 + \hat{l}_z^2 \tag{B69}$$

which can be regarded as the operator for the square of the magnitude of the orbital angular momentum of a particle.

Using the commutation relations (B52) of coordinates and linear momenta, we can easily verify the following commutation relations of the angular momentum operators

$$\begin{aligned}
&[\hat{l}_x, x] = 0, \quad [\hat{l}_x, y] = \mathrm{i}\hbar z, \quad [\hat{l}_x, z] = -\mathrm{i}\hbar y \\
&[\hat{l}_y, y] = 0, \quad [\hat{l}_y, z] = \mathrm{i}\hbar x, \quad [\hat{l}_y, x] = -\mathrm{i}\hbar x \\
&[\hat{l}_z, z] = 0, \quad [\hat{l}_z, x] = \mathrm{i}\hbar y, \quad [\hat{l}_z, y] = -\mathrm{i}\hbar z
\end{aligned} \tag{B70}$$

For example,

$$\begin{aligned}
\hat{l}_x y - y\hat{l}_x &= (y\hat{p}_z - z\hat{p}_y)y - y(y\hat{p}_z - z\hat{p}_y) \\
&= -z(\hat{p}_y y - y\hat{p}_y) = \mathrm{i}\hbar z
\end{aligned}$$

One can easily show that similar commutation relations hold between the angular momentum and linear momentum operators, i.e.

$$[\hat{l}_x, \hat{p}_x] = 0, \quad [\hat{l}_x, \hat{p}_y] = \mathrm{i}\hbar\hat{p}_z, \quad [\hat{l}_x, \hat{p}_z] = -\mathrm{i}\hbar\hat{p}_y \quad \text{etc.} \tag{B71}$$

Using the above formulae we establish the following commutation relations of $\hat{l}_x$, $\hat{l}_y$, $\hat{l}_z$ with one another

$$[\hat{l}_x, \hat{l}_y] = \mathrm{i}\hbar\hat{l}_z, [\hat{l}_y, \hat{l}_z] = \mathrm{i}\hbar\hat{l}_x, [\hat{l}_z, \hat{l}_x] = \mathrm{i}\hbar\hat{l}_y \tag{B72}$$

For example,

$$\begin{aligned}
\hat{l}_x\hat{l}_y - \hat{l}_y\hat{l}_x &= \hat{l}_x(z\hat{p}_x - x\hat{p}_z) - (z\hat{p}_x - x\hat{p}_z)\hat{l}_x \\
&= (\hat{l}_x z - z\hat{l}_x)\hat{p}_x - x(\hat{l}_x\hat{p}_z - \hat{p}_z\hat{l}_x) \\
&= -\mathrm{i}\hbar y\hat{p}_x + \mathrm{i}\hbar x\hat{p}_y = \mathrm{i}\hbar\hat{l}_z
\end{aligned}$$

It follows from (B72) that the three components of the angular momentum do not have a common set of eigenfunctions and therefore, cannot simultaneously be known exactly. In this respect the angular momentum is fundamentally different from linear momentum whose three components have a common set (B39) of eigenfunctions.

Finally one can show, using (B72), that

$$[\hat{l}^2, \hat{l}_x] = 0, \quad [\hat{l}^2, \hat{l}_y] = 0, \quad [\hat{l}^2, \hat{l}_z] = 0 \tag{B73}$$

For example,

$$\begin{aligned}
[\hat{l}_x^2, \hat{l}_z] &= \hat{l}_x[\hat{l}_x, \hat{l}_z] + [\hat{l}_x, \hat{l}_z]\hat{l}_x \\
&= -\mathrm{i}(\hat{l}_x\hat{l}_y + \hat{l}_y\hat{l}_x) \\
[\hat{l}_y^2, \hat{l}_z] &= \mathrm{i}(\hat{l}_x\hat{l}_y + \hat{l}_y\hat{l}_x) \\
[\hat{l}_z^2, \hat{l}_z] &= 0
\end{aligned}$$

and by adding the above three equations, we obtain $[\hat{l}^2, l_z] = 0$; and we can prove the other two relations of (B73) in the same way. These relations show that we can write a common set of eigenfunctions for the square of the magnitude of the orbital angular momentum and one of its components. In what follows we choose this to be the z-component which implies that the square of the magnitude of the angular momentum and its z-component can be known exactly at the same time. It remains for us to find this common set of eigenfunctions and the corresponding eigenvalues of $\hat{l}^2$ and $\hat{l}_z$. It will be easier to do so if we write $\hat{l}^2$ and $\hat{l}_z$ in spherical coordinates.

Introducing the spherical coordinates, r, θ, φ in accordance with the relations (see Fig. 2.1)

$$x = r\sin\theta\cos\varphi,\ y = r\sin\theta\sin\varphi,\ z = r\cos\theta$$

we obtain after some calculation the following expressions for $\hat{l}_x$, $\hat{l}_y$, $\hat{l}_z$ and $\hat{l}^2$:

$$\hat{l}_z = \frac{\hbar}{\mathrm{i}}\frac{\partial}{\partial\varphi} \tag{B74}$$

Instead of $\hat{l}_x$, $\hat{l}_y$ it is often more convenient to use the following combinations of them

$$\hat{l}_x + \mathrm{i}\hat{l}_y = \hbar\,\mathrm{e}^{\mathrm{i}\varphi}(\partial/\partial\theta + \mathrm{i}\cot\theta\,\partial/\partial\varphi) \tag{B75a}$$

$$\hat{l}_x - \mathrm{i}\hat{l}_y = \hbar\,\mathrm{e}^{-\mathrm{i}\varphi}(-\partial/\partial\theta + \mathrm{i}\cot\theta\,\partial/\partial\varphi) \tag{B75b}$$

We find the expression for $\hat{l}^2$ as follows. Using the first of the relations (B72) we obtain

$$\begin{aligned}(\hat{l}_x + \mathrm{i}\hat{l}_y)(\hat{l}_x - \mathrm{i}\hat{l}_y) &= \hat{l}_x^2 + \hat{l}_y^2 - \mathrm{i}(\hat{l}_x\hat{l}_y - \hat{l}_y\hat{l}_x)\\ &= \hat{l}_x^2 + \hat{l}_y^2 + \hbar\hat{l}_z\end{aligned} \tag{B76}$$

Adding $\hat{l}_z^2$ to the left and right hand sides of the above equation we find

$$\hat{l}^2 = (\hat{l}_x + \mathrm{i}\hat{l}_y)(\hat{l}_x - \mathrm{i}\hat{l}_y) + \hat{l}_z^2 - \hbar\hat{l}_z \tag{B77a}$$

Similarly

$$\hat{l}^2 = (\hat{l}_x - \mathrm{i}\hat{l}_y)(\hat{l}_x + \mathrm{i}\hat{l}_y) + \hat{l}_z^2 + \hbar\hat{l}_z \tag{B77b}$$

Substituting (B74) and (B75) into (B77) we obtain, after a little calculation, the following expression for $\hat{l}^2$

$$\hat{l}^2 = -\hbar^2\left[\frac{\partial^2}{\partial\theta^2} + \frac{\cos\theta}{\sin\theta}\frac{\partial}{\partial\theta} + \frac{1}{\sin^2\theta}\frac{\partial^2}{\partial\varphi^2}\right] \tag{B78}$$

To find the eigenvalues and the corresponding (common) set of eigenfunctions of $\hat{l}^2$ and $\hat{l}_z$ we proceed as follows. It is evident from (B74) that the eigenfunctions of $\hat{l}_z$ will have the form: $f(r,\theta)\Phi(\varphi)$, where $f(r,\theta)$ is an arbitrary function of r and θ, and $\Phi(\varphi)$ satisfies the equation

$$\frac{\hbar}{\mathrm{i}}\frac{\mathrm{d}\Phi}{\mathrm{d}\varphi} = m\hbar\Phi(\varphi) \tag{B79}$$

where we have chosen, without any loss of generality, to seek the eigenvalues of $\hat{l}_z$ in the form: $m\hbar$. The solution of (B79) is

$$\Phi(\varphi) = A\,\mathrm{e}^{\mathrm{i}m\varphi} \tag{B80}$$

If the above function is to be single-valued, we must have $\Phi(\varphi + 2\pi) = \Phi(\varphi)$, which is possible only if m is zero or an integer (positive or negative). Therefore the eigenvalues of $\hat{l}_z$ are

$$m\hbar,\ m = 0, \pm 1, \pm 2, \pm 3, \ldots \tag{B81}$$

and the corresponding eigenfunctions are given by

$$\Phi_m(\varphi) = (2\pi)^{-1/2}\,\mathrm{e}^{\mathrm{i}m\varphi} \tag{B82}$$

where we have put $A = (2\pi)^{-1/2}$. One can easily verify that

$$\int_0^{2\pi} \Phi_m^*(\varphi)\Phi_{m'}(\varphi)\,d\varphi = 1, \quad \text{if } m = m' \\ = 0, \quad \text{if } m \neq m' \tag{B83}$$

The expression (B78) allows us to seek the eigenfunctions of $\hat{l}^2$ in the following form: $R(r)X(\theta)\Phi_m(\varphi)$, where $R(r)$ is an arbitrary function of r. In this way, the eigenvalue equation for $\hat{l}^2$

$$\hat{l}^2 R(r)X(\theta)\Phi_m(\varphi) = \hbar^2 l(l+1)R(r)X(\theta)\Phi_m(\varphi) \tag{B84}$$

reduces to the following equation for $X(\theta)$

$$\frac{d^2X}{d\theta^2} + \frac{\cos\theta}{\sin\theta}\frac{dX}{d\theta} - \frac{m^2}{\sin^2\theta}X + l(l+1)X = 0 \tag{B85}$$

where we have chosen, without any loss of generality, to seek the eigenvalued of $\hat{l}^2$ in the form: $\hbar^2 l(l+1)$. (B85) is a well known equation in mathematical physics. It has solutions which are finite, continuous and single-valued, only for positive integer values of $l \geqslant |m|$. They are denoted by (we assume that $m \geqslant 0$)

$$X_{lm} = N_{lm} P_l^m(\cos\theta) \tag{B86}$$

where N_{lm} is a normalisation constant and P_l^m are the so-called associated Legendre polynomials defined by

$$P_l^m(\cos\theta) = \frac{1}{2^l l!}(\sin\theta)^m \frac{d^{l+m}}{(d\cos\theta)^{l+m}}(\cos^2\theta - 1)^l \tag{B87}$$

The associated Legendre polynomials satisfy the relations

$$\int_0^{\pi} [P_l^m(\cos\theta)]^2 \sin\theta\,d\theta = \frac{2}{2l+1}\frac{(l+m)!}{(l-m)!} \tag{B88}$$

and

$$\int_0^{\pi} P_l^m(\cos\theta)P_{l'}^m(\cos\theta)\sin\theta\,d\theta = 0, \quad \text{if } l \neq l' \tag{B89}$$

which tells us that associated Legendre polynomials with different l and the same m are orthogonal. Using (B88) we can normalise X_{lm} so that

$$\int_0^{\pi} [X_{lm}(\theta)]^2 \sin\theta\,d\theta = 1 \tag{B90}$$

We obtain

$$X_{lm} = (-1)^m[\tfrac{1}{2}(2l+1)(l-m)!/(l+m)!]^{1/2} P_l^m(\cos\theta) \tag{B91}$$

where $m \geqslant 0$. The factor of $(-1)^m$ in (B91) is a matter of convention. We note that (B85) depends on the square of m and that therefore the same solution is obtained for m and $-m$. For conventional reasons we need not go into here, we write the solutions corresponding to negative values of m as

$$X_{l-|m|}(\theta) = (-1)^m X_{l|m|}(\theta) \tag{B92}$$

We conclude from the above that the eigenfunctions of $\hat{l}^2$ and $\hat{l}_z$ are given by

$$Y_{lm}(\theta, \varphi) \equiv X_{lm}(\theta)\Phi_m(\varphi) \tag{B93}$$

where $l = 0, 1, 2, 3, \ldots$; and $m = -l, -l+1, \ldots, l-1, l$. The corresponding eigenvalues of $\hat{l}^2$ and $\hat{l}_z$ are $\hbar^2 l(l+1)$ and $m\hbar$ respectively. We have

$$\begin{aligned} \hat{l}^2 R(r) Y_{lm}(\theta, \varphi) &= \hbar^2 l(l+1) R(r) Y_{lm}(\theta, \varphi) \\ \hat{l}_z R(r) Y_{lm}(\theta, \varphi) &= m\hbar R(r) Y_{lm}(\theta, \varphi) \end{aligned} \tag{B94}$$

where $R(r)$ is an arbitrary function of r. $Y_{lm}(\theta, \varphi)$ are the so-called spherical harmonics introduced in Section 2.2. We note their following property

$$\begin{aligned} \int_0^{2\pi}\int_0^{\pi} Y^*_{l'm'}(\theta, \varphi) Y_{lm}(\theta, \varphi) \sin\theta \, d\theta \, d\varphi &= 1, \text{ if } l = l' \text{ and } m = m' \\ &= 0 \text{ otherwise} \end{aligned} \tag{B95}$$

which follows from equations (B83), (B89) and (B90). They are also a complete set of functions: any function $f(\theta, \varphi)$, defined over $0 \leqslant \theta \leqslant \pi$ and $0 \leqslant \varphi < 2\pi$ can be written as

$$f(\theta, \varphi) = \sum_{l=0}^{\infty} \sum_{m=-l}^{l} c_{lm} Y_{lm}(\theta, \varphi) \tag{B96}$$

where

$$c_{lm} = \int_0^{2\pi}\int_0^{\pi} Y^*_{lm}(\theta, \varphi) f(\theta, \varphi) \sin\theta \, d\theta \, d\varphi$$

More about the spherical harmonics and their properties can be found in many books of mathematics and theoretical physics because of their many applications in a great variety of problems.

B.6.2 SPIN AND SPIN-DEPENDENT OPERATORS

We shall consider only fermions of spin $s = \frac{1}{2}$. A spin-state is described by a spinor (see Section 2.10)

$$\underline{u} = \begin{pmatrix} u(1) \\ u(2) \end{pmatrix} \tag{B97}$$

which has the form of a column matrix (a matrix of one column and two rows); $u(1)$ and $u(2)$, the spin-up and spin-down components of the spinor, are in general complex numbers. 'Operating' on a spinor transforms it into another spinor; the operator can be written as a 2×2 matrix[6]

$$\underline{\underline{A}} = \begin{pmatrix} A_{11} & A_{12} \\ A_{21} & A_{22} \end{pmatrix} \tag{B98}$$

and the operation can be effected by using the rules of matrix multiplication as follows

$$\underline{v} = \underline{\underline{A}}\,\underline{u} = \begin{pmatrix} A_{11}u(1) + A_{12}u(2) \\ A_{21}u(1) + A_{22}u(2) \end{pmatrix} \tag{B99}$$

We often write down the components of $\underline{v}$ as follows

$$v(i) = A_{ij} u(j), \quad i = 1, 2 \tag{B100}$$

where a sum over the repeated index (j) is implied (we read (B100) as: $v(i) = A_{i1}u(1) + A_{i2}u(2)$). The indices i and j are often referred to as spin indices.

$\underline{\underline{B}}\,\underline{\underline{A}}\,\underline{u}$, where $\underline{\underline{A}}$ and $\underline{\underline{B}}$ are given operators (matrices), is obtained by two successive operations, first of $\underline{\underline{A}}$ on $\underline{u}$ and then of $\underline{\underline{B}}$ on $\underline{\underline{A}}\,\underline{u}$. The operation of $\underline{\underline{B}}\,\underline{\underline{A}}$ on any $\underline{u}$ can be described by a single matrix $\underline{\underline{C}}$ (the product of $\underline{\underline{B}}$ and $\underline{\underline{A}}$ in the given order) defined by

$$C_{ij} = B_{ik}A_{kj} \tag{B101}$$

We always assume a summation over repeated indices, e.g., $C_{12} = B_{11}A_{12} + B_{12}A_{22}$.

If two operators (matrices) $\underline{\underline{A}}$ and $\underline{\underline{B}}$ are such that $\underline{\underline{A}}\,\underline{\underline{B}} = \underline{\underline{B}}\,\underline{\underline{A}}$ we say that they commute. When $\underline{\underline{A}}\,\underline{\underline{B}} \neq \underline{\underline{B}}\,\underline{\underline{A}}$, which is more often the case, we say that they do not commute. The theorem stated in section (B4), namely that if two operators commute, they have a common set of eigenfunctions and vice-versa, applies here as well.

In the present case the eigenvalue-equation (B53) takes the form

$$\underline{\underline{A}}\,\underline{u}_\lambda = a_\lambda \underline{u}_\lambda \tag{B102}$$

where a_λ (a real number) is the eigenvalue and $\underline{u}_\lambda$ (a spinor) the corresponding eigenfunction. We can write (B102) explicitly as follows

$$\begin{aligned} (A_{11} - a_\lambda)u_\lambda(1) + A_{12}u_\lambda(2) &= 0 \\ A_{21}u_\lambda(1) + (A_{22} - a_\lambda)u_\lambda(2) &= 0 \end{aligned} \tag{B103}$$

The above system of two algebraic equations has non-trivial (non-zero) solutions only when a_λ is equal, respectively, to one or the other of the two roots, a_λ, $\lambda = 1, 2$, of the following equation

$$(A_{11} - a_\lambda)(A_{22} - a_\lambda) - A_{12}A_{21} = 0 \tag{B104}$$

The solutions a_λ, $\lambda = 1, 2$ of (B104), give the eigenvalues of $\underline{\underline{A}}$. To each one corresponds an eigenfunction, $\underline{u}_\lambda$, $\lambda = 1, 2$, which is obtained by solving (B103) with a_λ equal to the given eigenvalue. The two spinor-eigenfunctions, which are orthogonal to each other in the following sense

$$(\underline{u}_\lambda, \underline{u}_{\lambda'}) \equiv u^*_\lambda(1)u_{\lambda'}(1) + u^*_\lambda(2)u_{\lambda'}(2) = 0, \quad \text{if } \lambda \neq \lambda', \tag{B105}$$

can be normalised as follows

$$(\underline{u}_\lambda, \underline{u}_\lambda) \equiv u^*_\lambda(1)u_\lambda(1) + u^*_\lambda(2)u_\lambda(2) = 1 \tag{B106}$$

It follows from the above that the most general spinor state can be written as

$$\begin{aligned} \underline{u} &= c_1\underline{u}_1 + c_2\underline{u}_2 \\ c_\lambda &= (\underline{u}_\lambda, \underline{u}), \quad \lambda = 1, 2 \end{aligned} \tag{B107}$$

where $(\underline{u}, \underline{v})$ is defined, quite generally, by

$$\begin{aligned} (\underline{u}, \underline{v}) &= u^*(i)v(i) \\ &= u^*(1)v(1) + u^*(2)v(2) \end{aligned} \tag{B108}$$

We complete the above introduction to the matrix representation of spinors and operators acting on them, with the definition of the mean value over $\underline{u}$ of a physical quantity represented by $\underline{\underline{A}}$. This is given by

$$\begin{aligned}\langle A\rangle_{\underline{u}} &= (\underline{u}, \underline{\underline{A}}\underline{u}) \\ &= u^*(i)A_{ij}u(j) \qquad \text{(B109)} \\ &= u^*(1)(A_{11}u(1) + A_{12}u(2)) + u^*(2)(A_{21}u(1) + A_{22}u(2))\end{aligned}$$

We can now proceed with the definition of the operators $\hat{\sigma}_x$, $\hat{\sigma}_y$, $\hat{\sigma}_z$ of the components of the spin angular momentum $\boldsymbol{\sigma}$ introduced in Section 2.10. They are defined as follows

$$\hat{\sigma}_x \equiv \frac{\hbar}{2}\begin{pmatrix}0 & 1\\ 1 & 0\end{pmatrix}, \quad \hat{\sigma}_y \equiv \frac{\hbar}{2}\begin{pmatrix}0 & -\mathrm{i}\\ \mathrm{i} & 0\end{pmatrix}, \quad \hat{\sigma}_z \equiv \frac{\hbar}{2}\begin{pmatrix}1 & 0\\ 0 & -1\end{pmatrix} \qquad \text{(B110)}$$

and from these one obtains

$$\hat{\boldsymbol{\sigma}}^2 = \hat{\sigma}_x^2 + \hat{\sigma}_y^2 + \hat{\sigma}_z^2 = \frac{3\hbar^2}{4}\begin{pmatrix}1 & 0\\ 0 & 1\end{pmatrix} \qquad \text{(B111)}$$

which can be regarded as the operator of the square of the magntiude of the spin angular momentum. Using some matrix algebra, one can easily verify the following commutation relations

$$[\hat{\sigma}_x, \hat{\sigma}_y] = \mathrm{i}\hbar\hat{\sigma}_z, [\hat{\sigma}_y, \hat{\sigma}_z] = \mathrm{i}\hbar\hat{\sigma}_x, [\hat{\sigma}_z, \hat{\sigma}_x] = \mathrm{i}\hbar\hat{\sigma}_y \qquad \text{(B112)}$$

For example,

$$\begin{aligned}\hat{\sigma}_x\hat{\sigma}_y - \hat{\sigma}_y\hat{\sigma}_x &= \begin{pmatrix}0 & w\\ w & 0\end{pmatrix}\begin{pmatrix}0 & -\mathrm{i}w\\ \mathrm{i}w & 0\end{pmatrix} - \begin{pmatrix}0 & -\mathrm{i}w\\ \mathrm{i}w & 0\end{pmatrix}\begin{pmatrix}0 & w\\ w & 0\end{pmatrix} \\ &= \begin{pmatrix}\mathrm{i}w^2 & 0\\ 0 & -\mathrm{i}w^2\end{pmatrix} - \begin{pmatrix}-\mathrm{i}w^2 & 0\\ 0 & \mathrm{i}w^2\end{pmatrix} \\ &= \begin{pmatrix}\mathrm{i}\hbar w & 0\\ 0 & -\mathrm{i}\hbar w\end{pmatrix} = \mathrm{i}\hbar\hat{\sigma}_z\end{aligned}$$

where $w \equiv \hbar/2$. One can also easily verify that

$$[\hat{\boldsymbol{\sigma}}^2, \hat{\sigma}_x] = 0, \quad [\hat{\boldsymbol{\sigma}}^2, \hat{\sigma}_y] = 0, \quad [\hat{\boldsymbol{\sigma}}^2, \hat{\sigma}_z] = 0 \qquad \text{(B113)}$$

We note that the commutation relations (B112) and (B113) are the same, respectively, with the commutation relations (B72) and (B73) of the orbital angular momentum operators. The consequences of these relations are also the same. (B112) tells us that $\hat{\sigma}_x$, $\hat{\sigma}_y$ and $\hat{\sigma}_z$ do not commute, and so do not have a common set of eigenfunctions. Therefore, we cannot, simultaneously, know exactly two different components of the spin. On the other hand, according to (B113), we can obtain a common set of eigenfunctions of $\hat{\boldsymbol{\sigma}}^2$ and one of $\hat{\sigma}_x$, $\hat{\sigma}_y$, $\hat{\sigma}_z$; we choose $\hat{\sigma}_z$. We can readily see that the common set of eigenfunctions of $\hat{\boldsymbol{\sigma}}^2$ and $\hat{\sigma}_z$ consists of the following two spinors

$$\underline{u}_+ = \begin{pmatrix}1\\ 0\end{pmatrix} \quad \text{and} \quad \underline{u}_- = \begin{pmatrix}0\\ 1\end{pmatrix} \qquad \text{(B114)}$$

We have

$$\begin{aligned}\hat{\boldsymbol{\sigma}}^2\underline{u}_+ &= (3\hbar^2/4)\underline{u}_+, \quad \hat{\sigma}_z\underline{u}_+ = (\hbar/2)u_+ \\ \hat{\boldsymbol{\sigma}}^2\underline{u}_- &= (3\hbar^2/4)\underline{u}_-, \quad \hat{\sigma}_z\underline{u}_- = (-\hbar/2)\underline{u}_-\end{aligned}$$

We see that the square of the magnitude of the spin has only one eigenvalue $3\hbar^2/4$, in accordance with Dirac's theory of the electron (see Section 2.10). We remember that the most general spin-state will be a linear sum of $\underline{u}_+$ and $\underline{u}_-$, which means that the most general state will also be an eigenstate of $\hat{\boldsymbol{\sigma}}^2$, with the same eigenvalue $3\hbar^2/4$. The magnitude of the spin is always precisely known.

The two eigenstates of $\hat{\sigma}_z$, described by (B114), are the spin-up and spin-down states introduced in Section 2.11.

The reader can easily verify that the mean values of $\hat{\sigma}_y$ and $\hat{\sigma}_z$ over $\underline{u}_+$ and $\underline{u}_-$ equal zero. For example

$$\hat{\sigma}_y \underline{u}_+ = \frac{\hbar}{2}\begin{pmatrix} 0 & -\mathrm{i} \\ \mathrm{i} & 0 \end{pmatrix}\begin{pmatrix} 1 \\ 0 \end{pmatrix} = \frac{\hbar}{2}\begin{pmatrix} 0 \\ \mathrm{i} \end{pmatrix}$$

and therefore $\langle \hat{\sigma}_y \rangle_{\underline{u}_+} = (\underline{u}_+, \hat{\sigma}_y \underline{u}_+) = 0$.

Finally, we note that the orbital angular momentum operators $\hat{\boldsymbol{l}}^2$, $\hat{l}_z$ and the spin operators $\hat{\boldsymbol{\sigma}}^2$, $\hat{\sigma}_z$ commute with each other; this follows directly from the fact that the former act on the orbital part of the wavefunction and the latter on the spin part. They have a common set of eigenfunctions defined by

$$R(r)Y_{lm}(\theta, \varphi)\underline{u}_s \quad \begin{aligned} & l = 0, 1, 2, \ldots \\ & m = -l, -l+1, \ldots, l-1, l \\ & s = +, - \end{aligned} \tag{B115}$$

where $R(r)$ is an arbitrary function of r. We have

$$\left.\begin{matrix} \hat{\boldsymbol{l}}^2 \\ \hat{l}_z \\ \hat{\sigma}_z \end{matrix}\right\} R(r)Y_{lm}(\theta, \varphi)\underline{u}_s = \left.\begin{matrix} \hbar^2 l(l+1) \\ m\hbar \\ s\dfrac{\hbar}{2} \end{matrix}\right\} R(r)Y_{lm}(\theta, \varphi)\underline{u}_s \tag{B116}$$

We can write the most general spinor wavefunction as follows (see Section 2.11)

$$\underline{f}(\boldsymbol{r}) = \begin{pmatrix} f(\boldsymbol{r}; 1) \\ f(\boldsymbol{r}; 2) \end{pmatrix} = f(\boldsymbol{r}; 1)\begin{pmatrix} 1 \\ 0 \end{pmatrix} + f(\boldsymbol{r}; 2)\begin{pmatrix} 0 \\ 1 \end{pmatrix} \tag{B117}$$

or more simply as: $f(\boldsymbol{r}; i)$ with the understanding that the spin index i takes the values 1, 2. [f may depend of course on time, but in what follows we need not denote this explicitly.)

Let $\Omega(\boldsymbol{r}, \hat{\boldsymbol{p}})$ be an operator which does not involve the spin coordinates; its effect on $\underline{f}$ is described by

$$\underline{g} \equiv \underline{\underline{\Omega}}(\boldsymbol{r}, \hat{\boldsymbol{p}})\underline{f} = \begin{pmatrix} \Omega(\boldsymbol{r}, \hat{\boldsymbol{p}}) & 0 \\ 0 & \Omega(\boldsymbol{r}, \hat{\boldsymbol{p}}) \end{pmatrix}\begin{pmatrix} f(\boldsymbol{r}; 1) \\ f(\boldsymbol{r}; 2) \end{pmatrix} = \begin{pmatrix} \Omega(\boldsymbol{r}, \hat{\boldsymbol{p}})f(\boldsymbol{r}; 1) \\ \Omega(\boldsymbol{r}, \hat{\boldsymbol{p}})f(\boldsymbol{r}; 2) \end{pmatrix} \tag{B118}$$

or, more simply, by: $g(\boldsymbol{r}; i) = \Omega(\boldsymbol{r}, \hat{\boldsymbol{p}})f(\boldsymbol{r}; i)$.

Let $\underline{\underline{A}}$ be an operator, such as $\hat{\sigma}_x$, $\hat{\sigma}_y$ etc., which acts only on the spin coordinates; its effect on $\underline{f}$ is described by

$$\underline{g} = \underline{\underline{A}}\underline{f} = \begin{pmatrix} A_{11} & A_{12} \\ A_{21} & A_{22} \end{pmatrix}\begin{pmatrix} f(\boldsymbol{r}; 1) \\ f(\boldsymbol{r}; 2) \end{pmatrix} = \begin{pmatrix} A_{11}f(\boldsymbol{r}; 1) + A_{12}f(\boldsymbol{r}; 2) \\ A_{21}f(\boldsymbol{r}; 1) + A_{22}f(\boldsymbol{r}; 2) \end{pmatrix} \tag{B119}$$

or, more simply, by: $g(\boldsymbol{r}; i) = A_{ij}f(\boldsymbol{r}; j) = A_{i1}f(\boldsymbol{r}; 1) + A_{i2}f(\boldsymbol{r}; 2)$.

The most general operator, however, will have the form

$$\underline{\underline{\Omega}}(\boldsymbol{r}, \hat{\boldsymbol{p}}) = \begin{pmatrix} \Omega_{11}(\boldsymbol{r}, \hat{\boldsymbol{p}}) & \Omega_{12}(\boldsymbol{r}, \hat{\boldsymbol{p}}) \\ \Omega_{21}(\boldsymbol{r}, \hat{\boldsymbol{p}}) & \Omega_{22}(\boldsymbol{r}, \hat{\boldsymbol{p}}) \end{pmatrix} \tag{B120}$$

where $\Omega_{ij}(\boldsymbol{r}, \boldsymbol{p})$ is some function of $\boldsymbol{r}$ and $\boldsymbol{p}$ as in (B49), and its effect on $\underline{f}$ will be described by

$$\underline{g} = \underline{\underline{\Omega}}(\boldsymbol{r}, \hat{\boldsymbol{p}})\underline{f} = \begin{pmatrix} \Omega_{11}(\boldsymbol{r}, \hat{\boldsymbol{p}})f(\boldsymbol{r}; 1) + \Omega_{12}(\boldsymbol{r}, \hat{\boldsymbol{p}})f(\boldsymbol{r}; 2) \\ \Omega_{21}(\boldsymbol{r}, \hat{\boldsymbol{p}})f(\boldsymbol{r}; 1) + \Omega_{22}(\boldsymbol{r}, \hat{\boldsymbol{p}})f(\boldsymbol{r}; 2) \end{pmatrix} \tag{B121}$$

or, simply, by: $g(\boldsymbol{r}; i) = \Omega_{ij}(\boldsymbol{r}, \hat{\boldsymbol{p}})f(\boldsymbol{r}; j)$.

The interaction between the orbital and spin motions of an electron in a one-electron atom (see Sections 2.10 and 2.13) is described by the following operator

$$\begin{aligned} U_{ls}(r) &= \frac{A}{r^3}\{\hat{l}_x\hat{\sigma}_x + \hat{l}_y\hat{\sigma}_y + \hat{l}_z\hat{\sigma}_z\} \qquad\qquad \text{(B122)} \\ &= \frac{A}{r^3}\left\{\frac{\hbar}{2}\begin{pmatrix} 0 & \hat{l}_x \\ \hat{l}_x & 0 \end{pmatrix} + \frac{\hbar}{2}\begin{pmatrix} 0 & -\mathrm{i}\hat{l}_y \\ \mathrm{i}\hat{l}_y & 0 \end{pmatrix} + \frac{\hbar}{2}\begin{pmatrix} \hat{l}_z & 0 \\ 0 & -\hat{l}_z \end{pmatrix}\right\} \\ &= \frac{A}{r^3}\frac{\hbar}{2}\begin{pmatrix} \hat{l}_z & \hat{l}_x - \mathrm{i}\hat{l}_y \\ \hat{l}_x + \mathrm{i}\hat{l}_y & -\hat{l}_z \end{pmatrix} \\ A &\equiv Ze^2/(8\pi\epsilon_O m^2c^2) \end{aligned}$$

which obviously has the form (B120).

The mean value of $\underline{\underline{\Omega}}(\boldsymbol{r}, \hat{\boldsymbol{p}})$ over the state $\underline{f}$ is, accordingly, defined by

$$\begin{aligned} \langle\Omega\rangle &= (\underline{f}, \hat{\Omega}\underline{f}) \\ &= \int f^*(\boldsymbol{r}; i)\Omega_{ij}(\boldsymbol{r}, \hat{\boldsymbol{p}})f(\boldsymbol{r}; j)\,\mathrm{d}V \end{aligned} \tag{B123}$$

where $\mathrm{d}V = \mathrm{d}x\,\mathrm{d}y\,\mathrm{d}z$ and a summation over i and j $(i, j = 1, 2)$ is implied as usual. Quite often we write the above expression as follows

$$\langle\Omega\rangle = \int f^*(q)\hat{\Omega}(q)f(q)\,\mathrm{d}q \tag{B124}$$

where q stands for the spatial and spin coordinates and $\mathrm{d}q$ includes integration over all space of the spatial coordinates (x, y, z) and summation over the spin coordinates (indices).

More generally, we may write

$$(\underline{f}, \underline{g}) = \int f^*(q)g(q)\,\mathrm{d}q \equiv \sum_{i=1}^{2}\int f^*(\boldsymbol{r}; i)g(\boldsymbol{r}; i)\,\mathrm{d}V \tag{B125}$$

$$(\underline{f}, \hat{\Omega}\underline{g}) \equiv \int f^*(q)\hat{\Omega}(q)g(q)\,\mathrm{d}q = \sum_{i=1}^{2}\sum_{j=1}^{2}\int f^*(\boldsymbol{r}; i)\Omega_{ij}(\boldsymbol{r}; \hat{\boldsymbol{p}})g(\boldsymbol{r}; j)\,\mathrm{d}V \tag{B126}$$

B.7 Motion in a spherically symmetric field

The Hamiltonian of a particle of mass m in a spherically (centrally) symmetric potential field has the form

$$\hat{H} = -\frac{\hbar^2}{2m}\nabla^2 + U(r) \tag{B127}$$

which, using equation (B5) and (B78), we can write

$$\hat{H} = -\frac{\hbar^2}{2m}\left(\frac{\partial^2}{\partial r^2} + \frac{2}{r}\frac{\partial}{\partial r}\right) + \frac{2}{2mr^2}\hat{l}^2 + U(r) \tag{B128}$$

One can readily verify that the above Hamiltonian commutes with $\hat{l}^2$, $\hat{l}_z$ and with $\hat{\boldsymbol{\sigma}}^2$, $\hat{\sigma}_z$ (if the particle has spin $s = \frac{1}{2}$), which implies that its eigenfunctions will have the form (B115); the energy eigenvalues and corresponding eigenfunctions will be obtained by solving the following equation (in the case of a spinless particle $\underline{u}_s$ is dropped):

$$\hat{H}R(r)Y_{lm}(\theta, \varphi)\underline{u}_s = ER(r)Y_{lm}(\theta, \varphi)\underline{u}_s \tag{B129}$$

which, by using (B116), reduces to

$$-\frac{\hbar^2}{2m}\left(\frac{d^2}{dr^2} + \frac{2}{r}\frac{d}{dr}\right)R + \left(\frac{\hbar^2 l(l+1)}{2mr^2} + U(r)\right)R = ER(r) \tag{B130}$$

an ordinary differential equation involving only the radial part $R(r)$ of the eigenfunction. We note that only the quantum number l enters the above equation, which implies that the energy eigenvalues will have a $2(2l + 1)$-degeneracy deriving from the $(2l + 1)$ values of m $(-l, -l + 1, \ldots, l)$ and the two orientations of the spin $(s = +, -)$ allowed by (B115) for a given l.

Certain properties of the eigenvalue-problem (B130) are made apparent by the substitution

$$R(r) = u(r)/r \tag{B131}$$

which transforms (B130) into the following equation

$$-\frac{\hbar^2}{2m}\frac{d^2u}{dr^2} + U_l(r)u(r) = Eu(r) \tag{B132}$$

where

$$U_l(r) = U(r) + \frac{\hbar^2 l(l+1)}{2mr^2} \tag{B133}$$

We note that, since $R(r)$ must be everywhere finite, $u(r)$ must vanish at the origin, i.e.

$$u(0) = 0 \tag{B134}$$

We observe that (B132) is formally identical with Schrödinger's time-independent equation for a particle moving in an one-dimensional potential field defined by (B133). We have already stated, in Section 1.4.2, a theorem which tells us that the energy levels of bounded in space one-dimensional motion are not degenerate. From this is follows that the discrete energy levels to be had from the solution of (B130) can be listed as follows: E_{nl}, $n = 1, 2, 3, \ldots$ in order of increasing energy, n being the quantum number which defines the energy and the radial part $R_{nl}(r)$, a real function[7] of r, of the corresponding eigenfunction. The normalisation of this (a bound) eigenfunction is effected by

$$\int_0^{2\pi}\int_0^{\pi}\int_0^{\infty} [R_{nl}(r)]^2 |Y_{lm}(\theta, \varphi)|^2 r^2\, dr \sin\theta\, d\theta\, d\varphi = 1 \tag{B135}$$

which, using (B95), reduces to

$$\int_0^\infty [R_{nl}(r)]^2 r^2 \, dr = \int_0^\infty [u_{nl}(r)]^2 \, dr = 1 \tag{B136}$$

B.7.1 FREE MOTION (SPHERICAL COORDINATES)

The energy-eigenfunctions of a free particle, which are also eigenfunctions of the square of the orbital angular momentum and its z-component, have the form (apart from a spin-factor, if the particle has spin):

$$R_{El}(r) Y_{lm}(\theta, \varphi) \tag{B137}$$

where E, the energy of the particle, can take any value between zero and infinity. The radial part of the wavefunction must satisfy (B130) with $U(r) = 0$, i.e.

$$\left(\frac{d^2}{dr^2} + \frac{2}{r}\frac{d}{dr}\right) R(r) + \left(k^2 - \frac{l(l+1)}{r^2}\right) R(r) = 0 \tag{B138}$$

where $k = (2mE)^{1/2}/\hbar$. The above, a well known equation of mathematical physics, has, for given k and l, two linearly independent solutions: the first one, the so-called spherical Bessel function, denoted by $j_l(kr)$, is finite at the origin ($r = 0$); while the second, known as the spherical Hankel function and denoted by $h_l^+(kr)$, is not. Obviously, the sought eigenfunction is the physically acceptable solution, given by

$$j_l(kr) Y_{lm}(\theta, \varphi) \tag{B139}$$

in agreement with (2.5).

We have seen (in Section 2.3.2) that, although

$$h_l^+(kr) Y_{lm}(\theta, \varphi) \tag{B140}$$

do not, on their own, constitute physically acceptable wavefunctions, they are very useful in the description of scattering states.

Another set of useful functions is given by

$$h_l^+(iqr) Y_{lm}(\theta, \varphi) \tag{B141}$$

where $h_l^+(iqr)$, the Hankel functions of imaginary argument, are solutions of (B138) with $k = iq$, $0 < q < \infty$ which decay exponentially as $r \to \infty$.

The reader can verify the above for the Bessel and Hankel functions listed in Table 2.2.

B.7.2 THE SPHERICAL POTENTIAL WELL

This is defined by

$$\begin{aligned} U(r) &= -U_O, \quad r < a \\ &= 0, \qquad r > a \end{aligned} \tag{B142}$$

The discrete energy levels of a particle of mass m in the above field and the corresponding bound eigenstates are to be found by solving the eigenvalue problem

(B130). In the energy region $-U_O < E < 0$, where the above levels are to be found, equation (B130) takes the form

$$\left(\frac{d^2}{dr^2} + \frac{2}{r}\frac{d}{dr}\right)R(r) + \left(k^2 - \frac{l(l+1)}{r^2}\right)R(r) = 0, \quad r < a \qquad \text{(B143a)}$$

$$\left(\frac{d^2}{dr^2} + \frac{2}{r}\frac{d}{dr}\right)R(r) + \left((iq)^2 - \frac{l(l+1)}{r^2}\right)R(r) = 0, \quad r > a \qquad \text{(B143b)}$$

where

$$k = [2m(U_O + E)]^{1/2}/\hbar \quad \text{and} \quad q = (2m|E|)^{1/2}/\hbar \qquad \text{(B144)}$$

According to the results of the previous section, the solution of (B143a) which is finite at the origin is

$$R(r) = Aj_l(kr), \quad r < a \qquad \text{(B145a)}$$

and the solution of (B143b) which has the right asymptotic form (it decays exponentially as $r \to \infty$) is

$$R(r) = Bh_l^+(iqr), \quad r > a \qquad \text{(B145b)}$$

where A and B are constants to be determined. The continuity of $R(r)$ and its derivative at $r = a$, leads to the following equation

$$[j_l(ka)]'/j_l(ka) = [h_l^+(iqa)]'/h_l^+(iqa) \qquad \text{(B146)}$$

where $[f(pa)]'$ gives the value of $df(pr)/dr$ at $r = a$. The energies (if any) for which (B146) is satisfied are the energy-eigenvalues for the given l, which we can present schematically as in Fig. 2.8. The radial parts of the corresponding eigenfunctions are given by (B145); B and A are determined by the continuity of R at $r = a$ ((B146) secures the continuity of dR/dr) and the normalisation condition (B136).

Let us investigate (B146) for $l = 0$. Substituting in (B146) the corresponding spherical Bessel and Hankel functions (see Table 2.2) we obtain

$$k \cot ka = -q = -(2mU_O/\hbar^2 - k^2)^{1/2} \qquad \text{(B147)}$$

Squaring both sides of the above equation, using the identity $\sin^2 x + \cos^2 x = 1$, and taking the square root of both sides at the end of the manipulation, we can write the above equation as follows

$$[\hbar^2/(2ma^2 U_O)]^{1/2} ka = \pm \sin ka \qquad \text{(B148)}$$

and we must take the roots of (B148) for which $\cot ka < 0$, according to (B147). The energy eigenvalues for $l = 0$ are obtained from the above roots (k_1, k_2, etc.) through the equation (B144) for k. Let us determine the lowest of these energy eigenvalues. In Fig. B2 we plot the left and right-hand sides of (B148) as functions of $x \equiv ka$. The parts of the $y = \pm\sin x$ curves where $\cot x < 0$ are shown by continuous lines. It is evident from this figure that, if the depth U_O of the well is small enough, there will be no negative energy levels; the well will not hold the particle. We shall obtain the first such level when with increasing U_O the slope of the straight line in the diagram becomes small enough to intersect the $y = \sin x$ curve at $x = \pi/2$, (for $x \geqslant \pi/2$, $\cot x < 0$ as it must). Putting $[\hbar^2/(2ma^2 U_O)]^{1/2}\pi/2 = 1$, we obtain

$$U_{O,\min} = \pi^2\hbar^2/(8ma^2) \qquad \text{(B149)}$$

We note that as the radius a of the well beocmes smaller, its depth must increase in

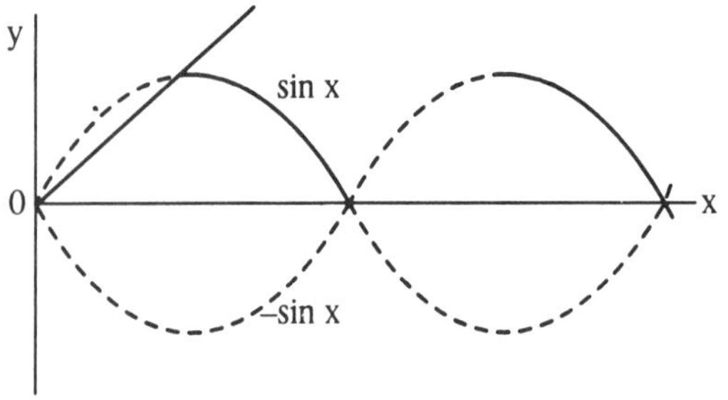

Figure B2. Graphical solution of equation (B148)

order to hold the particle in a bound state. The first energy level, when it first appears, has the value $E_1 = -U_{O,\text{min}} + \hbar^2 k_1^2/2m = -U_{O,\text{min}} + (\hbar^2/2m)\ (\pi/2a)^2 = 0$. As the depth of the well increases, the E_1 level descends and a second level appears when $U_O = 9U_{O,\text{min}}$.

One can easily show that the lowest energy level with $l > 0$ lies higher than that of $l = 0$ calculated above (see Fig. 2.8). This comes about because the effective one-dimensional potential field described by (B133) for $l > 0$ is obtained by adding a positive term to that for $l = 0$; a positive term which increases with l. We may summarise as follows. A three-dimensional spherical well will hold a particle (in a bound state) only when its depth exceeds a certain value; this is very different from what happens in an one-dimensional rectangular well which has at least one-discrete energy level (one bound state) however small its depth (see Section B.2.1). The first (bound) state to appear as the depth of the well increases, which by descending, becomes the ground state of the particle in the given well as the depth of the well increases further, is an s-state (it has $l = 0$).

We conclude this section with a brief reference to the scattering states of the spherical well. These correspond to a continuous spectrum of energy levels ($E > 0$) and can be calculated in essentially the same manner as those, of Section 2.3.2, for the hard-sphere barrier (see problem 2.2).

B.8 Addition of angular momenta

The classical orbital momentum of a system of N particles is defined by

$$\boldsymbol{L} = \sum_{j=1}^{N} \boldsymbol{l}_j = \sum_{j=1}^{N} \boldsymbol{r}_j \times \boldsymbol{p}_j \qquad \text{(B150)}$$

where $\boldsymbol{r}_j$ and $\boldsymbol{p}_j$ denote respectively the position vector and the linear momentum of the jth particle. We know that when the only forces acting on the particles are internal forces, which appear in pairs of two opposite forces of the same magnitude (as will be the case, for example, in a system of charged particles), the orbital angular momentum of the system will be a constant of the motion: a conserved quantity. The same will be true if the particles move in a centrally symmetric field (the same for all particles). The above remains true in quantum mechanics; which implies that the operator for the total orbital angular momentum of the system, defined by

$$\hat{\boldsymbol{L}} = \sum_{j=1}^{N} \hat{\boldsymbol{l}}_j = \sum_{j=1}^{N} \boldsymbol{r}_j \times \hat{\boldsymbol{p}}_j \tag{B151}$$

commutes with the Hamiltonian

$$\hat{H} = -\sum_{j=1}^{N} \frac{\hbar^2}{2m_j} \nabla_j^2 + U(\boldsymbol{r}_1, \ldots, \boldsymbol{r}_N) \tag{B152}$$

of the system of the N particles.

Using (B68) we can write the components of $\hat{\boldsymbol{L}}$ along the x, y and z directions as follows

$$\begin{gathered} \hat{L}_x = \sum_{j=1}^{N} \frac{\hbar}{\mathrm{i}}\left(y_j \frac{\partial}{\partial z_j} - z_j \frac{\partial}{\partial y_j}\right), \quad \hat{L}_y = \sum_{j=1}^{N} \frac{\hbar}{\mathrm{i}}\left(z_j \frac{\partial}{\partial x_j} - x_j \frac{\partial}{\partial z_j}\right), \\ \hat{L}_z = \sum_{j=1}^{N} \frac{\hbar}{\mathrm{i}}\left(x_j \frac{\partial}{\partial y_j} - y_j \frac{\partial}{\partial x_j}\right) \end{gathered} \tag{B153}$$

The operator of the square of the magnitude of the total orbital angular momentum is, accordingly, defined by

$$\hat{\boldsymbol{L}}^2 = \hat{L}_x^2 + \hat{L}_y^2 + \hat{L}_z^2 \tag{B154}$$

One can show that $\hat{L}_x$, $\hat{L}_y$, $\hat{L}_z$ and $\hat{\boldsymbol{L}}^2$ satisfy the same commutation relations, (B72) and (B73), as the corresponding operators for one particle. We have

$$[\hat{L}_x, \hat{L}_y] = \mathrm{i}\hbar\hat{L}_z,\ [\hat{L}_y, \hat{L}_z] = \mathrm{i}\hbar\hat{L}_x,\ [\hat{L}_z, \hat{L}_x] = \mathrm{i}\hbar\hat{L}_y \tag{B155}$$

$$[\hat{\boldsymbol{L}}^2, \hat{L}_x] = 0,\ [\hat{\boldsymbol{L}}^2, \hat{L}_y] = 0,\ [\hat{\boldsymbol{L}}^2, \hat{L}_z] = 0 \tag{B156}$$

When the particles of the system have spin (we shall assume that they are fermions with spin $s = \frac{1}{2}$), the system will have a total spin, $\boldsymbol{S}$, as well. The operators of the x, y and z components of the total spin are defined by

$$\hat{S}_x = \sum_{j=1}^{N} \hat{\sigma}_{jx},\ \hat{S}_y = \sum_{j=1}^{N} \hat{\sigma}_{jy},\ \hat{S}_z = \sum_{j=1}^{N} \hat{\sigma}_{jz} \tag{B157}$$

where $\hat{\sigma}_{jx}$, $\hat{\sigma}_{jy}$, $\hat{\sigma}_{jz}$, defined by (B110), operate on the spin coordinates of the jth particle. The operator of the square of the magnitude of the total spin is given by

$$\hat{\boldsymbol{S}}^2 = \hat{S}_x^2 + \hat{S}_y^2 + \hat{S}_z^2 \tag{B158}$$

And we can show that the above operators satisfy the same commutation relations, (B112) and (B113), as the corresponding operators for one particle. We have

$$[\hat{S}_x, \hat{S}_y] = \mathrm{i}\hbar\hat{S}_z,\ [\hat{S}_y, \hat{S}_z] = \mathrm{i}\hbar\hat{S}_x,\ [\hat{S}_z, \hat{S}_x] = \mathrm{i}\hbar\hat{S}_y \tag{B159}$$

$$[\hat{\boldsymbol{S}}^2, \hat{S}_x] = 0,\ [\hat{\boldsymbol{S}}^2, \hat{S}_y] = 0,\ [\hat{\boldsymbol{S}}^2, \hat{S}_z] = 0 \tag{B160}$$

Finally, we define the total angular momentum of the system

$$\boldsymbol{J} \equiv \boldsymbol{L} + \boldsymbol{S} \tag{B161}$$

The corresponding quantum-mechanical operators are given by

$$\hat{J}_x = \hat{L}_x + \hat{S}_x,\ \hat{J}_y = \hat{L}_y + \hat{S}_y,\ \hat{J}_z = \hat{L}_z + \hat{S}_z \tag{B162}$$

$$\hat{\boldsymbol{J}}^2 = \hat{J}_x^2 + \hat{J}_y^2 + \hat{J}_y^2 \tag{B163}$$

which can be shown to satisfy the same commutation relations as $\hat{\boldsymbol{L}}$ and $\hat{\boldsymbol{S}}$; we have

$$[\hat{J}_x, \hat{J}_y] = i\hbar\hat{J}_z,\ [\hat{J}_y, \hat{J}_z] = i\hbar\hat{J}_x,\ [\hat{J}_z, \hat{J}_x] = i\hbar\hat{J}_y \tag{B164}$$

$$[\hat{\boldsymbol{J}}^2, \hat{J}_x] = 0,\ [\hat{\boldsymbol{J}}^2, \hat{J}_y] = 0,\ [\hat{\boldsymbol{J}}, \hat{J}_z] = 0 \tag{B165}$$

Now, when the particles of the system under consideration are charged particles of spin $s = \frac{1}{2}$ (e.g., electrons), the Hamiltonian of the system will, in general, contain a spin-orbit interaction term. For the lighter atoms (those listed in Table 2.5) this takes the form (a generalisation of (B122))

$$U_{LS} \propto \{\hat{L}_x\hat{S}_x + \hat{L}_y\hat{S}_y + \hat{L}_z\hat{S}_z\} \tag{B166}$$

Because of the above term, no component of $\hat{\boldsymbol{L}}$ or $\hat{\boldsymbol{S}}$ will commute with the Hamiltonian of the system but the components of $\hat{\boldsymbol{J}}$ do, and so does $\hat{\boldsymbol{J}}^2$. From the above and equations (B164) and (B165) we conclude that $\hat{H}$, $\hat{\boldsymbol{J}}^2$ and $\hat{J}_z$ will have a common set of eigenfunctions and therefore the energy, the square of the total angular momentum of the system and its z-component can be known exactly at the same time. The above holds generally, even when the spin–orbit interaction is not given by (B166) but by a more complicated formula. Its validity derives in the final analysis from the isotropicity of space. Occasionally, for example when the spin–orbit interaction of an atom can be written as in (B166), $\hat{\boldsymbol{L}}^2$ and $\hat{\boldsymbol{S}}^2$ (but not their components) will also commute with $\hat{H}$, $\hat{\boldsymbol{J}}^2$ and $\hat{J}_z$ and then the stationary states of the system can be eigenstates of these quantities also, but this, we should emphasise, is not always the case (e.g., it is not true for the heavier atoms).

The commutation relations obeyed by the angular momentum operators are important also in another respect. It can be shown (the proof of this theorem lies beyond the scope of the present book) that, because $\hat{J}_x$, $\hat{J}_y$, $\hat{J}_z$ and $\hat{\boldsymbol{J}}^2$ satisfy the commutation relations (B164) and (B165) the eigenvalues of $\hat{\boldsymbol{J}}^2$ are given by: $J(J+1)\hbar^2$, where J can be an integer or a half integer ($J = 0, 1/2, 1, 3/2, \ldots$); and for given J, the eigenvalues of $\hat{J}_z$ are given by: $M_j\hbar$, $M_j = -J, -J+1, \ldots, J-1, J$. Since the total orbital angular momentum and the total spin angular momentum operators obey the same commutation relations, the corresponding eigenvalues of $\hat{\boldsymbol{L}}^2$ and $\hat{L}_z$ and those of $\hat{\boldsymbol{S}}^2$ and $\hat{S}_z$ are given by similar formulae. The eigenvalues of $\hat{\boldsymbol{L}}^2$ are given, as we shall see, by $L\ (L+1)\hbar^2$ where L can be any positive integer ($L = 0, 1, 2, \ldots$), and those of $\hat{L}_z$, for given L, are given by $M\hbar$, $M = -L$, $-L+1, \ldots, L-1, L$. (The orbital angular momentum of a single particle is a special case of the above). The eigenvalues of $\hat{\boldsymbol{S}}^2$ are given, as we shall see, by $S(S+1)\hbar^2$ where S can be an integer or half-integer ($S = 0, 1/2, 2, 3/2, \ldots$) and those of $\hat{S}_z$, for given S, are given by $M_s\hbar$, $M_s = -S, -S+1, \ldots, S-1, S$. (The spin of a single particle is a special case of the above).

In many applications we would like to know the eigenstates of the total angular momentum of a system in terms of angular momentum states of the particles which make up the system. Let us see how this works. Let us for the sake of clarity consider, to begin with, two non-interacting distinguishable spinless particles, denoted by '1' and '2', moving in the same centrally symmetric potential field.

Because the particles are distinguishable and non-interacting, an eigenfunction of the energy of the system of the two particles can be written as

$$|n_1, l_1, n_2, l_2; m_1, m_2\rangle = R_{n_1l_1}(r_1)R_{n_2l_2}(r_2)Y_{l_1m_1}(\theta_1, \varphi_1)Y_{l_2m_2}(\theta_2, \varphi_2) \quad \text{(B167)}$$

The corresponding energy-eigenvalue of the system is $E_{n_1l_1} + E_{n_2l_2}$, where $E_{n_1l_1}$ is the energy of '1' in the state $R_{n_1l_1}Y_{l_1m_1}$ and $E_{n_2l_2}$ is the energy of '2' in the state $R_{n_2l_2}Y_{l_2m_2}$. We note also that (B167) is an eigenfunction of $\hat{l}_1^2$ and $\hat{l}_{1z}$ (the corresponding eigenvalues: $l_1(l_1+1)\hbar^2$ and $m_1\hbar$), and of $\hat{l}_2^2$ and $\hat{l}_{2z}$ (the corresponding eigenvalues: $l_2(l_2+1)\hbar^2$ and $m_2\hbar$). For given n_1l_1, n_2l_2, we obtain $(2l_1+1)(2l_2+1)$ states (B167) corresponding to the $(2l_1+1)$ possible values of m_1 $(-l_1, -l_1+1, \ldots, l_1)$ and the $(2l_2+1)$ possible values of m_2 $(-l_2, -l_2+1, \ldots, l_2)$. These $(2l_1+1)(2l_2+1)$ states have the same energy $E_{n_1l_1} + E_{n_2l_2}$, which is therefore $(2l_1+1)(2l_2+1)$-fold degenerate.

Next we note that (B167) is an eigenfunction of the z-component of the total orbital angular momentum of the system. Given that $\hat{L}_z = \hat{l}_{1z} + \hat{l}_{2z}$, we obtain

$$\hat{L}_z|n_1, l_1, n_2, l_2; m_1, m_2\rangle = (m_1 + m_2)\hbar|n_1, l_1, n_2, l_2; m_1, m_2\rangle \quad \text{(B168)}$$

It is also obvious that (B167) cannot, in general, be an eigenfunction of $\hat{L}^2$. We know, however, that $\hat{L}^2$ and $\hat{L}_z$ have a common set of eigenfunctions, which implies that we can obtain linear combinations of the $(2l_1+1)(2l_2+1)$ functions defined by (B167) for given n_1l_1, n_2l_2 which are eigenfunctions of $\hat{L}_z$ *and* $\hat{L}^2$.

One can readily see that the maximum eigenvalue of $\hat{L}^2$ compatible with given l_1 and l_2 is $L(L+1)\hbar^2$ with $L = l_1 + l_2$; we argue as follows: for given L we obtain $(2L+1)$ states corresponding to different eigenvalues $M\hbar$, $M = -L, -L+1, \ldots, L-1, L$, of $\hat{L}_z$; M has to be an integer (it is the sum of two integers m_1 and m_2) and it cannot exceed the maximum value of $m_1 + m_2$ which equals $l_1 + l_2$; and this obviously determines the maximum value of the magnitude of the total orbital angular momentum too. The complete argument (which we shall not give here) tells us that for given l_1 and l_2, the square of the magnitude of the total orbital angular momentum of the two particles obtains the following eigenvalues:

$$\begin{aligned} L(L+1)\hbar^2,\ L &= l_1 + l_2 \\ &= l_1 + l_2 - 1 \\ &\ \ \vdots \\ &= |l_1 - l_2| \end{aligned} \quad \text{(B169)}$$

For given L, we obtain $(2L+1)$ states corresponding to different eigenvalues $M\hbar$ $(M = -L, -L+1, \ldots, L)$ of $\hat{L}_z$. The number of states obtained in this way for the totality of (B169) is: $(2(l_1+l_2)+1) + (2(l_1+l_2-1)+1) + \ldots + (2|l_1-l_2|+1) = (2l_1+1)(2l_2+1)$, as it should be; it is equal to the number of states, for the given l_1 and l_2, described by (B167). The eigenfunction corresponding to certain values of L and M can be written as follows

$$|n_1, l_1, n_2, l_2; L, M\rangle = \sum_{m_1=-l_1}^{l_1} \sum_{m_2=-l_2}^{l_2} C_{LM}(m_1, m_2)|n_1, l_1, n_2, l_2; m_1, m_2\rangle \quad \text{(B170)}$$

where the sum is taken over all states (B167). The coefficients $C_{LM}(m_1, m_2)$, known as the Clebsh–Gordan coefficients, depend only on the angular momentum quantum numbers (they do not depend on n_1 and n_2) and therefore can be determined quite generally. This has been done, and tables of these coefficients can be found in a number of books. It is worth noting (this follows from (B168)) that $C_{LM}(m_1, m_2)$ is different from zero only when $m_1 + m_2 = M$.

The advantage of (B170) over (B167) lies in the following observation. When the two particles do not interact with each other, all states (B167) corresponding to given n_1, l_1, n_2, l_2 have the same energy: $E_{n_1l_1} + E_{n_2l_2}$. The same is true of the states (B170). Introducing a weak interaction between the particles splits the above energy level (see Section B.10). The energy levels of the perturbed system will depend on L (but not on M), and the corresponding eigenfunctions will be given, to a good approximation, by (B170). The eigenfunctions of the Hamiltonian (including the interaction) are also eigenfunctions of $\hat{L}^2$ and $\hat{L}_z$. Moreover, when the perturbation is sufficiently weak, they will also be eigenfunctions of $\hat{l}_1^2$ and $\hat{l}_2^2$ (but not of $\hat{l}_{1z}$ and $\hat{l}_{2z}$), as implied by (B170). When the perturbation is strong, (B170) will not be good enough; the correct energy-eigenfunctions will also be eigenfunctions of $\hat{L}^2$ and $\hat{L}_z$, but they will not be eigenfunctions of $\hat{l}_1^2$ and $\hat{l}_2^2$.

Let us demonstrate the result (B169) by applying it to the case of two particles in p-states ($l_1 = l_2 = 1$). In this case L takes the values: $L = 2, 1, 0$. We call a state of the system with $L = 0, 1, 2, 3$, etc., an S-state, a P-state, a D-state, an F-state, etc. The corresponding states can be written down using (B170) and standard tables of Clebsh–Gordan coefficients. We shall not bother to do so here.

Finally, we can obtain the total orbital angular momentum (the eigenvalues and corresponding eigenfunctions of $\hat{L}^2$ and $\hat{L}_z$) of a system of three particles with certain n_1l_1, n_2l_2 and n_3l_3 respectively, as follows: After evaluating the possible states for the two particles according to equations (B168) and (B170), we combine the eigenfunctions of given L (for the two particles) with those of the third particle to obtain the eigenvalues and corresponding eigenfunctions of $\hat{L}^2$ and $\hat{L}_z$ for the three particles. In so doing we treat the eigenfunctions corresponding to a given value of L (for the two particles) in the same way as we treated those of the one particle in our derivation of equations (B169) and (B170). In this way, combining the $(2(l_1 + l_2) + 1)$ eigenfunctions of the two particles corresponding to $L = l_1 + l_2$, with the $(2l_3 + 1)$ eigenfunctions of the third particle, we obtain eigenfunctions of $\hat{L}^2$ (for the three particles) corresponding to the following set of eigenvalues

$$\begin{aligned} L(L+1)\hbar^2,\ L &= (l_1 + l_2) + l_3 \\ &= (l_1 + l_2) + l_3 - 1 \\ &\ \vdots \\ &= |(l_1 + l_2) - l_3| \end{aligned} \tag{B171}$$

and, of course, to every one of the above values of L correspond $(2L + 1)$ states with different eigenvalues $M\hbar$ $(M = -L, -L + 1, \ldots, L)$ of $\hat{L}_z$. And what we have done for $(l_1 + l_2)$, we repeat for every one of the allowed values of L listed in (B169). The process is laborious but straightforward, and can be extended to a system of any number of particles.

Let us next consider the addition of two spin angular momenta. We assume that the particles are fermions with spin s_1 and s_2 respectively. Assuming that the two spins do not interact with each other or with the orbital motion, we can write the spin part of an eigenstate of the system in the following form

$$\begin{gathered} |s_1, s_2; m_{s1}, m_{s2}\rangle = |s_1, m_{s1}\rangle|s_2, m_{s2}\rangle \\ m_{s1} = -s_1, -s_1 + 1, \ldots, s_1;\ m_{s2} = -s_2, -s_2 + 1, \ldots, s_2 \end{gathered} \tag{B172}$$

In particular, when $s_1 = s_2 = 1/2$, as in the case of two electrons, we have:

$$|\tfrac{1}{2}, \tfrac{1}{2}; \tfrac{1}{2}, \tfrac{1}{2}\rangle = \begin{pmatrix}1\\0\end{pmatrix}_1 \begin{pmatrix}1\\0\end{pmatrix}_2 \quad |\tfrac{1}{2}, \tfrac{1}{2}; -\tfrac{1}{2}, \tfrac{1}{2}\rangle = \begin{pmatrix}0\\1\end{pmatrix}_1 \begin{pmatrix}1\\0\end{pmatrix}_2$$

$$|\tfrac{1}{2}, \tfrac{1}{2}; \tfrac{1}{2}, -\tfrac{1}{2}\rangle = \begin{pmatrix}1\\0\end{pmatrix}_1 \begin{pmatrix}0\\1\end{pmatrix}_2 \quad |\tfrac{1}{2}, \tfrac{1}{2}; -\tfrac{1}{2}, -\tfrac{1}{2}\rangle = \begin{pmatrix}0\\1\end{pmatrix}_1 \begin{pmatrix}0\\1\end{pmatrix}_2$$

where the subindices 1 and 2 refer to particles '1' and '2' respectively. (B172) is an eigenfunction of $\hat{\boldsymbol{\sigma}}_1^2$ (eigenvalue: $s_1(s_1+1)\hbar^2$), of $\hat{\sigma}_{1z}$ (eigenvalue: $m_{s_1}\hbar$), and of $\hat{\boldsymbol{\sigma}}_2^2$ (eigenvalue: $s_2(s_2+1)\hbar^2$) and $\hat{\sigma}_{2z}$ (eigenvalue: $m_{s_2}\hbar$). We note that (B172) is also an eigenfunction of $\hat{S}_z$; given that $\hat{S}_z = \hat{\sigma}_{1z} + \hat{\sigma}_{2z}$, we obtain

$$\hat{S}_z|s_1, s_2; m_{s1}, m_{s2}\rangle = (m_{s1} + m_{s2})\hbar|s_1, s_2; m_{s1}, m_{s2}\rangle \tag{B173}$$

As in the case of orbital angular momenta, we wish to obtain those linear combinations of (B172), which are eigenfunctions of $\hat{S}_z$ and $\hat{S}^2$. The rules for doing so are exactly the same as those for the addition of orbital angular momenta. We find that for given s_1 and s_2, $\hat{S}^2$ obtains the following eigenvalues:

$$S(S+1)\hbar^2, \quad \begin{aligned} S &= s_1 + s_2 \\ &= s_1 + s_2 - 1 \\ &= |s_1 - s_2| \end{aligned} \tag{B174}$$

For given S, we obtain $(2S+1)$ states corresponding to different eigenvalues $M_s\hbar$ ($M_s = -S, -S+1, \ldots, S$) of $\hat{S}_z$. The number of states obtained in this way for the totality of (B174) equals $(2s_1+1)(2s_1+1)$ as it should; it is equal to the number of states, for the given s_1 and s_2, defined by (B172). The corresponding eigenfunctions are given by linear sums of the functions (B172), as for the case of orbital angular momenta. The case of $s_1 = s_2 = \frac{1}{2}$ is easy to work out; we obtain two eigenvalues of $\hat{S}^2$: $S(S+1)\hbar^2$, $S = 0, 1$. For $S = 0$, we have just one state

$$|s_1, s_2; S, M_s\rangle = |\tfrac{1}{2}\tfrac{1}{2}; 0, 0\rangle = \frac{1}{\sqrt{2}}\left\{\begin{pmatrix}1\\0\end{pmatrix}_1\begin{pmatrix}0\\1\end{pmatrix}_2 - \begin{pmatrix}1\\0\end{pmatrix}_2\begin{pmatrix}0\\1\end{pmatrix}_1\right\} \tag{B175}$$

One can easily verify that

$$\hat{S}^2|\tfrac{1}{2}, \tfrac{1}{2}; 0, 0\rangle = 0|\tfrac{1}{2}, \tfrac{1}{2}; 0, 0\rangle$$
$$\hat{S}_z|\tfrac{1}{2}, \tfrac{1}{2}; 0, 0\rangle = 0|\tfrac{1}{2}, \tfrac{1}{2}; 0, 0\rangle$$

For $S = 1$ we have three states

$$|s_1, s_2; S, M_s\rangle = |\tfrac{1}{2}, \tfrac{1}{2}; 1, 1\rangle = \begin{pmatrix}1\\0\end{pmatrix}_1\begin{pmatrix}1\\0\end{pmatrix}_2 \tag{B176a}$$

$$|s_1, s_2; S, M_s\rangle = |\tfrac{1}{2}, \tfrac{1}{2}; 1, 0\rangle = \frac{1}{\sqrt{2}}\left\{\begin{pmatrix}1\\0\end{pmatrix}_1\begin{pmatrix}0\\1\end{pmatrix}_2 + \begin{pmatrix}1\\0\end{pmatrix}_2\begin{pmatrix}0\\1\end{pmatrix}_1\right\} \tag{B176b}$$

$$|s_1, s_2; S, M_s\rangle = |\tfrac{1}{2}, \tfrac{1}{2}; 1, -1\rangle = \begin{pmatrix}0\\1\end{pmatrix}_1\begin{pmatrix}0\\1\end{pmatrix}_2 \tag{B176c}$$

and again one can easily verify that

$$\hat{S}^2|s_1, s_2; S, M_s\rangle = S(S+1)\hbar^2|s_1, s_2; S, M_s\rangle$$
$$\hat{S}_z|s_1, s_2; S, M_s\rangle = \hbar M_s|s_1, s_2; S, M_s\rangle$$

We can obtain the total spin angular momentum (the eigenvalues and corresponding eigenfunctions of $\hat{S}^2$ and $\hat{S}_z$) of a system of three or more particles proceeding in exactly the same way as in the evaluation of the total orbital angular momentum of the system. It is worth noting that, in general, the eigenvalues of $\hat{S}^2$ are given by: $S(S+1)\hbar^2$, where S may be an integer or half-integer ($S = 0, 1/2, 1, 3/2, 2, \ldots$); the corresponding values of $\hat{S}_z$ are in every case given by $M_s\hbar$ ($M_s = -S, -S+1, \ldots, S$).

Finally we can combine the eigenfunctions of the total orbital angular momentum with those of the total spin angular momentum to find those of the total angular momentum (of $\hat{J}^2$ and $\hat{J}_z$). Let us demonstrate this for a system of two electrons. The eigenfunctions of the total orbital angular momentum are given by (B170); in what follows we shall write them in the simpler form: $|L, M\rangle$, where L takes the values (B169) and, for given L, $M = -L, -L+1, \ldots, L$. The eigenfunctions (B175) and (B176) of the total spin angular momentum will also be written in the simpler form: $|S, M_s\rangle$, where S takes the values (B174), i.e. $S = 0, 1$ and, for given S, $M_s = -S, -S+1, \ldots, S$. We introduce the products of these wavefunctions:

$$|L, S; M, M_s\rangle = |L, M\rangle|S, M_s\rangle \tag{B177}$$

(B177) is an eigenfunction of $\hat{L}^2$ (eigenvalue: $L(L+1)\hbar^2$), of $\hat{L}_z$ (eigenvalue: $M\hbar$), of $\hat{S}^2$ (eigenvalue: $S(S+1)\hbar^2$) and of $\hat{S}_z$ (eigenvalue: $M_s\hbar$). It is also an eigenfunction of the z-component of the total angular momentum defined by (B162), with eigenvalue: $(M + M_s)\hbar$, as one can easily verify. As we have done before, when adding spin or orbital angular momenta, we wish to know those combinations of (B177) which are eigenfunctions of $\hat{J}_z$ and $\hat{J}^2$. We find that for given L and S, $\hat{J}^2$ obtains the following eigenvalues

$$\begin{aligned} J(J+1)\hbar^2, \quad J &= L + S \\ &= L + S - 1 \\ &\ \vdots \\ &= |L - S| \end{aligned} \tag{B178}$$

For given J, we obtain $(2J+1)$ states corresponding to different eigenvalues $M_j\hbar$ ($M_j = -J, -J+1, \ldots, J$) of J_z. The number of states obtained in this way for the totality of (B178) equals $(2L+1)(2S+1)$, i.e. it is equal to the number of states (B177) for given L, S. The corresponding eigenfunctions can be written as

$$|L, S; J, M_j\rangle = \sum_{M=-L}^{L} \sum_{M_s=-S}^{S} C_{JM_j}(M, M_s)|L, S; M, M_s\rangle \tag{B179}$$

where $C_{JM_j}(M, M_s)$ are well known (tabulated) Clebsh–Gordan coefficients (we note that $C_{JM_j}(M, M_s)$ is different from zero only when $M + M_s = M_j$). We have already pointed out the physical significance of (B179) (see text following (B166)): the energy eigenstates of the system under consideration will, in general, have the form (B179) and not that of (B177) (see also Section B.10).

We note that (B179) is valid quite generally, i.e. for systems of N particles, where

$N \geqslant 2$. But of course when $N > 2$, the calculation of $|L, S; M, M_s\rangle$ becomes more laborious.

The simplest application of (B178) and (B179) we have already met in the description of spin-orbit coupling and its effect on the energy spectrum and corresponding states of the hydrogen atom, where L and S refer to the orbital and spin angular momentum, respectively, of a single electron (see Section 2.13). We have noted then that the energy levels of the hydrogen atom do not depend on M_j, for that would mean that the energy of the atom depended on the arbitrary choice of the z-axis which, obviously, cannot be the case. This is valid quite generally; the energy eigenvalue corresponding to the state (B179) may depend on L, S and J but it cannot depend on M_j, which means that this eigenvalue is $(2J + 1)$-fold degenerate.

In dealing with systems of N identical particles (e.g., atoms of two or more electrons) a further complication arises from the requirement that the eigenfunctions of the system must be antisymmetric (for fermions) and symmetric (for bosons) when the spatial and spin coordinates of any two particles are permuted with each other (see Section 2.14). We shall consider only fermions here (an analogous treatment applies to bosons).

We note that the angular momentum operators of N particles are invariant (they do not change) under permutation of the coordinates of any two particles. This means that the function obtained by permuting the coordinates of two particles in any of the cited eigenfunctions of such an operator (e.g., (B170) or (B179)) will also be an eigenfunction, corresponding to the same eigenvalue, of this operator. Now, the antisymmetrised version of an eigenfunction of such an operator is a linear sum of functions obtained from the given eigenfunction by permuting the coordinates of pairs of particles as described in Section (2.14), and as such is an eigenfunction of the given operator corresponding to the same eigenvalue. Therefore, when dealing with systems of N fermions (say N electrons) we must replace (B179) by its antisymmetrised version calculated in the manner of Section 2.14. In what follows we shall denote the antisymmetrised version of (B179) by the same symbol; in other words we shall assume that (B179) has been antisymmetrised. Now it may happen that the antisymmetrised version of (B179) vanishes identically; for example the following eigenfunction of two particles

$$R_{nl}(r_1)Y_{lm}(\theta_1, \varphi_1)R_{nl}(r_2)Y_{lm}(\theta_2, \varphi_2)\begin{pmatrix}1\\0\end{pmatrix}_1\begin{pmatrix}1\\0\end{pmatrix}_2$$

will vanish when antisymmetrised. This means that the corresponding state does not exist. In other respects, all our conclusions (reached by assuming distinguishable particles) remain valid. The requirement of antisymmetry has, nevertheless, a profound effect on the energy spectrum of the system as can be seen from the following example.

The atom of carbon has four electrons in closed s-shells and two electrons in its open p-shell. One can easily show that the contributions of a closed shell to the total orbital and total spin angular momentum of an atom vanish. One can see this as follows: in a closed shell, for every electron with a z-component of orbital angular momentum equal to $m\hbar$ there is another with a z-component equal to $-m\hbar$, therefore the z-component of the total (for the shell) orbital angular momentum vanishes identically. This implies that the magnitude of the total (for the shell) orbital angular momentum is zero too. The same argument applies to the spin angular momentum. Therefore the lower energy eigenstates of the carbon atom (those obtainable from the configuration $(1s^2)2s^22p^2$, listed in Table 2.5) will have the form (B179) where L may take one of the following values: $L = 0$ (S-state), $L = 1$

(P-state), $L = 2$ (D-state) and $S = 0$, 1. The corresponding non-vanishing antisymmetrised states are:

$$\begin{aligned}
|L, S; J, M_j\rangle &= |0, 0; 0, 0\rangle && \equiv {}^1\mathrm{S}_0 \\
&= |2, 0; 2, M_j = 2, 1, 0, -1, -2\rangle && \equiv {}^1\mathrm{D}_2 \\
&= |1, 1; 2, M_j = 2, 1, 0, -1, -2\rangle && \equiv {}^3\mathrm{P}_2 \\
&= |1, 1; 1, M_i = 1, 0, -1\rangle && \equiv {}^3\mathrm{P}_1 \\
&= |1, 1; 0, 0\rangle && \equiv {}^3\mathrm{P}_0
\end{aligned} \tag{B180}$$

where the notation of these states introduced in Section 2.15.2 is also given. The letter specifies the value of L as explained above; the right lower index gives the value of J and the upper left index equals $(2S + 1)$. The corresponding energy levels of the atom when the residual electrostatic interaction (repulsion) between the electrons referred to in Section 2.15.2 and the spin–orbit interaction (B166) are taken into account, are shown schematically in Fig. 2.23.

For a qualitative interpretation of the spectrum shown in Fig. 2.23 we need only consider the two electrons of the outer p-shell of the atom. We write the eigenfunctions (B180) as sums of products of an orbital part and a spin part as in (B179). The spin part is given by (B175) when $S = 0$ (i.e. for the ${}^1\mathrm{S}_0$ and ${}^1\mathrm{D}_2$ states), and by (B176) when $S = 1$ (i.e. for the states ${}^3\mathrm{P}_2$, ${}^3\mathrm{P}_1$, ${}^3\mathrm{P}_0$). Next we note that (B175) is antisymmetric with respect to the spin-coordinates permutation $(1 \rightleftarrows 2)$, which implies (since the complete wavefunction must be antisymmetric) that the orbital parts of the ${}^1\mathrm{S}_0$ and ${}^1\mathrm{D}_2$ states must be symmetric with respect to the operation $\boldsymbol{r}_1 \rightleftarrows \boldsymbol{r}_2$. On the other hand, (B176) are symmetric with respect to the spin-coordinates permutation $(1 \rightleftarrows 2)$ which implies that the orbital parts of the ${}^3\mathrm{P}_{2,1,0}$ states must be antisymmetric with respect to the operation $\boldsymbol{r}_1 \rightleftarrows \boldsymbol{r}_2$. Now an antisymmetric orbital wavefunction has the property $f(\boldsymbol{r}_1, \boldsymbol{r}_2) = -f(\boldsymbol{r}_2, \boldsymbol{r}_1)$, which implies that $f(\boldsymbol{r}_1 = \boldsymbol{r}, \boldsymbol{r}_2 = \boldsymbol{r}) = 0$, which implies, by a continuity argument, that the particles cannot be found close together in space, at least not with a high probability, and since the residual electrostatic interaction between two electrons (a positive potential energy term) is inversely proportional to the distance between the electrons, we expect the mean value of this quantity to be larger for the ${}^1\mathrm{S}_0$ and ${}^1\mathrm{D}_2$ states than for the ${}^3\mathrm{P}_{2,1,0}$ states, and therefore we expect the energy levels of ${}^1\mathrm{S}_0$ and ${}^1\mathrm{D}_2$ to lie above those of the ${}^3\mathrm{P}_{2,1,0}$ states, which is indeed the case according to Fig. 2.23. The above example demonstrates the considerable effect that the antisymmetrisation of the wavefunction has on the energy spectrum of the system under consideration. It also shows quite clearly the origin of Hund's rule cited in Section 2.15.2.

B.9 Eigenstates of an electron in a periodic potential field

The potential field (4.16), being periodic, can be written as follows (see Section 4.2):

$$U(\boldsymbol{r}) = \sum_{\boldsymbol{K}_g} v(\boldsymbol{K}_g) \exp(\mathrm{i}\boldsymbol{K}_g \cdot \boldsymbol{r}) \tag{B181}$$

where

$$\begin{aligned}
\boldsymbol{K}_g &= g_1\boldsymbol{b}_1 + g_2\boldsymbol{b}_2 + g_3\boldsymbol{b}_3 \\
g_1, g_2, g_3 &= 0, \pm 1, \pm 2, \pm 3, \ldots
\end{aligned} \tag{B182}$$

are the vectors of the reciprocal lattice, and $v(\boldsymbol{K}_g)$ are given coefficients. In practice $v(\boldsymbol{K}_g) = 0$ when $|\boldsymbol{K}_g| > K_o$, where K_o is an appropriate constant.

Let us assume that the space lattice under consideration is face-centred cubic, in which case the reciprocal lattice is body-centred cubic: $\boldsymbol{b}_1$, $\boldsymbol{b}_2$, $\boldsymbol{b}_3$ are given by (4.15). We can enumerate $\boldsymbol{K}_g$ in order of increasing magnitude as shown in Table B1. The magnitude $|\boldsymbol{K}_g|$ of a given $\boldsymbol{K}_g$ is noted in the fifth column of the table and the corresponding coefficient $v(\boldsymbol{K}_g)$ in the sixth column. We have assumed that $v(\boldsymbol{K}_g) = 0$ when $|\boldsymbol{K}_g|a/2\pi \geqslant 2$. It is worth noting that, because of symmetry, the coefficients $v(\boldsymbol{K}_g)$, $g = 2, \ldots, 9$ have the same value (v_1).

The eigenstates of an electron corresponding to a given value of the wavevector $\boldsymbol{k}$ (within the BZ) are written, according to (4.17)–(4.18), as follows

$$\psi_{\boldsymbol{k}\alpha} = \sum_{\boldsymbol{K}_h} C(\boldsymbol{K}_h) \exp[\mathrm{i}(\boldsymbol{k} + \boldsymbol{K}_h) \cdot \boldsymbol{r} - \mathrm{i}E_\alpha(\boldsymbol{k})t/\hbar] \tag{B183}$$

and we shall assume, for the sake of simplicity, that $C(\boldsymbol{K}_h) = 0$ when $|\boldsymbol{K}_h|a/2\pi > 2$, which means that the sum (B183) contains 15 terms, corresponding to the 15 reciprocal vectors listed in Table B1.

The energy eigenvalues $E_\alpha(\boldsymbol{k})$ and the coefficients $C(\boldsymbol{K}_h)$ which determine the corresponding eigenfunctions (B183) are calculated as follows (the dependence of $C(\boldsymbol{K}_h)$ on $\boldsymbol{k}$, α is not denoted explicitly to simplify the formulae). The eigenfunctions (B183) satisfy the equation

$$\left\{-\frac{\hbar^2}{2m}\nabla^2 + U(\boldsymbol{r})\right\}\psi_{\boldsymbol{k}\alpha} = E_\alpha(\boldsymbol{k})\psi_{\boldsymbol{k}\alpha} \tag{B184}$$

Substituting (B181) and (B183) in the above equation we obtain

$$\sum_{\boldsymbol{K}'_h} \frac{\hbar^2(\boldsymbol{k} + \boldsymbol{K}'_h)^2}{2m} C(\boldsymbol{K}'_h) \exp(\mathrm{i}\boldsymbol{K}'_h \cdot \boldsymbol{r}) + \sum_{\boldsymbol{K}'_h}\sum_{\boldsymbol{K}_g} v(\boldsymbol{K}_g) C(\boldsymbol{K}'_h) \exp[\mathrm{i}(\boldsymbol{K}_g + \boldsymbol{K}'_h) \cdot \boldsymbol{r}] =$$

$$E_\alpha(\boldsymbol{k}) \sum_{\boldsymbol{K}'_h} C(\boldsymbol{K}'_h) \exp(\mathrm{i}\boldsymbol{K}'_h \cdot \boldsymbol{r}) \tag{B185}$$

Table B.1. Components of a periodic potential field

$\boldsymbol{K}_g$	g_1	g_2	g_3	$\|\boldsymbol{K}_g\|a/2\pi$	$v(\boldsymbol{K}_g)$
$\boldsymbol{K}_1$	0	0	0	0	v_0
$\boldsymbol{K}_2$	1	0	0	$\sqrt{3}$	v_1
$\boldsymbol{K}_3$	0	1	0	$\sqrt{3}$	v_1
$\boldsymbol{K}_4$	0	0	1	$\sqrt{3}$	v_1
$\boldsymbol{K}_5$	−1	0	0	$\sqrt{3}$	v_1
$\boldsymbol{K}_6$	0	−1	0	$\sqrt{3}$	v_1
$\boldsymbol{K}_7$	0	0	−1	$\sqrt{3}$	v_1
$\boldsymbol{K}_8$	1	1	1	$\sqrt{3}$	v_1
$\boldsymbol{K}_9$	−1	−1	−1	$\sqrt{3}$	v_1
$\boldsymbol{K}_{10}$	1	1	0	2	0
$\boldsymbol{K}_{11}$	1	1	1	2	0
$\boldsymbol{K}_{12}$	1	0	1	2	0
$\boldsymbol{K}_{13}$	−1	−1	0	2	0
$\boldsymbol{K}_{14}$	0	−1	−1	2	0
$\boldsymbol{K}_{15}$	−1	0	−1	2	0

From table B1 we have $\boldsymbol{K}_1 = 0$ and $v(\boldsymbol{K}_1) = v_0$, which allows us to rewrite (B185) as

$$\sum_{\boldsymbol{K}'_h}\left\{E_\alpha(\boldsymbol{k}) - v_0 - \frac{\hbar^2}{2m}(\boldsymbol{k}+\boldsymbol{K}'_h)^2\right\}C(\boldsymbol{K}'_h)\exp(\mathrm{i}\boldsymbol{K}'_h\cdot\boldsymbol{r}) - \sum_{\boldsymbol{K}'_h}\sum_{\boldsymbol{K}_g(\neq\boldsymbol{K}_1)} v(\boldsymbol{K}_g)C(\boldsymbol{K}'_h)\exp[\mathrm{i}(\boldsymbol{K}_g+\boldsymbol{K}'_h)\cdot\boldsymbol{r}] = 0 \quad \text{(B186)}$$

The theory of Fourier series tells us that

$$\int_{V_o}\exp[\mathrm{i}(\boldsymbol{K}_h - \boldsymbol{K}'_h)\cdot\boldsymbol{r}]\,\mathrm{d}V = V_o, \quad \text{if } \boldsymbol{K}'_h = \boldsymbol{K}_h$$
$$= 0, \quad \text{if } \boldsymbol{K}'_h \neq \boldsymbol{K}_h \quad \text{(B187)}$$

where V_o is the volume of a primitive cell. Therefore, multiplying (B186) with $\exp(-\mathrm{i}\boldsymbol{K}_h\cdot\boldsymbol{r})$, integrating over the volume of a primitive cell, and using (B187), we obtain

$$\left\{E_\alpha(\boldsymbol{k}) - v_0 - \frac{\hbar^2}{2m}(\boldsymbol{k}+\boldsymbol{K}_h)^2\right\}C(\boldsymbol{K}_h) - \sum_{\boldsymbol{K}'_h(\neq\boldsymbol{K}_h)} v(\boldsymbol{K}_h - \boldsymbol{K}'_h)C(\boldsymbol{K}'_h) = 0 \quad \text{(B188)}$$

We note that we have 15 equations of the above form: one for every $\boldsymbol{K}_h$ of table B1. Therefore (B188) constitutes a homogeneous system of 15 algebraic equations with 15 unknowns ($C(\boldsymbol{K}_h)$, $h = 1, 2, \ldots, 15$). We know that such a system has 15 non-trivial (different from zero) linearly independent solutions

$$\{C(\boldsymbol{K}_h)\} = \{C_{\boldsymbol{k}\alpha}(\boldsymbol{K}_h)\}$$
$$\alpha = 1, 2, \ldots, 15 \quad \text{(B189}$$

which are obtained for specific values of $E_\alpha(\boldsymbol{k})$, the energy-eigenvalues, given by the 15 roots

$$E = E_\alpha(\boldsymbol{k}), \ \alpha = 1, 2, \ldots, 15 \quad \text{(B190)}$$

of the following algebraic equation, of the 15th degree with respect to E,

$$\begin{vmatrix} \dfrac{\hbar^2k^2}{2m} + v_0 - E & v(\boldsymbol{K}_1 - \boldsymbol{K}_2) & \ldots & v(\boldsymbol{K}_1 - \boldsymbol{K}_{15}) \\ v(\boldsymbol{K}_2 - \boldsymbol{K}_1) & \dfrac{\hbar^2}{2m}(\boldsymbol{k}+\boldsymbol{K}_2)^2 + v_0 - E & \ldots & v(\boldsymbol{K}_2 - \boldsymbol{K}_{15}) \\ \vdots & \vdots & & \vdots \\ v(\boldsymbol{K}_{15} - \boldsymbol{K}_1) & & \ldots & \dfrac{\hbar^2}{2m}(\boldsymbol{k}+\boldsymbol{K}_{15})^2 + v_0 - E \end{vmatrix} = 0 \quad \text{(B191)}$$

obtained by putting the determinant of the coefficients of (B188) equal to zero. Only when the above determinant of the coefficients vanishes will the system of equations (B188) admit a non-zero solution. In our example, $v(\boldsymbol{K}_h - \boldsymbol{K}'_h) = v_1$ if $\boldsymbol{K}_h - \boldsymbol{K}'_h$ is one of the vectors $\boldsymbol{K}_i$ ($i = 2, 3, \ldots, 9$) and zero otherwise (see Table B1).

The fifteen roots of (B191), which are all real numbers, can be calculated numerically. After evaluating the energy-eigenvalues in this way, we can easily obtain for each eigenvalue $E_\alpha(\boldsymbol{k})$ the corresponding eigenfunction, i.e., the coefficients (B189), by solving (B188). The solution of (B188) determines the coefficients (B189) apart from a multiplicative constant, which is determined by the normalisation condition

$$N\int_{V_o} |\psi_{k\alpha}(\boldsymbol{r})|^2 \,\mathrm{d}V = 1 \tag{B192}$$

where N is the number of primitive cells in the given crystal.

In conclusion: in our example, for given $\boldsymbol{k}$ we obtain fifteen energy-eigenvalues and a corresponding number of eigenstates. Some of the eigenvalues may be degenerate (they correspond to double or generally multiple roots of (B191)). It is obvious that a small change in the value of $\boldsymbol{k}$ will produce correspondingly small changes in the roots of (B191), and this leads to energy bands which are continuous functions of $\boldsymbol{k}$ as described in Section 4.3. In our example, at the end of the calculation we have just described, we shall have fifteen energy bands. Since the spin does not enter into the above calculation, the results will be the same for spin up and spin down: the energy-eigenvalues and the orbital parts of the corresponding eigenfunctions will be the same for the two orientations of the spin.

Finally we should point out that in a realistic calculation one needs to include a much larger number of terms in the sum (B183) than the fifteen terms we have used in our example and consequently, the number of roots of the corresponding determinantal equation will be greater and we shall therefore obtain at the end of such a calculation, a greater number of energy bands. We shall know whether we have included a sufficient number of terms in (B183), if by increasing the number of terms we still obtain the same lower-energy bands (those occupied at $T = 0$ and the lower of the empty ones). The higher-energy bands are, usually, of no interest (see also Section 4.11). The reader may wish to actually evaluate the energy bands for a simple (hypothetical) crystal with one electron per atom; if so, reasonable values for the parameters of Table B1 would be: $a = 3$ Å, $v_0 = 0$, $v_1 = -0.04$ Ryd.

The method of calculation of the energy bands and corresponding one-electron states we have described is not the only possible one. In fact, there are methods, available to us nowdays, which are computationally much faster than the one we described. We shall not be concerned with them in the present volume.

B.10 Time-independent perturbation theory

Quite often the Hamiltonian of a physical system can be written in the following form

$$\hat{H} = \hat{H}_0 + \hat{H}' \tag{B193}$$

where $\hat{H}'$ is a relatively small perturbation to the main (unperturbed) part of the Hamiltonian denoted by $\hat{H}_0$. We assume that the energy-eigenvalues and corresponding eigenfunctions of the unperturbed system are known; we have

$$\hat{H}_0 u_\alpha^{(0)}(q) = E_\alpha^{(0)} u_\alpha^{(0)}(q) \tag{B194}$$

where q stands for the spatial and spin coordinates of all the particles in the system under consideration, and α denotes, as usual, a set of quantum numbers. The addition of $\hat{H}'$ can make the solution of the energy-eigenvalue problem for $\hat{H}$ extremely difficult. In that case we can approximate the eigenvalues of $\hat{H}$, which cannot be very different from those of H_0 in view of the smallness of the perturbation, as follows. We distinguish between non-degenerate and degenerate energy levels.

(i) Non-degenerate levels. Let us assume that $E_1^{(0)}$ is such a level and let us denote the corresponding normalised eigenfunction of $\hat{H}_0$ by $u_1^{(0)}(q)$. We can, evidently, estimate the correction to this level due to the perturbation $\hat{H}'$ from the formula

$$\Delta E_1 = \int (u_1^{(0)}(q))^* \hat{H}' u_1^{(0)}(q)\,\mathrm{d}q \tag{B195}$$

which gives the mean value of $\hat{H}'$ over the unperturbed state which, since we are dealing with a weak perturbation, cannot be a bad approximation to the actual (perturbed) state of the system.

(The dq-integration in (B195) involves integration of the spatial coordinates of all the particles of the system over all space and a summation over spin-indices (if they have spin) of all particles. By an obvious extension of equations (B125) and (B126) we write (for a system of N particles of spin $s = \frac{1}{2}$):

$$\int f^*(q)g(q)\,\mathrm{d}q = \sum_{i_1=1}^{2} \cdots \sum_{i_N=1}^{2} \int \mathrm{d}^3r_1 \cdots \mathrm{d}^3r_N f^*(\boldsymbol{r}_1, i_1; \cdots; \boldsymbol{r}_N, i_N) g(\boldsymbol{r}_1, i_1; \cdots; \boldsymbol{r}_n, i_N) \tag{B196}$$

$$\int f^*(q)\hat{\Omega}g(q)\,\mathrm{d}q \equiv \sum_{i_1=1}^{2} \ldots \sum_{i_N=1}^{2} \sum_{j_1=1}^{2} \ldots \sum_{j_N=1}^{2} \int \mathrm{d}^3r_1 \ldots \mathrm{d}^3r_N \tag{B197}$$

$$\times f^*(\boldsymbol{r}_1, i_1; \ldots; \boldsymbol{r}_N, i_N)\hat{\Omega}_{i_1,\ldots,i_N,;j_1,\ldots,j_N}(\boldsymbol{r}_1, \boldsymbol{\nabla}_1; \ldots; \boldsymbol{r}_N, \boldsymbol{\nabla}_N) g(\boldsymbol{r}_1, j_1; \ldots; \boldsymbol{r}_N, j_N)$$

where $\mathrm{d}^3r \equiv \mathrm{d}V$).

We therefore write for the level E_1 of the perturbed system (corresponding to the $E_1^{(0)}$ level of the unperturbed system) the following first-order approximation formula

$$E_1 = E_1^{(0)} + \Delta E_1 \tag{B198a}$$

which can also be written as

$$E_1 = \int (u_1^{(0)}(q))^* \hat{H} u_1^{(0)}(q)\,\mathrm{d}q \tag{B198b}$$

The above equation follows directly from (B194) and (B195).

Let us, for the sake of clarity, write down the above equations for the simplest case: that of a spinless particle in a given potential field, $U_0(\boldsymbol{r}) + \Delta U(\boldsymbol{r})$. We have

$$\hat{H}_0 = -\frac{\hbar^2}{2m}\nabla^2 + U_0(\boldsymbol{r}) \tag{B199}$$

and let us assume that $u_1^{(0)}(\boldsymbol{r})$ is an eigenfunction of $\hat{H}_0$ corresponding to a non-degenerate energy level $E_1^{(0)}$, i.e., $\hat{H}_0 u_1^{(0)} = E_1^{(0)} u_1^{(0)}$. Then, provided $\Delta U(\boldsymbol{r})$ is sufficiently weak, the particle in the given field, $U_0(\boldsymbol{r}) + \Delta U(\boldsymbol{r})$, has a corresponding energy level given, to first-order approximation, by

$$\begin{aligned} E_1 &= E_1^{(0)} + \Delta E_1 \\ \Delta E_1 &= \int (u_1^{(0)}(\boldsymbol{r}))^* \Delta U(\boldsymbol{r}) u_1^{(0)}(\boldsymbol{r})\,\mathrm{d}V \end{aligned} \tag{B200}$$

(ii) Degenerate levels. Let us assume that the unperturbed Hamiltonian $\hat{H}_0$ of (B193) has an n-fold degenerate level $E^{(0)}$ with corresponding eigenfunctiosn $u_i^{(0)}(q)$, $i = 1, 2, \ldots, n$. We have

$$\hat{H}_0 u_i^{(0)}(q) = E^{(0)} u_i^{(0)}(q), \quad i = 1, \ldots, n \tag{B201}$$

and we assume that

$$\int (u_i^{(0)}(q))^* u_j^{(0)}(q)\,\mathrm{d}q = 1, \quad \text{if } i = j \qquad \qquad = 0, \quad \text{otherwise} \tag{B202}$$

We note immediately that any linear combination of the $u_i^{(0)}(q)$

$$u(q) = \sum_{i=1}^{n} c_i u_i^{(0)}(q) \tag{B203}$$

where c_i are coefficients to be determined, is also an eigenfunction of $\hat{H}_0$ with the same eigenvalue $E^{(0)}$. And there are n linearly independent such combinations which we can choose in a variety of ways. We shall choose them in such a way that they approximate, as far as possible, corresponding states of the perturbed Hamiltonian. We proceed as follows. We wish to have

$$\hat{H}u(q) = Eu(q) \tag{B204}$$

where E remains to be determined. Substituting (B203) in the above equation and using (B201) we obtain

$$(E^{(0)} - E)\sum_{i=1}^{n} c_i u_i^{(0)}(q) + \sum_{i=1}^{n} c_i \hat{H}' u_i^{(0)}(q) = 0 \tag{B205}$$

Multiplying the above equation from the left with $(u_j^{(0)}(q))^*$, integrating with respect to q, and using (B202) we obtain

$$(E^{(0)} - E)c_j + \sum_{i=1}^{n} H'_{ji} c_i = 0 \tag{B206}$$

where

$$H'_{ji} \equiv \int (u_j^{(0)}(q))^* \hat{H}' u_i^{(0)}(q)\,\mathrm{d}q \tag{B207}$$

We obtain n equations (B206) corresponding to $j = 1, 2, \ldots, n$. Taken together, they constitute a homogeneous system of n algebraic equations with n unknowns c_i, $i = 1, 2, \ldots, n$. This system of equations has non-trivial solutions only for specific values of the energy

$$E = E_\alpha, \; \alpha = 1, 2, \ldots, n \tag{B208}$$

which are given by the roots of the characteristic equation

$$\begin{vmatrix} E^{(0)} + H'_{11} - E & H'_{12} & \cdots & H'_{1n} \\ H'_{21} & E^{(0)} + H'_{22} - E & \cdots & H'_{2n} \\ \vdots & \vdots & \vdots & \vdots \\ H'_{n1} & H'_{n2} & \cdots & E^{(0)} + H'_{nn} - E \end{vmatrix} = 0 \tag{B209}$$

obtained by putting the determinant of the coefficients of (B206) equal to zero. The roots (B208) of the above equation determine, to first order approximation, the energy-eigenvalues of the perturbed system corresponding to the n-fold degenerate level $E^{(0)}$ of the unperturbed system. Putting $E = E_\alpha$ in (B206), we determine, apart from a normalisation constant, a wavefunction (B203) $\{c_i = c_{\alpha i}\}$ which, provided the perturbation is sufficiently weak, is a reasonably good approximation to the corresponding state of the perturbed system.

Unless the n-roots of (B209) have all the same value, the degeneracy of the unperturbed level will be, at least partly, lifted. Usually, the perturbation reduces the symmetry of the Hamiltonian and this splits the unperturbed level $E^{(0)}$ into a number of levels of lesser degeneracy.

The simplest example of the above is again obtained for a spinless particle in a given potential field: $U_0(\boldsymbol{r}) + \Delta U(\boldsymbol{r})$. Let us assume that

$$\hat{H}_0 u_i^{(0)}(\boldsymbol{r}) = E^{(0)} u_i^{(0)}(\boldsymbol{r}), \quad i = 1, 2 \tag{B210}$$

where $\hat{H}_0$ is given by (B199). In this case (B203) reduces to

$$u(\boldsymbol{r}) = c_1 u_1^{(0)}(\boldsymbol{r}) + c_2 u_2^{(0)}(\boldsymbol{r}) \tag{B211}$$

and the characteristic equation (B209) takes the form

$$\begin{vmatrix} E^{(0)} + H'_{11} - E & H'_{12} \\ H'_{21} & E^{(0)} + H'_{22} - E \end{vmatrix} = 0 \tag{B212}$$

where

$$H'_{ji} = \int (u_j^{(0)}(\boldsymbol{r}))^* \Delta U(\boldsymbol{r}) u_i^{(0)}(\boldsymbol{r})\, \mathrm{d}V$$

Let us, for the sake of simplicity, assume that $H'_{11} = H'_{22} = 0$ and $H'_{12} = H'_{21} = v$; in this case the two roots of (B212) are

$$E_1 = E^{(0)} + v, \quad E_2 = E^{(0)} - v \tag{B213}$$

which tells us that the doubly degenerate $E^{(0)}$ level of the unperturbed system is split by the perturbation into two levels separated by $2v$. One can easily show that the corresponding normalised eigenfunctions (B211) are given by

$$u_1(\boldsymbol{r}) = \frac{1}{\sqrt{2}}\{u_1^{(0)}(\boldsymbol{r}) + u_2^{(0)}(\boldsymbol{r})\} \tag{B214a}$$

and

$$u_2(\boldsymbol{r}) = \frac{1}{\sqrt{2}}\{u_1^{(0)}(\boldsymbol{r}) - u_2^{(0)}(\boldsymbol{r})\} \tag{B214b}$$

Quite often we can write down the correct eigenfunctions of the perturbed system (correct linear combinations (B203) of unperturbed eigenfunctions) using solely symmetry arguments. If that is the case, we can obtain the perturbed level corresponding to a given $u(q)$ by taking the mean of the perturbed Hamiltonian over this eigenfunction. For example, knowing the eigenfunctions (B214) from symmetry considerations, we could obtain (B213) from the following formula

$$E_i = \int (u_i(\boldsymbol{r}))^* (\hat{H}_0 + \Delta U(\boldsymbol{r})) u_i(\boldsymbol{r})\, \mathrm{d}V, \quad i = 1, 2 \tag{B215}$$

We have applied a version of this formula in our analysis of the electronic terms of the molecular hydrogen ion (see footnote 5 of Chapter 3). And our discussion of the one-electron levels of s-bands in crystals, where we have written the eigenfunctions of the electron in the crystal as Bloch sums of unperturbed atomic orbitals (see problem 4.3), is based on an extension of the same formula to the case of an N-fold degenerate level.

A treatment of p-bands (the same arguments apply to d-bands) can be given along similar lines. In this case, the eigenfunctions of given $\boldsymbol{k}$ are Bloch waves (4.27) where the φ_α are linear sums of the three atomic p-orbitals. We can put it differently. We can write three Bloch waves where the φ_α is given respectively by the first, second and third atomic p-orbital. We then seek the corresponding one-electron states in the crystal as linear sums of these three Bloch waves as in (B203) (we identify $u_i^{(0)}$, $i = 1$, 2, 3 with these waves). We then proceed as in the derivation of ((B206)), except that the procedure must be modified slightly, to take into account the fact that in the present case the $u_i^{(0)}$ do not satisfy (B202). The final result is a system of three homogeneous equations with three unknowns (as in B206), the solution of which determines the energy levels of the p-bands for the given $\boldsymbol{k}$, and, approximately, the corresponding eigenfunctions, i.e. the appropriate φ_α in the Bloch sums which describe these states (see Section 4.3.1).

The above examples refer to one-particle states in a given potential field, but similar arguments apply to the general case described by (B201). If we can write down the correct linear combinations (B203) of the unperturbed states from just symmetry considerations, we can obtain the corresponding energy levels, to first order approximation, from the formula

$$E_\alpha = \int (u_\alpha(q))^*(\hat{H}_0 + \hat{H}')u_\alpha(q)\,\mathrm{d}q \tag{B216}$$

where $u_\alpha(q)$ is given by (B203) with the $c_i \equiv c_{\alpha i}$ determined by symmetry. We have seen, for example, how symmetry considerations allow us to write down approximate eigenfunctions of an N-electron atom in terms of unperturbed states obtained in the mean-field approximation (see Section B.8). Having obtained these eigenfunctions we can obtain to first-order approximation, the corresponding energy levels of the atom using (B216); in this case $\hat{H}_0$ represents the mean-field Hamiltonian and $\hat{H}'$ includes the residual electrostatic interaction between the electrons, spin–orbit and other corrections.

B.11 Time-dependent perturbation theory

In the preceding sections of this appendix we have assumed that the Hamiltonian of the system under consideration is independent of the time, but this need not be always the case (see, e.g. Section 2.7.1). Let us then assume that the Hamiltonian of a physical system has the following form

$$\hat{H} = \hat{H}_0(q, \partial/\partial q) + \hat{H}'(q, \partial/\partial q; t) \tag{B217}$$

and let us also assume that the time-dependent part of the Hamiltonian, denoted by $\hat{H}'$, is much smaller than the time-independent part denoted by $\hat{H}_0$. Let us further assume that the eigenvalues and corresponding eigenfunctions of $\hat{H}_0$ are known. We have

$$\hat{H}_0 u_\alpha^{(0)}(q) = E_\alpha^{(0)} u_\alpha^{(0)}(q)$$

$$-\frac{\hbar}{\mathrm{i}}\frac{\partial}{\partial t} u_\alpha^{(0)}(q)\exp(-\mathrm{i}E_\alpha^{(0)}t/\hbar) = \hat{H}_0 u_\alpha^{(0)}(q)\exp(-\mathrm{i}E_\alpha^{(0)}t/\hbar) \tag{B218}$$

Because the eigenfunctions of $\hat{H}_0$ are a complete set of functions, we can write the wavefunction $u(q, t)$ which describes the state of the given system (whose Hamiltonian is that of (B217)), as follows

$$u(q, t) = \sum_{\alpha} c_\alpha(t) u_\alpha^{(0)}(q) \exp(-\mathrm{i}E_\alpha^{(0)} t/\hbar) \tag{B219}$$

If we know the wavefunction at some initial moment of time, to be denoted by t_0, we can determine the wavefunction at later times ($t > t_0$) by solving the Schrödinger equation

$$-\frac{\hbar}{\mathrm{i}}\frac{\partial u}{\partial t} = \hat{H}_0(q, \partial/\partial q)u(q, t) + \hat{H}'(q, \partial/\partial q; t)u(q, t) \tag{B220}$$

in accordance with what has been stated in Section B.1. Knowing the wavefunction means knowing the coefficients $c_\alpha(t)$ of (B219), so the above statement tells us that if we know $c_\alpha(t_0)$, we can determine $c_\alpha(t)$ for $t > t_0$, by solving (B220). We can rewrite (B220) in terms of these coefficients as follows. Substituting (B219) in (B220) and using (B218) we obtain

$$-\frac{\hbar}{\mathrm{i}}\sum_{\beta}\frac{\mathrm{d}c_\beta}{\mathrm{d}t}u_\beta^{(0)}(q)\exp(-\mathrm{i}E_\beta^{(0)}t/\hbar) = \sum_{\beta} c_\beta(t)\hat{H}'(q, \partial/\partial q; t)u_\beta^{(0)}(q)\exp(-\mathrm{i}E_\beta^{(0)}t/\hbar) \tag{B221}$$

Multiplying the above equation from the left with $(u_\alpha^{(0)}(q))^*$, integrating over q, and using the orthogonality property of the eigenfunctions of $\hat{H}_0$, i.e.

$$\int (u_\alpha^{(0)}(q))^* u_{\alpha'}^{(0)}(q)\,\mathrm{d}q = 1, \text{ if } \alpha = \beta$$
$$= 0, \text{ otherwise} \tag{B222}$$

we obtain

$$\frac{\mathrm{d}c_\alpha}{\mathrm{d}t} = -\frac{\mathrm{i}}{\hbar}\sum_{\beta} H'_{\alpha\beta}(t)\exp(\mathrm{i}\omega_{\alpha\beta}t)c_\beta(t) \tag{B223}$$

where

$$H'_{\alpha\beta}(t) = \int (u_\alpha^{(0)}(q))^*\hat{H}'(q, \partial/\partial q; t)u_\beta^{(0)}(q)\,\mathrm{d}q \tag{B224}$$

$$\omega_{\alpha\beta} = (E_\alpha^{(0)} - E_\beta^{(0)})/\hbar \tag{B225}$$

If we know $c_\alpha(t_0)$ we can determine $c_\alpha(t)$ for $t > t_0$ by solving (B223). We can formally integrate (B223) to obtain

$$c_\alpha(t) = c_\alpha(t_0) - \frac{\mathrm{i}}{\hbar}\sum_{\beta}\int_{t_0}^{t} H'_{\alpha\beta}(t')\exp(\mathrm{i}\omega_{\alpha\beta}t')c_\beta(t')\,\mathrm{d}t' \tag{B226}$$

which is an integral equation for the $c_\alpha(t)$, which incorporates the initial condition ($c_\alpha(t_0)$ are given quantities). So far we have not made use of our assumption that $\hat{H}'$ is a relatively small quantity and therefore (B226) is an exact equation valid quite generally.

When $\hat{H}'$ is relatively small, we can solve (B226), using time-dependent perturbation theory, as follows. Let us for the sake of simplicity assume that at $t = t_0$ the given system is described by a single unperturbed eigenfunction $u_{\alpha 0}(q)$, so that $c_\alpha(t_0) = \delta_{\alpha\alpha_0}(\delta_{\alpha\alpha_0} = 1$ if $\alpha = \alpha_0$ and equals zero otherwise). Because $\hat{H}'$ is small, we can assume that to a zeroth approximation $c_\beta(t) = \delta_{\beta\alpha_0}$. By substituting this quantity in the right hand side of (B226) we obtain $c_\alpha(t)$, to first order approximation, as follows

$$c_\alpha(t) = \delta_{\alpha\alpha_0} - \frac{\mathrm{i}}{\hbar}\int_{t_0}^{t} H'_{\alpha\alpha_0}(t')\exp(\mathrm{i}\omega_{\alpha\alpha_0}t')\,\mathrm{d}t' \tag{B227}$$

In many cases the above approximation is valid and in the applications that we shall consider we shall assume that this is indeed the case. In cases when the above is not sufficient we can obtain a higher order approximation as follows. We substitute (B227) in the right hand side of (B226) to obtain $c_\alpha(t)$ in second order approximation, and then we can repeat the process to obtain $c_\alpha(t)$ in third order approximation, and so on.

B.11.1 PERIODIC PERTURBATION—AN EXAMPLE

An electron moving in the field of a nucleus, when perturbed by an incident electromagnetic wave obeys the following Hamiltonian

$$\begin{aligned} \hat{H} &= \hat{H}_0 + \hat{H}'(z;t) \\ \hat{H}_0 &= -\frac{\hbar^2}{2m}\nabla^2 + U(r) \end{aligned} \tag{B228}$$

where $U(r)$ is given by (2.1) and $\hat{H}'(z;t)$ can be written, according to (2.15), as follows

$$\hat{H}'(z;t) = ezF_z^O\cos\omega t = ezF_z^O(\mathrm{e}^{\mathrm{i}wt} + \mathrm{e}^{-\mathrm{i}wt})/2 \tag{B229}$$

For computational purposes we shall assume that the perturbation is turned on gradually and write

$$\begin{aligned} \hat{H}'(z;t) &= \{ezF_z^O(\mathrm{e}^{\mathrm{i}wt} + \mathrm{e}^{-\mathrm{i}wt})/2\}s(t) \\ s(t) &= \exp(\delta t),\ t < 0 \\ &= 1,\ t > 0 \end{aligned} \tag{B230}$$

where δ is a very small positive quantity.

In the present case the unperturbed states (B218) are the eigenstates of the free atom:

$$u_\alpha^{(0)}(q)\exp(-\mathrm{i}E_\alpha^{(0)}t/\hbar) \Rightarrow \psi_{nlm} = u_{\mathrm{nlm}}(\mathbf{r})\exp(-\mathrm{i}E_n t/\hbar) \tag{B231}$$

where $u_{\mathrm{nlm}}(r) = R_{nl}(r)Y_{lm}(\theta,\varphi)$ are given in Table 2.3 and E_n are the corresponding energy-eigenvalues given by (2.12). In the presence of the perturbing field (B230), and when the initial (at $t = -\infty$) state of the atom is its unperturbed ground state, the wavefunction of the atom will be given by (2.16) with the coefficients in that formula determined, according to (B227), by

$$c_{100}(t) = 1 - \frac{\mathrm{i}}{\hbar}\int_{-\infty}^{t} H'_{100;100}(t')\,\mathrm{d}t' \simeq 1 \tag{B232}$$

$$c_{nlm}(t) = -\frac{\mathrm{i}}{\hbar}\int_{-\infty}^{t} H'_{nlm;100}(t')\,\mathrm{e}^{\mathrm{i}\omega_{n1}t'}\,\mathrm{d}t';\ n,\ l,\ m \neq 1, 0, 0 \tag{B233}$$

$$\omega_{n1} = (E_n - E_1)/\hbar \tag{B234}$$

$$H'_{nlm;100}(t) = (eF_z^O/2)\,Z_{nlm;100}(\mathrm{e}^{\mathrm{i}\omega t} + \mathrm{e}^{-\mathrm{i}\omega t})s(t) \tag{B235}$$

$$Z_{nlm;100} = \int (u_{nlm}(r))^* z u_{100}(r)\,\mathrm{d}V \tag{B236}$$

Substituting (B235) in (B233) we obtain, for $t > 0$,

$$c_{nlm}(t) = -(eF_z^O/2\hbar)\,Z_{nlm;100}\left\{\frac{\mathrm{e}^{\mathrm{i}(\omega_{n1}+\omega)t}}{\omega_{n1}+\omega} + \frac{\mathrm{e}^{\mathrm{i}(\omega_{n1}-\omega)t}}{\omega_{n1}-\omega}\right\} \tag{B237}$$

The induced electric dipole moment defined in Section 2.7.1 is given by

$$d_z(t) = -e\int \psi^*(\mathbf{r}, t) z \psi(\mathbf{r}, t)\,\mathrm{d}V \tag{B238}$$

$$\psi(r, t) \simeq u_{100}(\mathbf{r})\exp(-\mathrm{i}E_1 t/\hbar) + \sum_{nlm\neq 100} c_{nlm}(t) u_{nlm}(\mathbf{r})\exp(-\mathrm{i}E_n t/\hbar) \tag{B239}$$

Substituting (B237) and (B239) in (B238), dropping second order terms (proportional to $(F_z^O)^2$), and noting that

$$\int u_{100}^*(\mathbf{r}) z u_{100}(\mathbf{r})\,\mathrm{d}V = 0$$

because of the spherical symmetry of $|u_{100}(\mathbf{r})|^2$, we finally obtain

$$d_z(t) = (2e^2/\hbar)\sum_{nlm\neq 100}\frac{\omega_{n1}|Z_{nlm;100}|^2}{\omega_{n1}^2 - \omega^2}F_z^O\cos\omega t \tag{B240}$$

which is identical with (2.17) of Section 2.7.1. Obviously (B240) is not valid if, for some value of n, $\omega_{n1} = \omega$. We know that when $\omega = \omega_{n1}$, the atom can be excited with certain probability from its ground state to an excited state with energy $E_n = E_1 + \hbar\omega$. In what follows we shall calculate the so-called transition rate of such a transition.

B.11.2 TRANSITIONS

We now take the initial time to be $t = 0$, and we describe the perturbing interaction by (B229). For $t > 0$ the wavefunction of the atom is given by (2.16) and the coefficients are given by the following expression (which replaces (B233))

$$\begin{aligned} c_{nlm}(t) &= -\frac{\mathrm{i}}{\hbar}\int_0^t H'_{nlm;100}(t')\,\mathrm{e}^{\mathrm{i}\omega_{n1}t'}\,\mathrm{d}t' \\ &= -(eF_z^O/2\hbar)\,Z_{nlm;100}\left\{\frac{\mathrm{e}^{\mathrm{i}(\omega_{n1}+\omega)t} - 1}{\omega_{n1}+\omega} + \frac{\mathrm{e}^{\mathrm{i}(\omega_{n1}-\omega)t} - 1}{\omega_{n1}-\omega}\right\} \end{aligned} \tag{B241}$$

where $n, l, m \neq 1, 0, 0$. The probability that the atom will be found at time $t > 0$ in the (nlm)-state is given by $|c_{nlm}(t)|^2$. When $\omega \simeq \omega_{n1}$ the second term in the brackets of (B241) dominates and we can write

$$|c_{nlm}(t)|^2 = (eF_z^O/2\hbar)^2 |Z_{nlm;100}|^2 f(\omega_{n1} - \omega;\, t) \tag{B242}$$

$$f(\omega_{n1} - \omega;\, t) = \left|\frac{e^{i(\omega_{n1}-\omega)t} - 1}{\omega_{n1} - \omega}\right|^2 = \left(\frac{\sin\frac{1}{2}(\omega_{n1} - \omega)t}{(\omega_{n1} - \omega)/2}\right)^2 \tag{B243}$$

One can easily see that $f(\omega_{n1} - \omega;\, t)$ peaks at $\omega_{n1} - \omega = 0$, and oscillates with diminishing amplitude as $\omega_{n1} - \omega$ moves away from this point. Moreover, for sufficiently large values of t, one finds that

$$\int_{-\infty}^{\infty} f(\omega_{n1} - \omega;\, t)\, d(\omega_{n1} - \omega) = 2\pi t \tag{B244}$$

and that the contribution to the above integral comes from a very narrow range of $\omega_{n1} - \omega$, of approximately $2\pi/t$ about zero. Therefore in the limit $t \to \infty$, a transition occurs only if $\omega_{n1} = \omega$.

If we now recall that the excited energy level of the atom is broadened to some degree by the interaction with the electromagnetic field (see Section 2.9), we may conclude that

$$|c_{nlm}(t)|^2 = (2\pi/\hbar)|(eF_z^O/2) Z_{nlm;100}|^2 t g_{nlm}(E_1 + \hbar\omega) \tag{B245}$$

gives the probability that the atom will be found, at time t, with energy $E_1 + \hbar\omega$; where $g_{nlm}(E_1 + \hbar\omega)\, dE$ is the fraction of the broadened level (band) between $E_1 + \hbar\omega$ and $E_1 + \hbar\omega + dE$.

The transition rate, defined as the transition probability per unit time, corresponding to (B245) is given by

$$w_{nlm;100} = (2\pi/\hbar)|(eF_z^O/2) Z_{nlm;100}|^2 g_{nlm}(E_1 + \hbar\omega) \tag{B246}$$

In the above example the initial state of the atom was its ground state, but one can easily see that the same procedure applies for any initial state of the atom.

Time-dependent perturbation theory can be applied to any physical system as long as the perturbing interaction is sufficiently weak. The induced polarisation in a crystal, given by (4.133) and (4.135), is obtained in essentially the same way as (B240). Therefore, these formulae are valid provided the external field is sufficiently weak. The description of the interaction of matter with the electromagnetic field presented in this book is based on this assumption. However, there are situations when one needs to go beyond first order perturbation theory. For example, an atom interacting with the intense field (F) of a laser beam acquires an induced dipole moment described by the following formula

$$d(F) = \chi_1 F + \chi_2 F^2 + \chi_3 F^3 + \cdots \tag{B247}$$

The first term in the above formula is the linear response term which is obtained by first order perturbation theory in the manner we have described; for the evaluation of the additional terms, which are proportional to F^2, F^3, and so on, one must go beyond first order perturbation theory. There are a number of very interesting phenomena associated with the non-linear terms of (B247), and the same applies in relation to non-linear effects in the polarisation of a crystal under the influence of an intense electric field, but their discussion lies beyond the scope of the present book.

Footnotes of Appendix B

1. We assume that the potential field does not depend on the spin of the particle (if the particle has spin). It is, therefore, sufficient to describe the time development of the orbital part of the wavefunction. The spin part of the wavefunction does not change with time.
2. It would be sufficient to say that the integral (B8), which extends over all space, has a finite value. The normalisation to unity is a matter of convention (see Section 1.3.1).
3. We can say that, by construction, the Hamiltonian of a physical system is such that its eigenvalues are real quantities. The same is true of the eigenvalues of other physical operators (see following sections).
4. The most general state of the particle is given by a linear sum which extends over the bound *and* scattering states of the particle. However, we shall assume that the particle is bounded in space, and that the corresponding wavefunction is a linear sum of the bound eigenstates only.
5. If to a given eigenvalue correspond more than one eigenstates, then the probability of that eigenvalue being materialised is given by the sum of the corresponding $|c_\alpha(t)|^2$. The way one obtains experimentally $|c_\alpha(t)|^2$ and $\langle \Omega \rangle_t$ has been described in Section 1.6.
6. It is assumed that the reader has some knowledge of matrix algebra. To read Section B.6.2 one needs to know no more than the rules of addition and multiplication. We have

   (i) $a\begin{pmatrix} A_{11} & A_{12} \\ A_{21} & A_{22} \end{pmatrix} = \begin{pmatrix} aA_{11} & aA_{12} \\ aA_{21} & aA_{22} \end{pmatrix}$

   (ii) $\begin{pmatrix} A_{11} & A_{12} \\ A_{21} & A_{22} \end{pmatrix} + \begin{pmatrix} B_{11} & B_{12} \\ B_{21} & B_{22} \end{pmatrix} = \begin{pmatrix} A_{11}+B_{11} & A_{12}+B_{12} \\ A_{21}+B_{21} & A_{22}+B_{22} \end{pmatrix}$

   where a, A_{ij}, B_{ij}, are in general complex numbers, and the same rules apply, of course, to column matrices. The multiplication of two matrices is defined in the text.
7. That $u(r)$ is a real function follows from the non-degeneracy of the energy-eigenvalues of the one-dimensional potential field (B133) and the fact that the coefficients in equation (B132) are real quantities. If $u(r)$ were complex, its real and imaginary parts would be independent solutions of (B132) and the corresponding energy-eigenvalue would be degenerate.

Appendix C
PHYSICAL CONSTANTS

Planck's constant: $\hbar = h/2\pi = (1.05459 \pm 0.000001) \times 10^{-34}$ J s

Speed of light: $c = (2.997925 \pm 0.000001) \times 10^{8}$ m/s
Electronic charge: $-e$
Proton charge: $+e$
$e = (1.6021892 \pm 0.0000046) \times 10^{-19}$ C
Electron mass: $m = (9.109534 \pm 0.000047) \times 10^{-31}$ kg
Proton mass: $M_p = (1.672648 \pm 0.000006) \times 10^{-27}$ kg
Permittivity of free space: $\epsilon_O = 8.85434 \times 10^{-12}$ F/m
Boltzmann's constant: $k_B = 1.38044 \times 10^{-23}$ J/K

Atomic unit of length (a.u.): $a_0 = 0.52917$ Å
Atomic unit of energy: Hartree = 2 Rydbergs (Ryd) = 27.21 eV
One electron-volt (eV) = 1.6×10^{-19} J
One angstrom (Å) = 10^{-10} m.

REFERENCES

In my own studies of the quantum theory of matter I benefited greatly from the books (cited below) of L. D. Landau and E. M. Lifshitz, R. B. Leighton, N. F. Mott and H. Jones, N. F. Mott and E. A. Davies, and J. C. Slater. I have of course learned many things from a number of other books. The list which follows includes some books which I found useful, and some which the reader might find useful for supplementary or further reading; they range in level from elementary to advanced.

A. Quantum mechanics

A. Beiser, *Concepts of Modern Physics*, (New York: McGraw-Hill), fifth edition, 1995.

B. H. Bransden and C. J. Joachain, *Introduction to Quantum Mechanics*, (Longman Scientific & Technical), 1994.

C. Cohen-Tannoudji, B. Diu and F. Laloë, *Quantum Mechanics*, (New York: Herman and John Wiley & Sons), 1977.

A. P. French and E. F. Taylor, *An Introduction to Quantum Physics*, (London: Chapman and Hall), 1979.

L. D. Landau and E. M. Lifshitz, *Quantum Mechanics—Non-Relativistic Theory*, (Oxford: Pergamon Press), third edition, 1977.

R. B. Leighton, *Principles of Modern Physics*, (New York: McGraw-Hill), 1959.

E. Merzbacher, *Quantum Mechanics*, (New York: John Wiley & Sons), second edition, 1970.

J. C. Polkinghorne, *The Quantum World*, (London: Longman), 1984.

J. J. Sakurai, *Modern Quantum Mechanics*, (Menlo Park: Benjamin), 1985.

L. I. Schiff, *Quantum Mechanics*, (New York: McGraw-Hill), second edition, 1955.

J. C. Slater, *Quantum Theory of Matter*, (New York: McGraw-Hill), second edition, 1968.

G. L. Squires, *Problems in Quantum Mechanics* (with solutions), (Cambridge University Press), 1995.

B. Atoms and molecules

B. H. Bransden and C. H. Joachain, *Physics of Atoms and Molecules*, (Longman Scientific & Technical), 1994.

A. Corney, *Atomic and Laser Spectroscopy*, (London: Oxford University Press), 1977.
C. A. Coulson, *Valence*, (London: Oxford University Press), 1961.
H. G. Kuhn, *Atomic Spectra*, (London: Longman), second edition, 1969.
M. A. Morrison, T. L. Estle and N. F. Lane, *Quantum States of Atoms, Molecules and Solids*, (New Jersey: Prentice Hall), 1976.
R. B. Leighton, *Principles of Modern Physics*, loc. cit.
J. C. Slater, *Quantum Theory of Matter*, loc. cit.
J. C. Slater, *The Self-consistent Field for Molecules and Solids: Quantum Theory of Molecules and Solids*, Volume 4, (New York: McGraw-Hill), 1974.

C. Solids

N. W. Ashcroft and N. D. Mermin, *Solid State Physics*, (Philadelphia: Saunders College Publishing), 1976.
J. S. Blakenmore, *Solid State Physics*, (Cambridge University Press), second edition, 1985.
N. Bloembergen, *Non-linear Optics*, (New York: Benjamin), second edition, 1982.
R. G. Chambers, *Electrons in Metals and Semiconductors*, (London: Chapman and Hall), 1990.
W. A. Harrison, *Solid State Theory*, (New York: Dover Publications), 1979.
W. A. Harrison, *Electronic Structure and the Properties of Solids*, (New York: Dover Publications), 1989.
J. R. Hook and H. E. Hall, *Solid State Physics*, (Chichester: John Wiley & Sons), second edition, 1991.
H. Ibach and H. Lüth, *Solid-State Physics*, (Berlin: Springer-Verlag), 1991.
C. Kittel, *Introduction to Solid State Physics*, (New York: John Wiley & Sons), sixth edition, 1986.
E. A. Lynton, *Superconductivity*, (London: Methuen), 1962.
A. Modinos, Field, *Thermionic and Secondary Electron Emission Spectroscopy*, (New York: Plenum), 1984.
L. Moruzzi, J. F. Janak and A. R. Williams, *Calculated Electronic Properties of Metals*, (New York: Pergamon), 1978.
N. F. Mott and H. Jones, *The Theory of the Properties of Metals and Alloys*, (New York: Dover Publications), 1958.
N. F. Mott and E. A. Davies, *Electronic Processes in Non-crystalline Materials*, (Oxford: Oxford University Press), 1979.
K. J. Pascoe, *Properties of Materials for Electrical Engineers*, (New York: John Wiley & Sons), 1973.
M. Prutton, *Introduction to Surface Physics*, (Oxford: Clarendon), 1994.
B. K. Ridley, *Quantum Processes in Semiconductors*, (Oxford: Clarendon), third edition, 1993.
J. C. Slater, *Symmetry and Energy Bands in Crystals*, (New York: Dover Publications), 1972.
J. C. Slater, *The Self-consistent Field for Molecules and Solids: Quantum Theory of Molecules and Solids, Volume 4*, loc. cit.
G. Vidali, *Superconductivity—The next revolution?*, (Oxford: Clarendon), 1993.
R. Zallen, *The Physics of Amorphous Solids*, (New York: Wiley Interscience), 1983.
J. Ziman, *Principles of Solid State Physics*, (Cambridge University Press), second edition, 1972.

D. Various

The extracts from the works of Galilei and Newton on pages 1 and 129 are cited by H. Weyl (*Philosophy of Mathematics and Natural Science* (New York: Atheneum), 1963). The attribution to Demokritus cited on page 45 is made by W. Heisenberg (*Physics and Philosophy* (London: Penguin), 1989).

INDEX

Absorption of radiation,
  by amorphous germanium, 279
  by atoms, 62, 68, 74
  by crystals, 254, 256
  by molecules, 145
  coefficient of, 256
Acoustic modes of vibration, 228
Adiabatic approximation, 131
Amorphous film, 269
Anderson transition, 276
Angular frequency, 4, 66
Angular momentum, addition of, 85, 324
  in classical mechanics, 47, 312, 324
  orbital, 47, 312
    eigenfunctions, 49, 315
    eigenvalues, 48, 316
    operators, 312–314
  spin, 77, 88
    eigenfunctions-spinors, 79, 82, 318
    eigenvalues, 77, 88, 319
    operators, 318
    total, 85, 106, 325
Atom, many-electron, 97
  one-electron, 45
  one-electron energy levels of, 104
  correlation energy of, 117
  self-consistent electronic structure of, 100
  size of, 114
  total energy of, 116
Atomic energy levels, 70, 76, 106, 109
Atomic number, 97, 102, 116
Atomic shells, 101
  closed, 101
  inner, 102, 112, 149, 176
  ionisation of, 103
  open, 101
  outer, 102, 112
  radii of, 111, 144
Atomic unit of length (a.u.), 133

Bessel functions, spherical, 53, 322
Binding, covalent, 144, 150
  ionic, 144
  or crystals, 180
  of the methane molecule, 150
Black-body radiation, 73
  Planck's law for, 73
Bloch waves, 170, 176, 241, 273, 339
Bloch's theorem, 170, 226, 244
Body-centred cubic lattice, 159, 233
Bohr magneton, 78
Bohr radius, 60
Boltzmann statistics, 96
Boltzmann's constant, 65
Born–Oppenheimer approximation, 131, 166
Bose–Einstein condensation, 92
Bose–Einstein distribution, 73, 91, 153
Bosons, 72, 88, 90, 153, 230
Bound states, 20, 38, 134, 300, 321
  and discreteness of energy spectrum, 11, 21
Bravais lattices, 163
Brillouin zone (BZ), 169, 239
  symmetry-lines of, 175

Brillouin zone (BZ) (*cont.*)
irreducible part of, 182
of the bcc lattice, 184
of the fcc lattice, 175

Centrally symmetric field, 46
Centre-of-mass (CM) system of coordinates, 118, 155
Chemical, potential, 91, 95, 194, 197
Clebsch–Gordan coefficients, 125, 327, 330
Cohesive energy of metals, 224
Commutation relations, 309, 313, 318, 326
Commuting operators, 310, 311, 317
Compensation, in semiconductors, 216, 275
Complex function, 3
Complex numbers, 296
Conduction band, of metals, 189
of semiconductors, 190
Conductivity,
hopping, 277
of amorphous semiconductors, 276, 279
of doped semiconductors, 213, 215
of intrinsic semiconductors, 209
of metals, 198, 201, 206, 209
Configuration of electrons in atoms, 108
Constant-energy surfaces, 184
Cooper pairs, 266
Copper crystal, 175
Core states, 179, 254
Correlation energy, of atoms, 117
of molecules, 144
of metals, 223
Coulomb, field, 45, 59
interaction, 97, 127, 257, 260
Critical exponent, 277
Critical temperature (T_c) for
Bose–Einstein condensation, 92
ferromagnetic transition, 288
superconducting transition, 263
Crystallographic plane, 232
Crystals, 159, 166
Curie temperature, 288
Current, electric, 201

De Broglie wave, 28
Debye, cut-off frequency, 230, 266
law of specific heat, 231
model of lattice vibrations, 229
temperature, 231
De-excitation of atoms, 69, 74
Defects in crystals, 202
Degeneracy of energy-eigenvalues, 14, 59
and symmetry, 15, 338
Density functional method, 224
Density of nomal modes, 228
Density of states, 17, 187
in amorphous solids, 274–276
in ferromagnetic nickel, 222
in semiconductors, 197, 284
in tungsten, 188
local, 247
of a free particle (in a box), 17, 41, 83
Diamond, unit cell of, 166
Dielectric function, 258
Dielectric constant, 212, 258
Dirac's theory of hydrogen atom, 78, 86
Dispersion, curves, 174
of a wavepacket, 32, 228
Dissociation energy, 117, 134, 143
Doping of semiconductors, 211, 275, 281

Effective mass, 185, 212
Eigenfunctions,
complete set of, 301, 302, 307, 309, 316
of energy (Hamiltonian), 300
of momentum, 29, 307
orthogonal set of, 301, 302, 307, 309, 316, 317
(see also eigenstates),
Eigenstates, stationary, 8, 311
of a free particle (Cartesian coordinates), 28, 308
(spherical coordinates), 52, 322
of a particle, in a box (see Particle in a box)
in a rectangular potential well, 22, 302, 304
in a spherical potential well, 58
in a spherically symmetric field, 49, 320
of an electron, in a Coulomb field, 60
in a periodic potential field, 170, 332
in a semi-infinite crystal, 242
in the non-periodic potential field of an amorphous solid, 274
of two or more bosons, 90
of two or more fiermions, 93

Eigenvalues, and measurement, 310
of energy (Hamiltonian), 300
of momentum, 307
(see also energy-eigenvalues)
Electric dipole moment, induced, 67, 341
intrinsic, 146
Electromagnetic field, 65
quantum description of, 70
interaction of atoms with, 66, 74
interaction of matter with, 69, 73, 75
Electron, 1
Electronic cloud, 67, 98, 111, 124, 134, 173, 178, 217, 224, 262
Electronic transitions,
in atoms, 68, 87, 109
in crystals, 251
in molecules, 148
Emission lines, 69, 75
Emission of light, 62, 69, 145
Energy, internal, 45, 129
kinetic, 5, 18, 26, 28, 299
potential, 18, 25, 299
total, 18, 299
translational, 45, 129
Energy bands, 137, 174, 240, 257, 335
in amorphous materials, 279
in copper, 175
in ferromagnetic nickel, 220
in gallium arsenide, 193
in germanium, 191
in potassium chloride, 193
in silicon, 193
in tungsten, 183
in zinc sulphide, 193
calculation of, 182, 218, 333
Energy-eigenvalues, continuous spectrum of, 7, 28, 301
discrete spectrum of, 6, 11, 21, 300
Energy gap, of a semiconductor, 190, 192, 279
of a superconductor, 267
Energy loss spectrum, 261
Exclusion principle, Pauli's, 94, 95, 99, 101, 108, 225
Exchange energy, 99, 103, 143, 219, 223
Exciton, 257

Face-centred cubic lattice, 163
Fermi–Dirac distribution, 95, 194
Fermi energy (level), 96, 176, 189, 194, 208, 235, 276
Fermi glass, 277
Fermi hole, 98, 117, 223
Fermi surface, 186
Fermions, 88, 93
Ferromagnetic metals, 219
Field emission, 250
Fine-structure constant, 86
Fowler–Nordheim equation, 250
Free-electron model of metals, 208

Gas in thermodynamic equilibrium, 64
Germanium, amorphous, 269
crystal, 166
Glasses, 270
Glass-transition temperature, 270

Hamiltonian operator, 299, 302
Hankel functions, spherical, 53, 55, 322
Harmonic oscillations, 141
coupled, 151, 226
Hartree unit of energy, 102
Hellmann–Feynman theorem, 144
High-temperature superconductors, 268
Holes in valence band, 197, 215
Hund's rule, 108, 127, 225, 332
Hybridisation, 149, 182
Hydrogen atom, 1
energy eigenvalues and eigenstates of, 59–64
fine structure of the energy levels of, 75, 86
hyperfine structure of the energy levels of, 125
spinors, of, 83, 87
transitions between states of, 68, 87
Hydrogen molecular ion, 132
electronic terms of, 135, 338

Identical particles, 89
wavefunction symmetry for, 90, 93
Image potential field, 235
Imaginary numbers, 295
Impurity states, 213, 214, 275
Infrared spectrum of diatomic molecules, 147
Insulators, 192
Ionisation energy, of atoms, 20, 103
of donor impurities, 212
Ionic crystals, 224
Isotopes, 116

Kronig–Penney model, 282
k-space, 168

Lattice, constant, 160
reciprocal, three-dimensional, 167, 169, 239
two-dimensional, 237
space, three-dimensional, 159, 163, 233
two-dimensional, 233
Legendre polynomials, associated, 315
Lifetime of excited state, 75, 251
Linear combination of atomic orbitals (LCAO) method, 132, 148, 176, 180, 182
Linear sum of functions, 4, 43, 301
Localisation due to disorder, 271
Localisation length, 274, 277
Low-energy electron diffraction (LEED), 234

Magnetic moment, 77, 78
Magnetisation, 219, 289
Martices, as operators, 316
Mean field approximation, 97, 106, 143, 148, 170, 216, 224
Mean free path, 207
inelastic, 252
Mean values, 25, 310
and measurement, 25, 310
Mechanics, classical, 1, 12, 18, 64, 130, 141
Meissner effect, 264
Metal, total energy of, 223
Metallic behaviour, 189
Methane molecule, 148, 149
Miller indices, 232
Mobility, of electrons, 210, 215
of holes, 210, 215
Mobility edge, 275
Molecules, diatomic,
dissociation energy of, 117, 134, 143, 144
electronic terms of, 130, 143
energy levels of, 139, 142, 145
rotational motion of, 138
total energy of, 130, 138, 142
transitions between states of, 146, 148
vibrational motion of, 139
Molecules, polyatomic, 148
Morse's curve, 154
Mott's law, 279
Multiplet splitting of atomic levels, 109

Normal frequencies, 4, 157, 227
Neutron, 116
Non-bound states, 27, 134, 300
Non-magnetic metals, 218
Normal modes, of vibrating lattice, 226
of vibrating string, 4
of vibration of polyatomic molecules, 151, 157
Normalisation of wavefunction, 9, 60, 81, 121, 132, 170, 176, 321

Ohm's law, 201
Operators, 299, 308
spin-dependent, 316
Optical modes of vibration, 228
Orbit, classical, 2, 47
Orbital, atomic, 113, 120, 132
molecular, 132, 137, 148
antibonding, 136
bonding, 136, 149
Oscillating dipole, 67
Oscillators, of the electromagnetic field, 71
Overlap of atomic orbitals, in molecules, 133, 135
in crystals, 176, 178, 181

Paramagnetism, 285, 289
Parity, 24, 303, 304
Particle in a box, 5
energy-eigenvalues and eigenstates of, 6, 10, 12, 38, 306
degeneracy of energy levels of, 14
density of states of, 17, 83
Periodic boundary conditions, 40, 172, 226
Periodic potential field, 170
Periodic table of elements, 115
Permittivity, 45
Perturbation theory,
time-dependent, 339
time-independent, 335
Phonons, 230
Photoemission, 251
Photons, 72
Planck's constant, 2, 73
Plane waves, electromagnetic, 66, 71
free-particle, 28, 54
Plasmons, 260
Polarisation, 71, 258

Potential barrier, rectangular, 32
  spherically symmetric, 54
  surface, 235
Potential field, 18
Potential well, rectangular, 22, 302
  spherical, 57, 322
Primitive unit cell, 160
Probability-current density, 34, 56, 62
Probability-density function, 8, 63
  classical, 8, 11
Proton, 1

Radial part of wavefunction, 58, 321
Real numbers, 295
Reciprocal lattice (see Lattice)
Reciprocal space, 168
Reciprocal vectors, 168
Reduced mass, 118, 156
Reduced **k**-zone, 172, 239
Reflection coefficient, for electrons, 34, 306
  for light, 287
Relativistic interaction, 78, 111
Relaxation time, 204
Residual interaction, 106, 223, 251, 260
Resistivity, 206, 216
Richardson's equation, 249
Rutherford's model of the atom, 1
Russell–Saunders approximation, 111
Rydberg unit of energy, 185

Scattering, 32, 54
  by impurities, 203, 215
  by surface barrier, 243
  cross section, 56, 57
  states, 52, 304
Schrödinger's equation, 9, 299, 302
Schrödinger's theory of hydrogen atom, 77
Selection rules, for many-electron atom, 109
  for molecules, 146
  for one-electron atom, 68, 87
Self-consistent field, for atoms, 102
  for crystals, 216
  for molecules, 143
Semiconductors, amorphous, 269, 279
  crystalline, 192
    doped, 211
    n-type, 213
    p-type, 214

Slater determinant, 126
Sommerfeld model, 208, 249
Sound velocity, 228
Specific heat, 192
  of metals, 195, 196
  of semiconductors, 198
Spherical coordinates, 45
Spherical harmonics, 49, 316
Spherical waves, 53, 56
Spin, electron, 77
  (see also Angular momentum)
Spin–orbit interaction, 78, 103, 109, 320, 326
Stationary states (see Eigenstates)
Stefan–Boltzmann law, 120
Superconductivity, 263
  BCS theory of, 264
  high-temperature, 268
Surface, potential barrier, 235
  reconstruction, 235
  states, 244
Surface Brillouin zone (SBZ), 238
Symmetry, 15, 24, 46, 162, 175, 338

Tetrahedral, configuration, 166
  orbitals, 149
Thermal expansion coefficient, 270
Thermionic emission, 249
Transition, Anderson, 275
Transition rate, 69, 341
Transitions, between atomic states, 68, 87, 109
  between molecular states, 146, 148
  electronic, in solids, 251
Transmission coefficient, 34, 37, 250, 306
Transport, electron, 198, 277
  velocity, 199, 228
Tunnelling, 37, 178, 249, 277

Uncertainty about the value of,
  a physical quantity, 311
  energy, 75, 252
  momentum, 2, 31
  position, 2, 32, 43
Uncertainty principle, 2, 10, 19, 29
Unit cells, 159, 164

Valence, band, 190
Vectors, 290
Vibrations, lattice, 166, 225
  of molecules, 139, 151, 157

Wave function, 3
  antisymmetric, 93
  symmetric, 90
Wave vector, 28, 71, 170, 226
Wave packet, 30, 198, 228
Wigner–Seitz cell, 163
  of the bcc lattice, 162
  of the fcc lattice, 165
Work function, 235
  of the different faces of tungsten, 237

X_α-model, 99, 126
X-ray diffraction, 167